ECOLOGICAL RISKS

Perspectives from Poland and the United States

Edited by

Wladyslaw Grodzinski
Ellis B. Cowling
Alicja I. Breymeyer

with

Anna S. Phillips
Stanley I. Auerbach
Ann M. Bartuska
Mark A. Harwell

Polish Academy of Sciences

National Academy of Sciences
of the United States of America

National Academy Press
Washington, D.C. 1990

National Academy Press • 2101 Constitution Avenue, N.W. • Washington, D.C. 20418

NOTICE: The project that is the subject of this report was approved by the Governing Board of the National Research Council, whose members are drawn from the councils of the National Academy of Sciences, the National Academy of Engineering, and the Institute of Medicine. The members of the committee responsible for this report were chosen for their special competences and with regard for appropriate balance.

This report has been reviewed by a group other than the authors according to procedures approved by a Report Review Committee consisting of members of the National Academy of Sciences, the National Academy of Engineering, and the Institute of Medicine.

The National Academy of Sciences is a private, nonprofit, self-perpetuating society of distinguished scholars engaged in scientific and engineering research, dedicated to the furtherance of science and technology and to their use for the general welfare. Upon the authority of the charter granted to it by the Congress in 1863, the Academy has a mandate that requires it to advise the federal government on scientific and technical matters. Dr. Frank Press is president of the National Academy of Sciences.

The National Academy of Engineering was established in 1964, under the charter of the National Academy of Sciences, as a parallel organization of outstanding engineers. It is autonomous in its administration and in the selection of its members, sharing with the National Academy of Sciences the responsibility for advising the federal government. The National Academy of Engineering also sponsors engineering programs aimed at meeting national needs, encourages education and research, and recognizes the superior achievements of engineers. Dr. Robert M. White is president of the National Academy of Engineering.

The Institute of Medicine was established in 1970 by the National Academy of Sciences to secure the services of eminent members of appropriate professions in the examination of policy matters pertaining to the health of the public. The Institute acts under the responsibility given to the National Academy of Sciences by its congressional charter to be an adviser to the federal government and, upon its own initiative, to identify issues of medical care, research, and education. Dr. Samuel O. Thier is president of the Institute of Medicine.

The National Research Council was organized by the National Academy of Sciences in 1916 to associate the broad community of science and technology with the Academy's purposes of furthering knowledge and advising the federal government. Functioning in accordance with general policies determined by the Academy, the Council has become the principal operating agency of both the National Academy of Sciences and the National Academy of Engineering in providing services to the government, the public, and the scientific and engineering communities. The Council is administered jointly by both Academies and the Institute of Medicine. Dr. Frank Press and Dr. Robert M. White are chairman and vice chairman, respectively, of the National Research Council.

Library of Congress Catalog Card No. 90-61630
International Standard Book Number 0-309-04293-3

Copies of this report are available from:
Office of Soviet and East European Affairs
National Research Council
2101 Constitution Avenue N.W.
Washington, D.C. 20418

Additional copies are for sale from:
National Academy Press
2101 Constitution Avenue, NW
Washington, D.C. 20418

S158

Printed in the United States of America

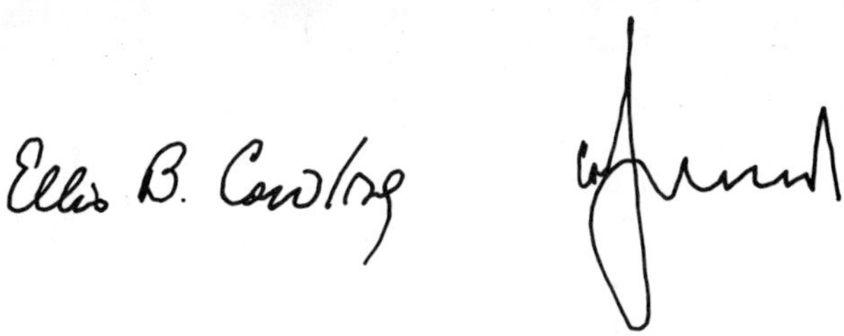

In memory of our good friend and colleague
Academician Wladyslaw Grodzinski

1934 - 1988

We will always remember
your creativity,
your intellectual energy,
your contagious enthusiasm,
your spirit of camaraderie,
and
your remarkable courage.

Editors' Note

During recent months, a major political revolution has taken place in Poland and in many other countries of Central Europe. In March of 1989, a great national debate took place among all political forces in Poland. These "Round Table Talks" were the first open discussions between Solidarity and the political parties then in power. An important part of the Round Table Talks focused on issues of ecology and the environment. A long list of final statements and demands were formulated and then signed jointly by leaders of both Solidarity and the government. Many of these demands had precise deadlines for fulfillment. In essence, demands for both political and ecological reform were formulated by the political opposition but were addressed to the government, which was to take responsibility for their realization.

Then, as in so many other aspects of history, the actual events of our lives changed the scenario created by our imagination. In June of 1989, the election for Polish Parliament was won by Solidarity. And in September of that same year, a new and freely elected government was formed—a government dominated by yesterday's opposition.

These remarkable political events were accompanied by equally dramatic increases in inflation and drastic economic reforms. While the statements and demands for reform resulting from the Round Table Talks still stand as goals for the nation, the responsibility for their fulfillment is changing. In the environmental field, several political forces are now competing and the so-called "green" parties are growing stronger and stronger. Polish society is now freely articulating its own goals and aspirations. And the new government is struggling to meet the immediate needs of the people and, at the same time, help the country make the necessary adjustments to a free-market economy.

Will ecological perspectives retain an important place in our thinking as Poland continues its struggle for economic viability and self- determination? Certainly, an exchange of experience and lessons deriving from comparative studies are of special value in this situation. Thus, we believe that this book on *Ecological Risks: Perspectives from Poland and the United States* is even more timely at the beginning of this new decade than it was at the end of the last decade—not only for Poland and the United States, but also for many other nations in this rapidly changing world.

ALICJA I. BREYMEYER
Warsaw, Poland

ELLIS B. COWLING
Raleigh, North Carolina

March 1990

Preface

This book is a product of cooperation between the Polish Academy of Sciences (PAN) and the National Academy of Sciences of the United States (NAS). In 1986, PAN and NAS agreed to organize a joint workshop on "Ecological Research and Environmental Protection." The meeting was held in October 1987 in the small village of Mogilany near Krakow, Poland. The enthusiasm generated by this first meeting led to a reciprocal workshop which was held in Washington, D.C. in November 1988. With few exceptions, the same Polish and American scientists from a wide range of scientific disciplines participated in both workshops. The names and addresses of the participants appear in Appendix 1; the summary memoranda prepared at each of the two workshops appear in Appendix 2.

At each workshop, the participants presented and debated concepts and impact studies which illustrate current challenges in ecological research and environmental protection. These ideas have value not only in Poland and the United States, but in many other countries around the world. Appendices 3 and 4 provide lists of collaborative research projects which were suggested during the two workshops, some of which have already come to fruition.

This book was written to stimulate interest in the assessment and management of ecological risks by five major groups of individuals:

- *leaders in* local, national, and multi-national *corporations and industries*;
- *leaders in* local, regional, and central *governments and international associations*;
- *educators* in colleges, universities, and secondary schools;

- *graduate students and other researchers* in many different aspects of biological, physical, engineering, and social sciences, natural resources and environmental studies;
- *volunteers in* such social movements as *political parties, churches, labor unions, trade associations, and environmental and ecological clubs.*

The ideas presented in the various chapters of this book are those of the authors. The book does not represent the official views of any organization or agency.

The participants in the two workshops are especially grateful to Glenn Schweitzer of NAS and Halina Obuchowicz and Zdzislaw Kaczmarek of PAN for suggesting the idea and then facilitating implementation of both workshops on ecological research and environmental problems. Financial support from the Rockefeller Brothers Fund and the Ford Foundation for the workshops and from the U.S. Environmental Protection Agency for this publication is also greatly appreciated.

The workshop participants were deeply saddened by the death in Poland of Academician Wladyslaw Grodzinski during the workshop in Washington. As co-chairman of these PAN-NAS workshops, Ladd provided vital leadership to assure their success. But his total contribution was much more profound—he inspired us all through his intellectual creativity, his spirit of camaraderie, and his remarkable courage to the end.

We dedicate our continuing collaborative efforts to his memory.

ALICJA I. BREYMEYER

ELLIS B. COWLING

Contents

HUMAN EFFECTS ON THE TERRESTRIAL ENVIRONMENT

AGRICULTURAL IMPACTS

IMPACTS ON AQUATIC ECOSYSTEMS

ENVIRONMENTAL MANAGEMENT CASE STUDIES

RECOMMENDATIONS

APPENDICES

Overview

Assessment and Management of Ecological Risks

ELLIS B. COWLING
North Carolina State University

WLADYSLAW GRODZINSKI
Jagiellonian University

ALICJA BREYMEYER
Institute of Geography and Spatial Organization

The purpose of this joint publication by the Polish Academy of Sciences (PAN) and the National Academy of Sciences of the United States (NAS) is to help our two countries learn more about the assessment and management of ecological risk. The interest of our two Academies to collaborate in this endeavor is an outgrowth of concern about the current state of our natural heritage.

IMPORTANCE OF ECOSYSTEMS FOR HUMANS

The prosperity and quality of life in every nation depend on a long list of natural endowments. These gifts from the evolutionary history of our planet vary greatly from continent to continent, country to country, locality to locality, and time to time (Conable, 1987; Brundtland, 1987).

In the cases of Poland and the United States, our natural heritage includes: a wide diversity of valuable plants, animals, and microorganisms; abundant air and water; productive soils, farmlands, forests, and aquatic ecosystems; beautiful mountains, lakes, streams, and rivers; valuable wilderness, coastal areas, wetlands, and groundwaters.

Throughout our history, every man, woman, and child has had good reason to care about the quality and stability of these natural resources. We care because so much of our economic and social well-being as individuals and as peoples is affected. But the history of our two countries shows that we still have a great deal to learn about how to care enough and to be wise enough to be good stewards of our natural resources—to fulfill the vision that seers such as Aldo Leopold (1968) had for a *land ethic*, that Van Rensselaer Potter (1959, 1987) had for a society committed to *ecological bioethics*, and that Brundtland (1987) had for a world society committed to *sustainable development*.

The processes of urbanization and industrialization have added greatly to the challenges of continuing the economic development in our societies while maintaining the quality of our environment, the productivity and health of our people, and the stability of the ecosystems on which our life depends. Successive generations of Poles and Americans have had progressively higher expectations than their predecessors for wholesome food, clean drinking water, material possessions, travel, education, satisfying jobs, enjoyable living and working conditions, and personal, social, and national security. All these aspirations have added to the demands of our people for an ever-increasing standard of living. Unfortunately, however, they also have added many waste materials which circulate through the air, land, surface waters, groundwaters, ecosystems, and all the living things with which we share our life on this planet.

Success in the conservation, wise use, and protection of ecosystems and other natural resources requires that scientists, industrial and political leaders, and citizens in every aspect of society learn a great deal more about:

- the distinctive features and dynamics of the ecosystems and other natural resources of our countries;
- the industrial, social, political, technical, and economic systems by which these resources are managed; and
- appropriate methods by which to assess the nature and magnitude of risks to ecosystems caused by both deliberate and inadvertent human activities. Such understanding is needed by citizens and industry leaders in every local community or district, as well as within the state and federal governments of our own and neighboring countries (Pillet and Murota, 1987).

These three types of knowledge provide the foundation for making wise (or unwise) choices about alternative means by which to:

- produce the goods and services our people need (or think they need); while
- maintaining the productivity and stability of the ecosystems on which the quality and abundance of our life depends (Breymeyer, 1986; Brundtland, 1987; NAS, 1986).

An especially important part of these scientific and educational challenges is to learn how to properly dispose of all the waste materials that our modern styles of living introduce into the air, waters, and soils of the cities, towns, villages, farms, forests, and the wilderness and recreational areas of our countries. As discussed more fully in the various chapters of this book, both Poland and the United States have had some satisfying successes and some spectacular failures in developing systems for proper disposal of wastes.

The purpose of this book is to review some of the concepts that societies must use to improve our assessments of risks and our management of ecosystems. Case histories are presented by authors from both countries to illustrate the advantages and limitations of several different methods. Before we begin, however, a few definitions are in order.

DEFINITIONS

Ecosystems

Ecosystems are composed of biotic (living) and abiotic (nonliving) components. The biotic components consist of three general types of living organisms:

- *producers*: green plants that capture the energy of the sun and produce organic matter;
- *consumers*: including humans and other animals that utilize as their energy sources the food materials stored by producers; and
- *decomposers*: mostly microorganisms that obtain their energy by breaking down and converting the dead bodies of organisms into simpler compounds.

The abiotic components of ecosystems include air, water, soil, nutrients, the geological substrata, sediment, and both particulate and dissolved dead organic matter.

W.L. Smith (1980) described the relationships among ecosystem components as follows. Ecosystems are energy-processing systems whose components have evolved together over a long period of time. The boundaries of natural ecosystems are determined by the environment, i.e., by what forms of life can be sustained by environmental conditions of a particular region. Plant and animal populations within the system represent the objects through which the system functions.

Ecosystems are also open systems. They receive materials from and contribute to the environment that surrounds them. The environment contributes gases, minerals, and energy. Ecosystems utilize these substances and, in turn, make their own contributions to the environment. Energy flows through the system unidirectionally while water, gases, and minerals are recycled and fed back into the system.

Stress occurs in all living organisms when some physical, chemical, or biological feature of the environment is outside the range that is optimal for development of that organism. Stress occurs in a whole ecosystem whenever some physical, chemical, or biological feature of the environment causes a significant alteration of the natural dynamics of (i.e., flows of energy and materials through) that system (Odum, 1985).

As ecosystems function, populations of organisms playing similar roles are joined in communities and trophic groups or levels. These large groups contain many species of organisms that act as ecological units and respond to stresses as ecological units. Some examples of the reactions of large groups of soil animals to environmental stress are discussed in Chapter 8 of this volume.

Ecological Risk

Ecological risk is a condition in which the normal functions of a population, ecosystem, or an entire landscape are threatened by external forces or stress factors which presently or in the future may diminish the health, productivity, genetic structure, economic value, or the aesthetic quality of the system (NAS, 1986, 1987; USEPA, 1983a, 1983b, 1986a). These external forces or stress factors can occur in the form of:
medskip

- excessive imputs of nutrients, toxic pollutant, pesticides, etc;
- disturbance of normal energy flows, such as by thermal pollution around a cooling tower or global warming due to so-called "greenhouse gases";
- drastic changes in the rates of ecosystem processes, such as by drying of peatlands or flooding of tropical grasslands, thus speeding up or slowing down decomposition processes;
- physical destruction of ecosystems, e.g., by compaction of soils in city parks, excessive grazing of rangelands, burning of forests or grasslands, or covering with heavy loads of volcanic or industrial dusts; or
- introduction of virulent pathogens such as parasitic, pathogenic, or predacious fungi, insects, bacteria, viruses, etc.

Ecological Risk Assessment

The assessment of ecological risk is the process of:

- quantifying the probability that adverse ecological effects may, or are, occurring as a result of exposure to one or more stress factors;
- determining the quantitative significance of such adverse effects; and
- determining how to manage the ecosystem or the sources of the stress factors so that effects can be maintained within limits that are acceptable to society (Andrews, 1987; Kates, 1978; U.S.EPA, 1987a, 1987b).

In practice, airborne and waterborne pollutant chemicals are among the most common stress factors of concern to society. Also, in most ecological risk assessments, the most important ecosystem components of concern are populations or communities of living organisms. In Chapter 8 of this volume, Breymeyer proposes production and decomposition of organic matter as measurable ecosystem processes at risk. Similarly, in Chapter 15, Ryszkowski proposes circulation of some nutrient elements in agricultural landscapes as a measurable process at risk. Thus, most ecological risk assessments involve identification of one or more of the following:

- the types and species of organisms that are at risk within a given ecosystem;
- the rate of production of organic matter and/or other ecosystem processes;
- the nature, concentrations, and timing of pollutant chemical exposures that occur within the system;
- the nature and magnitude of the response of the organisms or ecosystems to the stress imposed by pollutant chemicals (dose/response relationships);
- the physiological, biogeochemical, or ecological processes by which the pollutant chemicals induce their detrimental effects (mechanisms);
- the sources of pollutant chemicals that cause stress within the ecosystem; and
- the management procedures (e.g., changes in species composition, alterations of managed ecosystems, restructuring of landscapes, modifications of industrial or other processes, regulatory policies, emissions limitations, mitigative treatments, or other methods) by which the *exposure* of a given ecosystem to pollutant chemicals can be maintained within acceptable limits; and/or the *effects* of the pollutant chemical within the ecosystem can be mitigated (i.e., maintained within acceptable limits).

In many cases, the formal process of ecological risk assessment within a given organization (such as a federal, state, or provincial department of environmental protection, or a specific private industry or government enterprise) may include only one or a given set of the procedures listed above.

AN ILLUSTRATION OF CONFLICTS BETWEEN DESIRABLE GOALS IN SOCIETY

The assessment of ecological risks associated with the widespread use of pesticides in plant and animal agriculture presents a good example of the challenges in ecological risk assessment. Here the conflict between desirable goals of society is especially poignant because it involves a conflict between two taboos—the taboo against "bugs and filth" and the taboo against "poisons in our food and water."

The departments of agriculture in most countries are charged to provide wholesome food for the people. The food and drug laws of the United States, for example, state that food must not be "... filthy, putrid, or unfit for human consumption." Those are powerful words and agricultural scientists heard them loud and clear! They responded by developing chemical pesticides to help keep the bugs and filth out of our food. And soon, a grateful society gave Mueller a Nobel Prize for discovering DDT.

But then it was discovered that DDT and some other pesticides were not very biologically degradable. DDT tended to accumulate in the body fat of all the people who ate the food that was free of "bugs and filth." Furthermore, DDT could be dispersed in the air to many non-target organisms. Before long it was discovered that DDT accumulated in the body fat of penguins in Antartica where DDT was never used! Now the taboo against "poisons in our food" (and in innocent penguins near the South Pole) came into conflict with the taboo against "bugs and filth." Thus, a new assessment was called for and soon many countries around the world prohibited the use of DDT.

CONTRASTS AND SIMILARITIES BETWEEN POLAND AND THE UNITED STATES

The current environmental situation is quite different in Poland and the United States. In both countries there exist some areas of very high environmental quality and others of great environmental impact. Examples of the latter in the United States include the San Bernardino Valley of California, some parts of the Great Lakes, and the steel production areas near Gary, Indiana. In Poland they include Upper Silesia, the Krakow region, Turoszow, Pulawy, and several other areas. The essential difference, however, is that the percentage of heavily impacted areas is greater in Poland; such areas encompass practically the entire southwestern part of the country.

In Poland, both air and water quality have been decreasing over the past 30 to 40 years as the country has sought to use its substantial reserves of high-sulfur coal in heavy industry. In the United States, both air and water quality have improved in many parts of the country as the economy has shifted away from its earlier dependence on heavy industry and responded to the requirements of the Clean Air Act of 1970 and the Federal Water Pollution Control Act of 1972.

In Poland, the most severe air-quality problems are emissions of sulfur dioxide and industrial dusts and aerosols in the southwestern part of the

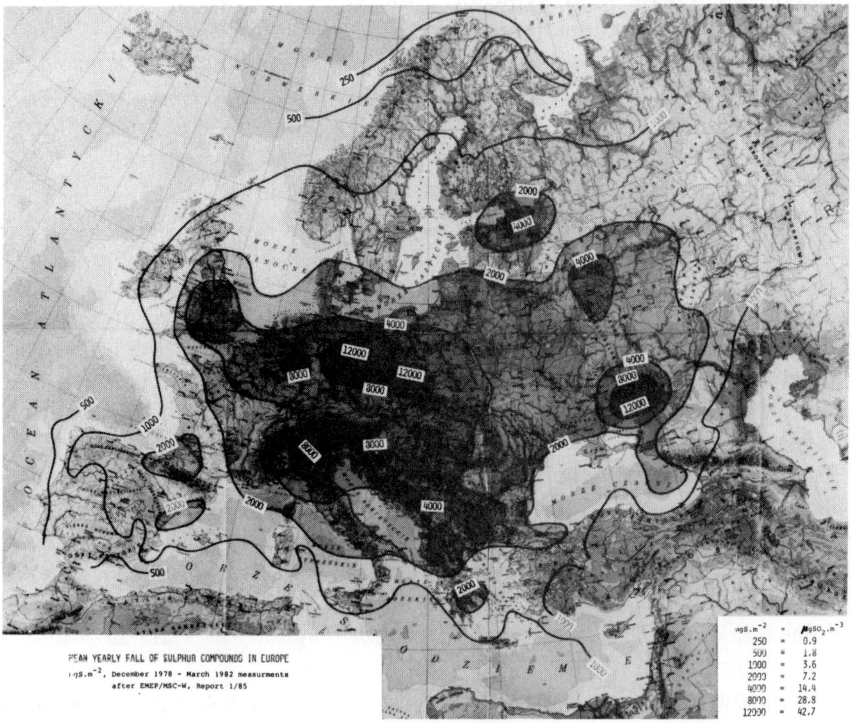

FIGURE 1 Distribution of sulfur pollution over Europe. Please note especially the so-called "black spot" of Europe which includes the southwestern part of Poland and portions of Czechoslovakia and the German Democratic Republic. The data shown are for the period December 1978 through March 1982.

country. In the eastern United States, by contrast, the most severe air quality problem is ozone and other photochemical oxidants which accumulate in the atmosphere mainly in the eastern half of the country and in southern California.

In Poland, the most important air pollutants are emitted directly from industrial smoke stacks in the form of gaseous sulfur dioxide and industrial dusts. In the eastern United States, by contrast, the most important air pollutants are not emitted directly from industrial sources; ozone, other photochemical oxidants, and acid deposition are formed in the atmosphere from a complex mixture of nitrogen oxides and volatile organic compounds released by both mobile and stationary sources.

In Poland, geographical gradients from major cities to the surrounding countryside are fairly steep, typically decreasing by a factor of ten over distances of 10 to 100 kilometers. In the United States, by contrast, geographical gradients in ozone pollution are fairly shallow, typically decreasing by a factor of two over distances of 100 to 500 kilometers.

In Poland, sulfur-dioxide and dust-pollution problems are more or less constant throughout the year. In the United States, by contrast, ozone pollution is a problem only during the summer months.

The maps in Figures 1 and 2 show a part of this story. The southwestern part of Poland lies in the so-called "black spot" of the European continent (see Figure 1). Here, heavy industries from three nations (Czechoslovakia, the German Democratic Republic, and Poland) are concentrated in a small area. This is the most polluted area in all of Europe. Most of the territory of Poland lies in the path of dominant southwesterly winds. Thus, much of the country receives a heavy dose of sulfur and dust pollution in all seasons of the year.

The area of highest ozone pollution in the eastern United States forms a broad band across the southern part of the country (see Figure 2). Here, many different mobile and stationary sources emit nitrogen oxides and volatile organic compounds which are precursors for photochemical oxidation reactions. High concentrations of ozone accumulate in the atmosphere mainly in the summer months when oxidation reaction rates are high and wind speeds are generally low.

These contrasts in air quality (see Chapters 9-12), and those in water quality (see Chapters 18 and 19) are caused in part by differences in the area and population density of the two countries. But they are also caused by differences in the history of industrialization, electrification, and urbanization, different energy resources and policies, and some geographical and political influences. The discovery of many similarities and differences in the ecological risks confronting a middle-sized European country and a very large North American country was very fascinating to the participants in the PAN-NAS workshops. We hope it will also be instructive and of general interest to the ecologists, economists, and environmental protection specialists for whom this book was written.

OVERVIEW OF THE BOOK

This book is organized in seven distinct sections:

- *Overview and Executive Summary* (Chapters 1 and 2);
- *Environmental Management Concepts* (Chapters 3-6);
- *Human Effects on the Terrestrial Environment* (Chapters 7-14);
- *Agricultural Impacts on Environmental Quality* (Chapters 15-16);
- *Impacts on Aquatic Ecosystems* (Chapters 17-19);

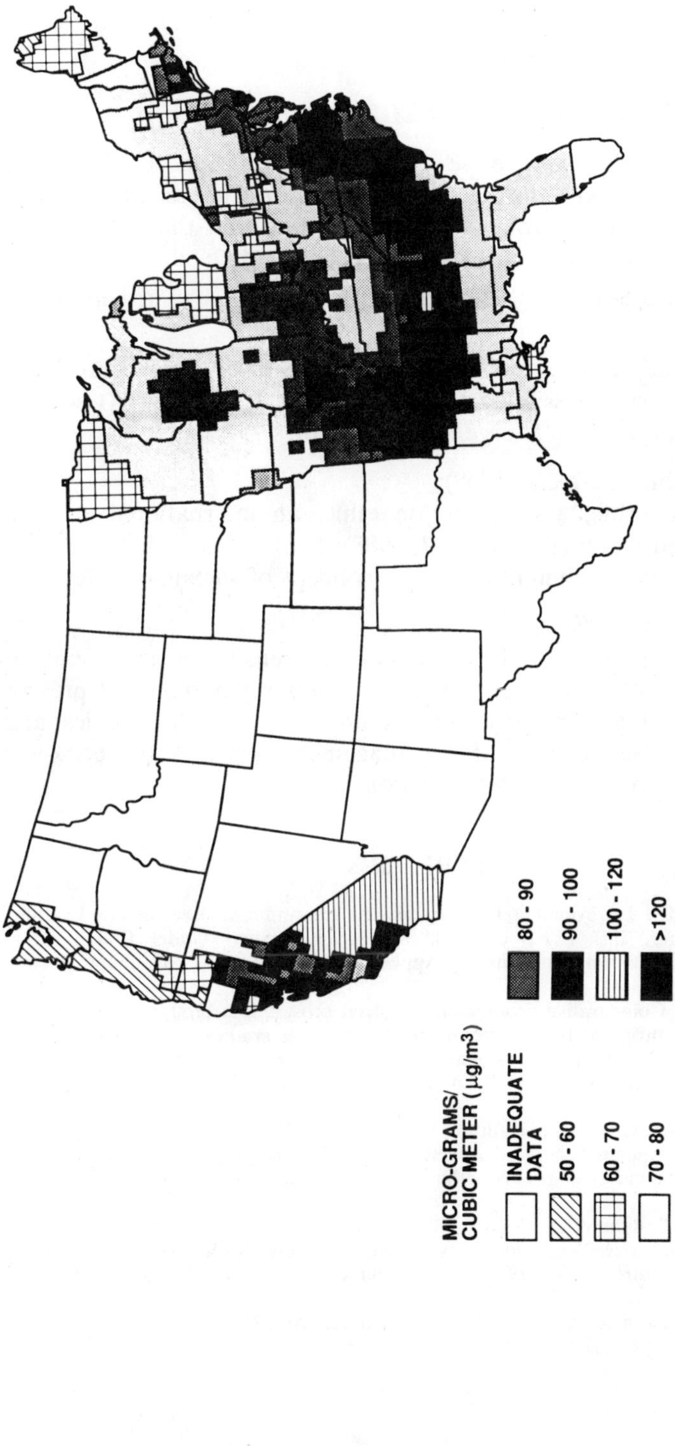

FIGURE 2 Ozone concentrations over the eastern and western United States. The monitoring data shown are based on seven-hour averages from April to October from 1978 through 1982. Note especially the very broad areas within the East Central United States and California where summertime ozone concentrations exceed 80 micro-grams per cubic meter. (This figure was prepared by Dr. David Shriner of the Oak Ridge National Laboratory.)

- *Environmental Management Case Studies* (Chapters 20-23);
- *Recommendations* (Chapter 24).

OUR HOPE FOR THE FUTURE

The editors of this book hope every reader will find the Executive Summary in Chapter 2 and the Recommendations in Chapter 24 especially valuable. But more importantly, we hope you find that the volume as a whole illustrates Oliver Wendell Holmes' conviction that:

> Man's mind, stretched to embrace a new idea, never returns to its original dimensions.

Preserving and restoring the natural heritage of our two countries is both an "evolutionary possibility and an ecological necessity" (Leopold, 1968). Mankind as a whole is urgently in need of:

- a land ethic (Leopold, 1968);
- a wildlife ethic, a consumption ethic, an international ethic, a geriatric ethic, and so on (Potter, 1987, 1988); and
- a world society committed to the concept of sustainable development (Brundtland, 1987).

We hope the next generation of scientists and citizens in our two countries and in every nation of the world will do a better job than the present generation in learning both the scientific foundations and the practical arts of ecological risk assessment and risk management. We hope this book will help bring that process closer to realization!

REFERENCES

Andrews, R.N.L. 1987. Environmental impact assessment and risk assessment: Learning from each other. Task Force on Risk and Policy Analysis Under Conditions of Uncertainty. International Institute for Applied Systems Analysis, Luxenburg, Austria. pp. 19.

Breymeyer, A. 1986. Comparative ecology of terrestrial ecosystems: Production/decomposition budget. A proposal for establishment of a SCOPE synthesis of data on organic matter budgets. Board on Environmental Studies and Toxicology, Commission on Physical Sciences, Mathematics, and Resources, National Research Council, Washington, D.C. pp. 6.

Brundtland, G.H. 1987. Our common future. Report of the World Commission on Environment and Development. Oxford University Press, Cambridge, England. pp. 238.

Conable, B.B. 1987. Sound ecology is good business. World Resources Institute, Washington, D.C. pp. 15.

Cowling, E.B. 1985. Pollutants in the air and acids in the rain: Influences on our natural environment and a challenge for every industrial society. XXV Horace M. Albright Conservation Lecture, College of Natural Resources, University of California, Berkeley. pp. 26.

Kates, R.W. 1978. Risk assessment of environmental hazard. SCOPE 8. New York: John Wiley and Sons, pp. 112.

Leopold, A. 1968. A sand county almanac. Oxford, England: Oxford University Press, pp. 226.

National Academy Sciences (NAS). 1986. Ecological knowledge and environmental problem solving: Concepts and case studies. Committee on the Applications of Ecological Theory to Environmental Problems, Commission on Life Sciences, National Research Council. Washington, D.C.: National Academy Press, pp. 388.

NAS. 1987. The mono basin ecosystem: Effects of changing lake level. Mono Basin Ecosystem Study Committee, Board on Environmental Studies and Toxicology, Commission on Physical Sciences, Mathematics, and Resources, National Research Council. Washington, D.C.: National Academy Press, pp. 270.

Odum, E.P. 1985. Trends expected in stressed ecosystems. Bioscience 35:419-422.

Pillet, G., and T. Murota. 1987. Environmental economics: The analysis of a major interface. Geneva, Switzerland: R. Leimgruber, pp. 320.

Potter, V.R. 1987. Aldo Leopold's land ethics revisited: Two kinds of bioethics. Perspectives in Biology and Medicine 30:157-169.

Potter, V.R. 1988. Global bioethics: Building on the Leopold legacy. East Lansing, Michigan: Michigan State University Press, pp. 203.

Smith, W.L. 1980. Air pollution and forest trees. New York: Academic Press, pp. 384.

U.S. Environmental Protection Agency (EPA). 1983a. Testing for environmental effects under the Toxic Substances Control Act. U.S. Environmental Protection Agency, Washington, D.C., pp. 55.

U.S. EPA. 1983b. Estimating "concern levels" for concentrations of chemical substances in the aquatic environment. Office of Environmental Processes and Effects Research, U.S. Environmental Protection Agency, Washington, D.C., pp. 46.

U.S. EPA. 1986a. Hazard Evaluation Division. Standard Evaluation Procedure. Ecological Risk Assessment. Office of Pesticide Programs, U.S. Environmental Protection Agency, Washington, D.C., pp. 96.

U.S. EPA. 1986b. Research plan for ecological risk assessment. Office of Environmental Processes and Effects Research, U.S. Environmental Protection Agency, Washington, D.C., p. 6.

U.S. EPA. 1987a. Unfinished business: A comparative assessment of environmental problems. Appendix III. Ecological Risk Work Group. Office of Research and Development, U.S. Environmental Protection Agency, Washington, D.C.

U.S. EPA. 1987b. EPA's current ecological risk research program and a strategy for a long-term research and related monitoring program. U.S. Environmental Protection Agency, Washington, D.C., p. 44.

Executive Summary

STANLEY I. AUERBACH
Oak Ridge National Laboratory

ALAN W. MAKI
Exxon Company, USA

ELLIS B. COWLING
North Carolina State University

The various chapters of this book were developed to illustrate the current state of knowledge about ecological risks in Poland and the United States. The scientists and engineers who prepared these chapters have very diverse backgrounds in terrestrial and aquatic ecology, environmental engineering, public administration, and economics. The authors of each chapter were challenged to represent their particular disciplines and experience during both the Polish and U.S. portions of the joint workshops sponsored by the Polish Academy of Sciences and the National Academy of Sciences of the United States. Through these interactions, the participants made some useful progress in developing a conceptual framework for science-based environmental policy which two of us (Auerbach and Maki) have presented in Chapter 24.

Because of the great differences in current environmental conditions in Poland and the United States, many of our workshop participants were surprised to discover many similarities in the environmental management goals and objectives in our two countries. The similarities were much greater than the differences. For this reason, the approaches that are needed in the future to ensure the conservation and wise use of natural resources, and the protection and enhancement of ecological values should also be similar.

Highlights from each major section and chapter of this book are summarized in the paragraphs that follow.

OVERVIEW OF THE CHAPTERS

The book begins with an overview chapter by Cowling, Grodzinski, and Breymeyer. These persons served as chairpersons for the workshops, both in Poland and the United States, and also as editors of the book itself. Their opening chapter sets the stage for all other parts of the book. They describe the importance of natural systems for human beings and then define what is meant by ecosystems, ecological risks, ecological risk assessment, and ecological risk management. They point up some interesting contrasts and similarities between ecological goals and conditions in Poland and the United States. The chapter ends with a statement of hope for a future in which ecological risks are assessed and management with greater wisdom—not only in Poland and the United States—but in many other countries of the world as well.

Chapter 2 is the Executive Summary of the book, which you are now reading.

ECOLOGICAL MANAGEMENT CONCEPTS

Chapters 3 through 6 deal with the fundamental concepts of ecological risk assessment and risk management. In Chapter 3, Russell proposes that all questions of ecological impact need to be answered in the context of human values—both economic and aesthetic or existential. In this context, the environment (including all terrestrial and aquatic ecosystems) are sources of materials and aesthetic or recreational experiences which contribute to the prosperity and quality of human life. In most societies, ecosystems are regarded as public or private property which should be managed for the benefit of humans rather than protected or conserved for their own sake.

Analysis of ecological impacts needs to take into account two major and differing social points of view concerning relations between human beings and the rest of nature. One extreme view is that society is an "interloper" and hence generally destructive to natural systems. The other, ethnocentric viewpoint is that ecological systems and all they encompass are factors of production and consumption in an economic sense and do not have intrinsic value. Under the latter viewpoint, changes in ecosystems are judged by their potential impact on the values and quality of human life. In evaluating changes in ecosystems (and hence potential ecological risk), one needs to take into account the dimensions of the problem (i.e., space involved), time (i.e., how much of the future needs to be taken into account), which groups of people have the most to gain or lose, and how differing perspectives should be considered in making social judgments or establishing environmental, industrial, economic, or social policy.

In summary, ecological impacts, whether avoided or encouraged, have both economic and social costs. In a human-centered value system, the social dilemma centers around whether those costs are worth bearing.

In Chapter 4, Marek and Kassenberg deal with the relationships between social planning and environmental protection. Although this chapter focuses on these issues in Poland, the concepts and ideas have much wider applicability. Because of their inherent complexity, issues of environmental protection should be addressed through a systems-analytic approach that encompasses all the interactions and interdependencies among social needs and requirements, economic activity, and environmental conditions. The scale of analysis should be at the level of major socio-economic subsystems that contain key social components and that have major environmental implications. An example of such a subordinate part of our social-economic system is the energy/environment complex. Research is needed in this (and many other) areas to determine the nature and magnitude of direct or indirect effects on humans and on ecosystems. In the environmental arena these include:

- identification and quantification of pollutants;
- improvement and verification of pollution dispersion models;
- understanding the fate of pollutants and their role in biogeochemical cycles; and
- estimation of changes in the environment caused by emissions of specific pollutants which act alone and in mixtures.

The concept of integrated planning provides an opportunity for formulating and implementing a socially acceptable strategy of development that integrates and harmonizes social, economic, and ecological interests.

In Chapter 5, Kabala analyzes the economic factors involved in Poland's contemporary environmental problems. He points out that the country is in the grip of a dual crisis of economic reversal and ecological degradation whose roots lie in the country's thirty-year policy of intense industrialization. The intensive use of energy and material resources in an economy biased toward heavy industry has come to produce ever smaller gains in material benefits and has begun to supplant opportunities for environmentally benign development. Efforts to modernize industry and invest in environmental protection are hampered by the country's high foreign debt. Events since the economic crisis of 1981 indicate that Poland will not be able to restore its economic vitality nor modify the structure of its economy without external assistance. The social, political, and ecological stresses resulting from this lack of internal capability are already manifest. A desirable first step in correcting the dual problem is investment of new foreign capital in ways that link financial considerations with environmental needs so that both problems are addressed simultaneously.

In Chapter 6, Maki and Slimak discuss the role of ecological risk assessment in environmental decision making. The formal assessment process used by many industries and by the Environmental Protection Agency in the United States consists of two parallel lines of investigation designed to relate observed effects to expected exposures. The evaluation includes consideration of the following factors:

- chemical and physical properties of the substances in question;
- how and in what quantities the material is to be used;
- expected concentrations in the environment;
- environmental fate tests;
- estimates of human and environmental exposure;
- tests for human health effects;
- tests for environmental effects;
- field monitoring and modeling;
- comparison of predicted concentrations and predicted effects; and
- decisions about the acceptability or unacceptability of the predicted risks.

Although quantitative methods are available for assessment of some of these factors, others require good judgment by industry and regulatory leaders and by workers or citizens who may come in contact with the substances in question.

HUMAN EFFECTS ON THE TERRESTRIAL ENVIRONMENT

Human effects on the terrestrial environment are discussed in Chapters 7-14. This section of the book begins with two chapters on concepts in stress ecology. It includes four chapters on air pollution impacts and ends with two chapters on environmental monitoring.

Concepts in Stress Ecology

In Chapter 7, Harwell, Harwell, Weinstein, and Kelley show how understanding stress ecology is essential to effective assessment and management of ecological risk. This requires understanding of:

- how components of ecosystems are exposed to human-induced stress;
- how these systems respond to stress; and
- how ecosystems recover once the stress is removed or ameliorated.

A scheme is required to separate important changes from less important ones. This scheme ultimately must relate to aspects of ecosystems that humans value and therefore are identified in social objectives or legal regulations such as protection of endangered species, prevention of erosion on construction sites, or maintenance of biological diversity. Ideally, these social objectives or legal requirements should be formulated in ecologically meaningful terms (sometimes called ecological endpoints) which are defined in terms of species, communities, or ecosystem processes. Changes in carefully selected ecological endpoints constitute changes that must be considered in ecological risk assessments. Harwell et al. argue that a complete assessment process should also identify those specific components of ecosystems (e.g., particular species, rates of particular processes, concentrations of particular chemicals) that need to be measured and monitored in order to detect socially significant ecological changes caused by stress.

Such indicators of ecological effects, properly measured and compared with non-stressed situations, can provide the basis for evaluating impacts from human activities and, through application of ecological understanding, can allow projections of future impacts. In this way, ecological risk assessment and management can be prospective and not just in reaction to unacceptable environmental damage already done.

In Chapter 8, Breymeyer asserts that analysis of risks to ecosystems should be predicated on understanding functional roles within the ecosystems themselves. For this purpose, ecosystems should be understood in terms of the amounts of organic matter produced and stored within the system and the fluxes of energy and nutrients flowing through the system. The sequence of basic processes of production, consumption, and decomposition are the key functional attributes essential for maintaining the stability and resiliency in ecosystems.

Ecosystem risk analysis can be facilitated by comparing the changes in processes under differing amounts of or gradients of stress. Examples of this approach include:

- comparing the natural range of ecosystem processes (including net primary productivity) and structures (e.g., above- and below-ground biomass) as a function of precipitation, climate, etc.;
- comparing rates of decomposition as a function of latitude and evapotranspiration;
- comparing the distribution of forests in relation to air pollution loadings; and
- analysis of the biomass and composition of soil fauna as a function of pollution stress.

Such comparative analyses of system-level attributes provide a sound conceptual foundation for ecological risk assessments.

Air Pollution Impacts

The known effects of air pollution and the possible effects of acid deposition on forests have evoked major public concern and a manyfold increase in research in North America in recent years. Earlier research on the effects of sulfur dioxide and fluoride near major point sources of these pollutants has given way to greatly increased research on the effects of regionally dispersed secondary pollutants such as ozone and acid deposition.

In Chapter 9, Bartuska describes the development of regional case studies of these effects in various parts of North America and the difficult challenges of:

- distinguishing the individual and combined effects of airborne pollutant chemicals from those of natural stress factors such as water stress, frost, and biotic pathogens; and
- determining effects on mature trees in whole landscapes rather than just the response of individual seedlings to particular stress agents in controlled chamber studies.

Improving this knowledge base is complicated by absence of visible symptoms in many cases. Research which links physiological or nutritional changes with growth appears to be a useful approach to evaluating air pollution impacts on trees in the short term. Ultimately, however, we need to know if the growth and productivity of whole forest ecosystems is affected. Such effects only can be detected with long-term measurements. The challenge will be to provide sufficient information to assist policy makers in making short-term decisions while emphasizing that only through long-term studies will uncertainty be reduced.

In Chapter 10, Godzik and Sienkiewicz describe recent changes in the health of forest in Central Europe (Poland, Czechoslovakia, and the German Democratic Republic). These changes have resulted in large part from a combination of regional and point-source pollution problems. Measurements of air pollutants in the forested areas of these countries suggest that sulfur dioxide is the pollutant most likely responsible for damage to forests. Other pollutants such as nitrogen oxides, fluoride, and heavy metals may be contributing or interacting factors in some cases, especially in local situations where large amounts of these substances are emitted from local sources. Part of the difficulty in attributing injury to forests to particular sources of pollution is that much of Central Europe has been subjected to many different kinds and sources of air pollution for a long time, and measurements of some pollutants such as ozone are generally lacking. For this reason, it is difficult to relate specific injuries to particular emission sources and insults.

In Chapter 11, Heck emphasizes that air pollution effects on agricultural plants have been studied for more than a century. Damage or injury from point sources were first recognized in the early 1900s. Research was directed at recognizing injury symptoms and assessing losses in productivity in accordance with the severity of symptom development. Table 1 in the chapter lists the airborne pollutant chemicals that have been shown to injure crops at some range of pollutant exposure expressed in terms of concentration and time. Much current research has centered on the impacts of ozone and acid precipitation on the growth and yield of crop plants. Ozone has been proven to affect the growth, reproduction, quality, and yield of many crops. Direct effects include changes in leaf surface structure, leaching of nutrients, and changes in metabolic function or reproductive processes. Conclusive evidence of injury in agricultural crops by acid deposition is very limited.

In the United States, ozone has by far the greatest impact on crop production. A number of regional and national crop-loss assessments have been made since 1980. Regional estimates range from about $30 million to $670 million per year. National estimates based on different groups of crops, ranged from $1.2 to $3 billion annually. In both regional and national estimates, corn and soybean were key components. A summary of economic estimates suggests that current seasonal ozone concentrations are causing an annual loss in excess of $3 billion in crop productivity. These assessments require three factors:

- response function that links yield to exposure;
- an air quality data base to link response to exposure on a geographic basis; and
- a crop census.

Sources of uncertainty that should be taken into account in future assessments include:

- development of a data base that is more fully representative of North America;
- effects of other biotic or abiotic stresses on crop response to air pollutants; and
- better control and measurement of experimental variables.

In distinct contrast to the situation in the United States, Godzik points out in Chapter 12 that air pollution effects on forests in Poland are better understood than effects on agricultural crops. Twenty-seven regions of Poland have been designated as *Areas of Ecological Hazard* (AEH) because of environmental pollution. Agricultural land constitutes about ten percent of these areas. The impacts of air pollution on a broad regional scale is currently not known because a country-wide survey has not been made. While investigations near sources of pollutants have shown significant decreases in crop yields, these impacted areas also have high soil concentrations of heavy metals which confound correlations with current air pollution.

Because of their implications for human health, Godzik stresses that yield reductions in crops are less important than the concentration of heavy metals and other contaminants in fruit and vegetables. At present it would be very difficult to determine air pollution losses in crops on a country-wide scale because of:

- lack of methods for determining yield reductions in already polluted areas;
- lack of data on air pollutant concentrations on a regional scale;
- lack of experimental results on decreases in crops yield in regions other than Upper Silesia (a heavily polluted region); and
- lack of financial resources.

Monitoring Ecological Stresses and Effects

In Chapter 13, Molski and Dmuchowski describe techniques for monitoring the distribution and intensity of toxic elements in the forest vegetation of Poland. They demonstrate that pine needles are useful bioindicators of the geographic distribution of pollutant elements within the country and show that the accumulation of particular elements within pine needles is a useful index of pollution impacts on vegetation.

These studies were initiated in the Bialowieza Forest District (a relatively clean area) and in the Panewnik Forest District near Katowice (among the most polluted areas in Poland). Later, the study was broadened to include a survey of all of Poland, using an 8 km by 8 km grid as the basis for a country-wide sampling scheme. Three hundred sites were selected from the grid for sampling using a random number process. The distribution patterns of the nine elements found in the needles were drawn on country-wide maps using a computer mapping program. Of the nine elements analyzed (sulfur, chromium, arsenic, iron, copper, zinc, cadmium, lead, and nickel), sulfur was the most widespread and abundant contaminant, with concentrations often exceeding the amounts needed for normal growth and development. Molski and Dmuchowski found that about 50 percent of the pine forest areas in Poland had sulfur contents exceeding 200 percent of the normal concentration, i.e., 1,200 ppm. Arsenic showed a bimodal type of distribution reflecting both long range transport from other countries and point-source contributions from industrial centers and coal-fired electric generating stations within Poland. The other elements also were limited in their quantities and distributions.

Molski and Dmuchowski point out that this type of research offers the possibility of determining the kinds and extent of plant resources that are being threatened by air pollution and identifying plant gene pools that may warrant protection against extinction by air pollution.

In Chapter 14, Grodzinska has demonstrated the value of mosses in national parks as long-term bioindicators of pollution in various parts of Poland. Her data show that both geographical and temporal trends can be determined by these methods. For example, mosses collected from national parks in the southern part of Poland are most contaminated with various heavy metals (including cadmium, chromium, nickel, lead, and zinc), whereas mosses collected in the least industrialized parts show much lower contaminations. Similarly, trends in concentration of certain metals in mosses over a series of years were well correlated with industrial activity in large areas of Poland.

Grodzinska also provides a useful summary of the morphological and physiological characteristics of mosses and the particular features of national parks that make them simple, rapid, inexpensive, and effective means by which to monitor long-term trends in pollution loadings.

AGRICULTURAL AND FORESTRY IMPACTS ON ENVIRONMENTAL QUALITY

In Chapters 15 and 16, Ryszkowski and Johnston give two remarkably contrasting views of the ecological impacts of agricultural and forestry practices on the environment. In Chapter 15, Ryszkowski describes the results of many decades of observation of landscapes in Poland containing many different mixtures of crops, pastures, and both natural and managed forests and grasslands growing on various types of soils, slopes, etc. He also describes the effects of different sizes of fields, natural areas, methods of cultivation, orientation of rows, and tillage, fertilization, irrigation, pesticide, and harvesting practices. These results show that optimum choices among all of these variables can make very large differences in the efficiency of agricultural and forest production in terms of yields, amount of purchased inputs, amounts of irrigation water used, amounts of energy needed, amounts of erosion, and both nutrient- and water-use efficiency. This chapter gives new meaning to the term "appropriate technology" and suggests that very large gains in efficiency of agricultural and forestry operations can be achieved through more holistic understanding of the land-use and production-management systems of whole landscapes. The innovative and familiar ecological guidelines for agriculture and silviculture contained in this paper offer great potential to farmers and foresters in a wide range of circumstances in many different countries throughout the world.

In Chapter 16, Johnston has summarized a series of ecological problems associated with agricultural development using some examples from the United States. Special attention is given to the twin problems of soil erosion and its impacts on both land quality and surface waters, and chemical contamination with fertilizers and pesticides and their impacts on both surface waters and groundwaters. Johnston distinguishes between on-site effects in which the principal impacts are on the farm itself, and off-site impacts in which the principal impacts are on down-stream water quality.

Off-site impacts can include both short-distance and sometimes very long-distance impacts. When the United States Congress passed the Federal Water Pollution Control Act in 1972, the nation's waters were threatened mainly by discharges from industry where it was fairly easy to identify a particular point from which the discharges came, hence the term "point sources." Five years later, the Soil and Water Resource Conservation Act of 1977 identified agriculture as the primary cause of changes in water quality resulting from many diffuse points of discharge, hence the term "non-point sources." Erosion and runoff from agricultural lands are leading sources of non-point pollutants.

A major advance in understanding losses of topsoil due to erosion was development of the universal soil loss equation (USLE) which expresses soil losses in tons of topsoil per acre of land per year. Average losses in the United States are about 4.8 tons per acre per year. Leaching of water soluble pesticide and fertilizer nutrient chemicals from agricultural lands is another serious problem in many parts of the United States. Johnston describes the development and use of a series of practices designed to decrease erosion and leaching of agricultural chemicals from farmland. These include conservation tillage, contour farming, strip cropping, terracing, land leveling, and planting of trees on highly erodable land. Practices aimed at decreasing chemical contamination of surface waters include decreased dependence on pesticides through increased use of crop rotations and genetically resistant varieties of crops and adoption of integrated pest management practices.

IMPACTS ON AQUATIC ECOSYSTEMS

In Chapters 17 through 19, Hillbricht-Ilkowska, Cooper, and Gromiec provide a comparative analysis of pollution impacts on aquatic ecosystems in Poland and the United States. Their collective experience shows that water quality problems in Poland resemble those that were familiar over two decades ago, before the United States undertook its massive water cleanup and sewage treatment programs. Currently, the majority of the major rivers in Poland have experienced serious degradation of water quality. If experience in the United States is typical, Poland can expect to discover it has major groundwater contamination problems in addition to its present surface-water problems. The monitoring and mitigation programs that will be needed to redress these problems will require a much more intensive investment in chemical methods for analysis and removal. High sulfur coal is the major domestic source of energy in Poland and western currencies are scarce. Because of this, Poland needs "soft energy" alternatives that are much more labor intensive rather than those utilized in the United States.

Based on the above assumptions, a number of recommendations are made by these authors. Their key points include the following:

- River basin management units for water quality regulation should be explored. The Tennessee Valley Authority might be considered as a possible model for institutional arrangements. Experimental watersheds in Poland must be utilized as prototype ecosystems to demonstrate the suitability of alternative technologies and to train managers on site. Emphasis should be given to discharges from non-point sources of nutrients, utilizing wetlands as pretreatment units, riparian vegetation strips as buffer zones, and *in situ* ecological manipulations to encourage aquatic ecosystem responses.
- Water resources in Poland are very scarce, and high quality water is even more limited. Legislatively based regulations and pricing policies should be evaluated with the focus on water conservation and water quality management.
- Appropriate technologies based on limited amounts of energy and western currencies for remediation and control of water quality must be identified. The infrastructure necessary to support these technologies must be developed in Poland.
- Consumer products that reduce environmental loadings must be encouraged. Biodegradable pesticides, low phosphate detergents, recyclable materials, and other means should be considered in terms of costs, benefits, and ecological risk assessments.
- Cooperative training programs between Poland and the United States that integrate environmental engineering and ecological risk assessment should be initiated for graduate students, postdoctoral students, and college faculties.
- A major new initiative is needed for monitoring of nitrates and other toxic substances in groundwater, including the detection and measurement of organic substances through the use of advanced analytical techniques such as gas chromatography/mass spectroscopy.

Poland must be considered a water-poor country, where conflicts over water-resource use and development are on the rise. Water resources are unevenly distributed among different parts of the country, and water supplies now appear inadequate in quantity and quality at the same time that demand is increasing in many regions of the country.

Lakes in Poland, especially in the northern region, are threatened by eutrophication. Recent trends in chemical analyses suggest that phosphorus loading from human sources is the primary cause. Unfortunately, there are no data on heavy metals and synthetic organic substances to evaluate the toxic chemical loadings to these lakes. The current lake-monitoring program should be expanded to include toxic chemicals in addition to cultural eutrophication. Eutrophication is likely to remain the chief cause of human disturbances of natural lakes and their environments even after the elimination of the point-source nutrients. It is the main cause of changes in the numerous small- and medium-sized natural lakes situated in arable regions (i.e., the North European postglacial lakelands) that serve as important tourist centers.

An integrated assessment and management system is needed to determine impacts on lake waters at all trophic levels. It should be based on the role of the watersheds in the supply and transport of nutrients, the natural resistance of lakes to eutrophication, and the response of lake biota and its internal processes along a trophic contiuum. A system of this sort was applied to 24 large lakes in the Masurian Landscape Protected Area and led to recommendations for different forms of protection for both watersheds and lakes in Poland.

Degradation of groundwater resources is a different class of environmental problems than cultural eutrophication of surface waters. Nitrate loadings from septic fields and agricultural practices are major problems. Contamination of potable water by halogenated solvents and pesticides is another class of emerging environmental problems for Poland. A monitoring program should be initiated for nitrates, synthetic organic substances, and metallo-organic substances in groundwater and ecological food chains.

ENVIRONMENTAL MANAGEMENT CASE STUDIES

In Chapter 20, Trojan presents the results of a two-year comparative analysis of the concepts and methods used in development of environmental policy in several of the socialist countries of Europe, including the USSR, the German Democratic Republic, Czechoslovakia, Hungary, and Poland. This study reveals substantial similarities in policy perspectives in these countries and in many democratic countries. These similarities are evident in the recommendations deriving from the study:

- Improve the organization of local and central government agencies concerned with environmental protection and natural resource.
- Increase collaboration between countries in developing equipment for environmental monitoring and protection.
- Develop legal, economic, and regional planning instruments that are consistent with ecological policy initiatives.
- Enhance educational and other social programs that will positively reinforce ecologically sound management decisions and actions.

In Chapter 21, Riordan reviews the progress in scientific understanding that was achieved and the stalemate in political negotiations that took place between Canada and the United States with regard to acid deposition and its effects during the years between 1982 and 1987. Both the scientific progress and the political stalemate that developed during this period illustrate how difficult it can be to achieve both a scientific and a political consensus on what should be done about a complex environmental problem like acid deposition.

Some scientists and industrial and political leaders in both countries were convinced that acid deposition posed a serious threat to sensitive aquatic ecosystems and that significant decreases in emissions of sulfur and nitrogen oxides should be undertaken immediately. Others were equally convinced that there was not an immediate need for decisions and that time was available for additional research to resolve important uncertainties. Still others stood somewhere between these extremes. The end result was a standoff which may be resolved by future changes in scientific consensus and/or industrial and political leadership in one country or the other. Whatever the future may hold, Riordan's analysis of the scientific and political issues provides a useful case study for future reference.

In Chapter 22, Kassenberg provides a detailed diagnosis of the current ecological crisis in Poland. His analysis begins with a presentation of monitoring data on air quality, water quality, changes in the condition of land, and the status of living resources in forests and landscape parks. It continues with a description of the principal factors which have led to changes which, in aggregate economic terms, represent more than 10 percent of the national economic product of Poland. He also describes a four-zone environmental classification system for Poland and outlines the environmental policies required for protection or restoration of each area. These four zones include:

Zone I: characterized by many contiguous or nearly contiguous areas of high ecological hazard; requires intensive efforts for restoration of living conditions and economic activities;

Zone II: includes two heavily polluted coastal areas; requires a significant decrease in human population density and intensity of use;

Zone III: several ecological hazard areas are dispersed; requires regional development that extends across natural barriers; and

Zone IV: relatively free of pollution problems; requires implementation of the principles of ecodevelopment.

In Chapter 23, Juda and Budzinski discuss the linkages between energy use and environmental problems. In Poland, as in many other countries, energy use is coupled with environmental issues especially where coal serves as the primary source of energy for generation of electricity. To understand the complexity of these relationships and interdependencies on a country-wide basis, complex system models are needed which incorporate both ecological and economic factors. Juda and Budzinski describe some aspects of these models which are predicated on optimum consumption. They outline a number of opportunities for application of these models and describe the sequential components of an overall systems analysis for Poland.

SYNTHESIS AND INTEGRATION

The book ends with a summary chapter by Auerbach and Maki. In Chapter 24, they have put forward a conceptual framework for a science-based environmental policy and restoration program for pollution-impacted ecosystems. This conceptual framework includes many ideas that are developed in the earlier chapters of this book.

The principal elements of this conceptual framework include the following steps in a continuous system of ecological risk, ecological assessment, and ecological management activities:

Step 1: Monitoring programs that provide objective data for analysis of current conditions and trends in the condition of ecosystems;

Step 2: Ecological risk assessments that define the scope of existing problems within ecosystems that are now being impacted or may be impacted in the future;

Step 3: Development of science-based policy recommendations that will lead to improvement in the condition of the ecosystems shown to be impacted or at risk in Steps 1 and 2 (above);

Step 4: Definition of environmental limits for continuing economic development and initiation of public education about these environmental limitations;

Step 5: Implementation of technologies that will allow for sustained use and enjoyment of the ecosystems that are at risk or of control technologies that will allow presently damaged ecosystems to recover;

Step 6: Reassessment and redefinition of needs for environmental monitoring in the light of experience which feeds back to Step 1 once again.

Environmental Management Concepts

Evaluating Ecological Impacts: A Conceptual Framework

MILTON RUSSELL
University of Tennessee
and
Oak Ridge National Laboratory

Editors' Note: This chapter differs from most of the others in this book. It is, in essence, a personal essay prepared at our request by the author. It presents a particular perspective on the evaluation of ecological impacts. We found it very stimulating and hope our readers will find it equally so.

A central thesis of ecology is that everything is connected to everything else; thus, any perturbation of an ecological system will/can have effects that alter conditions for life of all other components of the system, and also redound upon the perturbing event itself. Some of these linkages may be strong; others weak. Some may attenuate quickly over space and time, as do ripples in a pool; others may amplify and spread, as does the triggering event for an avalanche.

Human actions are among those perturbing events, and one of the reasons for the great interest in ecology is that it can provide useful insights into the fuller implications of human actions. Then, armed with greater understanding of these effects, decision makers can adjust behavior accordingly. They can seek to ameliorate or mitigate harmful effects or augment desirable ones. Note, however, that any influence on decisions of further insights on prospective outcomes depends not only on their size and nature. Outcomes will be serious or trivial, beneficial or harmful, or may just "be," depending on the *values* brought to the judgment, and on *who* is doing the judging.

The purpose of this essay is to explore a basis on which these judgments might be made. The range of potential views is large—in Chapter 4, Marek and Kassenberg provide a formulation very different from that presented below. The literature is also enormous, composed as it is of much of what has been written on the place of man in nature. This essay surveys neither that range of views nor that literature, but instead seeks to provide a sketch of an approach to valuing ecological change from a perspective that places humans at the center. Many may disagree with this formulation or find it too narrow or, at best, incomplete. However, this view deserves consideration if only because the humans who by action or inaction may alter the course of future events are the persons living now, and because the presumption that they seek to improve their welfare (defined broadly) rather than harm it seems reasonable.

In the human-centered view explored in this essay, the ecological system and its natural components are treated as factors of production and elements in consumption, and are not presumed to possess value in and of themselves. Of course, humans may value natural systems for their own sake—witness the affection for wilderness in the United Sates—and that fits well into the human-centered view. This essay knowingly avoids the more profound discussion of mankind as endogenous to natural systems and adopts an admittedly artificial division. Changes in the ecological system are judged by whether or not they add to or detract from the quality of life for the human population affected. Saying this, however, does not make decision processes much more straightforward. Key questions in valuing remain, some of which are identified below.

A TYPOLOGY OF CONSEQUENCES OF CHANGE

The discussion that follows starts by postulating an existing ecological situation, and then examines the major consequences to consider in deciding whether or not to change it. Such change could be of different sorts. For example, it could be to relieve existing stress on an ecological system by decreasing the amount of pollution that is affecting negatively some of its components. It could be to change relative species abundance through, for example, cultivating grasslands or logging forests. It could be to change the extent of human impact by enhancing access to a secluded area with a new road or, alternatively, by restricting it. It could be to dam a free-flowing river, or to drain and fill a swamp, or to convert wetlands to dry. Whatever the change, it will affect the ecological system and, in turn, those effects will impact humans in various ways.

The first way in which humans will be affected is through changes in the magnitude of economic contribution from the ecological system. That economic contribution can take many forms. One form is the yield of materials desired by humans; these materials are counted as part of the Gross National Product (GNP). Forests provide timber and habitat for animals and plants of direct use to humans. Rivers, lakes, and oceans provide fish and birds that are important to diets. Grasslands provide grazing and hay for domestic animals and habitat for useful wild ones.

Cultivated acreage is also part of the ecological system. Shifts in its condition that affect output—for example, due to pollution or to a change in population of beneficial insects—flow immediately to the GNP.

Indirect economic effects are also the consequence of changes affecting the ecological system. Change in cultivation patterns or amount or type of forest cover affect water flows with possible consequences of flooding, for example. These changed hydrologic patterns may require expenditures to offset an undesired effect, or they may have direct economic consequences through destruction of property. Wetlands have been shown to be efficient sinks for pollutants of various kinds, and their diminution may lead to declines in productivity of water bodies or to the need for capital expenditures to treat effluents or to prevent non-point pollution. Such indirect economic effects may or may not flow through the GNP, but in principle they are quantifiable and definable as factors in the overall output of the economy.

Other factors are not as easily evaluated in the same way. These are the indirect amenity values of the environment which add or detract from human satisfaction. An obvious example is the value of the recreational opportunities presented by forests, streams, and oceans. Still less direct, but no less real, are the benefits received by people in viewing a pleasing natural vista, in seeing or listening to wild birds, or in viewing animals in the wild.

Another effect felt by humans which must be considered here is the "existence value" which components of natural systems may have to those who never experience them directly. For example, many people would feel a deep sense of loss to hear that the giant panda of China or the elephant of Africa had been rendered extinct in the wild, even if they had never seen these animals there, nor expected to. Similarly, the non-fisherman may mourn the loss of a species from a river, or rejoice when salmon return to streams where they had been absent for a long time. This sense of responsibility and stewardship goes deep. It is not bounded by direct benefits or measured by an accountant's economic rationality. These values are difficult to measure as compared to, say, changes in the output of a fishery, but they have substantial importance where humans can express their interests in ecological outcomes.

Ecological change caused by human activities characteristically leads to less diversity in systems, and hence to greater vulnerability in the presence of shocks. Thus, monoculture increases the possibility of disastrous results from aberrant weather. Water control systems designed for 100-year floods can be overcome by 1,000-year episodes, with catastrophic results. Narrowed gene pools can leave important plant or animal populations susceptible to new disease. These risks are components of the human interest in ecological change which present misgivings, even when they are sufficiently vague as to defy explicit consideration.

Another risk factor is uncertainty, which comes in two forms. The first is lack of knowledge of what the second and higher order effects of a change will be, and how they will affect the human condition. The ecological havoc wrought by introduced plant and animal species are well known: rabbits in Australia; deer in New Zealand; the African bee in Latin America; the gypsy moth, English sparrow, and kudzu in the United States, to name a few. The harmful downstream effects of the Aswan Dam have been widely reported; some desertification in Africa and elsewhere has been traced to initially beneficial changes in agriculture practices. We have a tendency to remember the bad surprises and to ignore those which turned out well; for example, one can scarcely imagine agriculture without the massive exchange of plants and animals which has taken place. However, the point is that by definition there will be unforeseen effects, since those which are understood can be factored into a decision. Therefore, there will be an unquantifiable residuum of impacts for good or ill from any perturbation of an ecological system, and that fact alone offers reason for caution.

The second form of uncertainty is in how affected factors will be valued in the future and under changed conditions. The point is obvious when it comes to plant and animal species rendered extinct; their potential uses will never be realized. This form of uncertainty also works the other way. Some dams built to produce electric power and to foster navigation in the United States now have lake recreation as their most important contribution to human welfare. Such unexpected results are not due to the failure to comprehend how the ecological system would respond to perturbation, but also due to changes in how much humans care about what happens.

ISSUES IN EVALUATING CHANGES

Consequences of ecosystem changes must be considered by decision makers when making social choices. How might the effect on "the quality of life for the affected human population" be determined, since it is often the presumed basis for the decision?

Three issues come immediately to mind. The first is in the dimension of space—what is the scope of the human community whose well being is to be considered? The second is in the dimension of time—what consideration is to be given to those who will live in the future? The third is in whose wishes count—how will different wishes be considered in reaching a social judgment?

Take first the dimension of space. Ecological, economic, and social effects of a change may all attenuate rather quickly from the point of impact. When they do, the community affected is readily circumscribed and effects upon it relatively easy to discover and describe. For example, a single, small river may be affected by acid mine drainage. Decisions may still be controversial because their distributional effects are important, but the problem is at least not unmanageable due to failure of political and economic institutions to encompass all players.

In other cases, all of the meaningful effects of a change may reside within the boundaries of a nation-state. The scope may be large and the facts less clear, but in principle the institutions would be in place to consider the interests of competing parties. In yet other circumstances, however, significant effects cross national boundaries. The harmful effects of acid rain are one example; the beneficial effects of actions to preserve tropical forests or to protect symbolically important animals such as the wild elephant are others.

When ecological boundaries cross political ones, the potential lack of coincidence between the community with the ability to decide and the community that bears a share of the impact is clear. The human-centered, maximizing view would suggest that only those who share in both the origin *and* the impact of the change would have their wishes fully reflected, although those wishes could include a measure of altruism. However, this does not doom those outside to having no possible voice in the final outcome. If the issue is important enough, they can either attempt to achieve an agreement more to their liking through recourse to a superordinate power, as with an international treaty (e.g., the Montreal Protocol on Ozone Depletion), or seek to negotiate side payments or to offer threats that cause interests to coincide with effects (e.g., the proposed debt-for-forest swaps between parties in developed and in developing countries). In either case, interests are brought together albeit imperfectly because of the expense and difficulty of making bargains across institutional divisions. In doing so, evaluations of ecological effects by all those affected in the present are elicited, and in some way are the basis for action.

EVALUATING FUTURE CONSEQUENCES

The ecological effects of a human perturbation run forward in time as well as outward in space. They may attenuate readily: the fishery in the James River of Virginia recovered in a few years when paper and pulp pollution was controlled; and in just 50 years gross evidence of human habitation has disappeared in the Great Smoky Mountains National Park. On the other hand, effects may be as permanent as the extinction of a species or as near-irreversible as the desertification of portions of northwestern China, which occurred some 1,000 years ago. The evaluation of ecological change requires a decision as to whether persons in the near or distant future are to have their interests considered; if so, what those interests are; and finally, how those interests are going to be accommodated.

Whether (and how) to account for the wishes of future generations of humans is one of the most difficult conceptual problems in making decisions regarding environmental protection. At the limits the answer is easy: human society *does* value contributions to the welfare of future generations; but it *does not* care enough to embrace risk of its own suicide to contribute to that welfare. Away from these limits, however, no simple or absolute answers hold. And it is away from these limits that most important environmental questions exist.

With regard to the second question of identifying the interests of future generations, the human-centered view suggests that they will be something like those of present generations, i.e., there will be concern both for narrow economic output and for the broader values of natural systems as described above. Consequently, one objective should be to endow subsequent generations with a combination of the undifferentiated ability to satisfy economic wants with actions taken to limit the diminution of the diversity of ecological systems.

In achieving this, one approach is to assess future outcomes through the use of some positive discount rate. There are several possible value bases for applying this approach, but space permits discussion of only one. That basis would purport to leave an endowment no less rich than the one currently enjoyed. In that sense, future generations would be treated as if their wants had equal standing with present generations.

One predicate of this approach is that resources not consumed in the present will accumulate at a positive rate to be the basis for future consumption, including environmental protection. From this it follows that it is neither necessary nor wise to invest as much today in preventing future harm as the projected cost of that harm to future generations. It is not *necessary* in meeting the equal endowment goal because, assuming the correct discount rate is chosen, future generations will be at least as rich as those living today due to the return from investment taking place in the present. They will be able to afford to take care of their problems as well as we can. It is similarly not *wise*.

If investment in preventing future harm is not placed on the same discounted basis, as is investment in producing other goods and services, resources will be allocated inefficiently and total output will suffer. Future generations—as they judge it—will be *worse off* than they need to be. This is because by depleting resources to overcorrect problems now, we deny future generations the resources that could do the job at less sacrifice to them. This proposition has much to be said for it. In its most general form, it undergirds consumption and investment decisions of all modern economies. When it comes to ecological protection, however, caution should be exercised in its application.

One critical assumption inherent in this approach is that resources are fungible, so that future wants can be satisfied by whatever set of resources exist then, as long as enough of them are available. This assumption is brought into question when it comes to ecological protection because of the irreversible nature of some changes. Extinct species cannot be brought back, however rich in other assets a future society may be. Their loss may be seen as an irremediable decrease in future welfare. On the other hand, it might be deemed presumptuous for us, today, to determine the *pattern* of consumption of those who live in the future. While they would thank us for leaving them "better off" in the abstract, their values may be such that they would not prefer what we prefer, given our life experiences. Some humility may be in order, for example, before we decide to save some species at high cost in future production ostensibly for the benefit of *future* generations. The fact that those in the future are our wards (in that they have a vital interest in decisions they can't affect) means that we bear an obligation to them. But it is not obvious that fulfilling that obligation requires handing on a specific endowment.

Another assumption is that capital will continue to accumulate, or else that any future cataclysmic change would affect all capital equally. Human history has already provided evidence of wars and other events which have driven civilization backwards to greater reliance on natural systems, to a point from which progress could start anew. It could happen again. Perturbation of these systems to the detriment of their natural productivity may be of little consequence when man-made capital and technology are present to replace their output, but may be of an entirely different order of effect if the continuity of society is disrupted. For example, aquaculture may make natural fisheries unnecessary, but regrowth of a future civilization may be seriously handicapped if natural fisheries have been destroyed by pollution. Or, domestic hybrid animals and plants that are highly productive with intensive care may be unable to survive altered circumstances, and if native species are not available, there will be nothing on which people then living can depend.

These elements are joined by the fact that the time scale for ecological change is so long that decisions taken today affect not just the next generation but potentially affect generations far into the future. With any positive discount rate, the implication is that any actions today are of little importance to the well-being of distant generations. While there may be something to be said for this view in some circumstances, when stated baldly it gives many people pause, especially when it comes to ecological change.

A reasonable conclusion to be drawn from all this is that discounting is useful when assessing future effects, but only as a starting point. It must be used with caution and with full understanding of its assumptions and implications when long-term ecological change is at issue.

ECOLOGICAL VALUES AND OTHER VALUES

This essay thus far has treated ecological changes and their effects on human welfare in isolation from the motivation that brings them about. That motivation includes increased economic production and resulting social change.

In some cases, ecological perturbation is the direct and intended effect of actions to increase production. Dams are built to produce electric power and to prevent floods, and in the process drown valleys and change the flow of rivers. Swamps are drained and forests cleared to increase croplands. Noxious insects and animals are controlled to prevent disease and to lessen damage to economically valuable foodstuffs and materials. New species are introduced to yield higher returns.

In other cases, ecological perturbation is the unintended consequence of actions pursued to increase production. Acid water is pumped from coal mines to free deeper seams for exploitation, but it changes the character of streams. Effluents from factories producing goods desired by people introduce chemicals into water bodies which can destroy living organisms or change their relative abundance. Waste heat from power plants changes the biota in the receiving water body, as does the effluent from domestic sewage treatment plants. Air emissions from coal combustion changes the pH of rain. Indeed, virtually all production and consumption has an unintended effect on surrounding ecosystems, sometimes positive but often negative.

To avoid such effects, the activity must be eliminated or modified, and this has the first-order effect of decreasing the output of the economic goods sought. If the good is not produced or consumed, the resources devoted to it are available for transfer to an alternative use, but will produce output which by definition has lesser value or it would have been chosen in the first place. Alternately, the resources consumed in building and operating pollution control devices are not available to produce other desired goods and services.

In short, avoiding or fostering ecological change has costs. The question in a human-centered value system is whether those costs are worth bearing. The easy answer is that it depends—it depends on whether the total welfare is enhanced or reduced. This formulation gives the consequences of ecological change exactly the same standing as the benefits received from direct economic production. Trade-offs among alternative ways of meeting human wants can be made and indeed must be made.

Several cases may be distinguished when avoiding or fostering ecological change is contemplated. In some cases, the overall magnitude of economic output, even narrowly conceived, is increased. For example, the cost of reducing an effluent may be less than the added return from an enhanced fishery. Different persons gain and lose, but the total economic product is increased. In other cases, the losses in output are more than made up by ecological contributions to health or welfare of the sorts described above as indirect economic impacts. In still other cases, the loss of direct output is more than offset when existence value, uncertainty, or future effects are considered. There is, of course, also the class of cases where the cost of reducing an environmental insult is greater than the benefits derived, even when the broadest definition of benefits is used. In this class of cases, no control action is justified when tested against human-centered criteria.

Such a formulation of the structure of decisions gives rise to two problems. The first is identification of the boundaries within which the well-being of inhabitants are to be considered, a matter already discussed. Does this group only consist of persons within some political boundary and their heirs, or are the wishes of affected others to be considered? On the output side, the issue is similar.

The second problem has to do with the distribution of gains and losses. Only rarely will these distributions be coincident; therefore, either action or inaction will leave some better off and others worse off, notwithstanding the question of whether the change in overall outcome is positive or negative. The fact that compensation of the losers by the winners may be possible does not, in the view of some, alter the case as long as all individuals themselves do not judge the outcome as satisfactory.

How are these diverse views to be taken into account? And who decides? This intriguing and daunting question illuminates the complexity a social system faces in determining how to evaluate ecological impacts, and then in choosing what to do.

This discussion has proceeded along only two dimensions—ecological impacts and the sum of direct and indirect economic benefits, including such matters as existence values. However, the total decision matrix is far richer and other elements, including but not limited to distributional concerns, must be factored into any decision.

FROM CONCEPT TO PRACTICE

This essay has sought to describe a conceptual framework for how the products of ecological science might be used in making decisions that affect natural systems. The value system used on this essay puts human preferences at the center.

This essay does not consider the difficulty of obtaining data necessary to make wise decisions. These data are almost always expensive to acquire, will remain incomplete, and will be subject to change. Hence, any decision will have an unknown and *unknowable* outcome. This fact suggests that a premium should be placed on a decision that is robust in the presence of error, so that the possibility of irreversible damage is lessened.

The framework presented has also leaped over the problem of identifying human preferences, i.e., that set of outcomes which would indeed make people better off. The viewpoint is that of naive utilitarianism, with a benevolent entity seeking to maximize the sum of human welfare within a political boundary. That entity takes account of such matters as the distribution of benefits and costs. It takes an altruistic attitude towards those separated by space and time, but from the viewpoint of an outsider, not a participant.

Actually, the difficulty in doing this through any sort of political process is great enough when a circumscribed population is considered. It becomes greater still when small effects on large numbers of far-distant persons are considered. Thus, when concern for the values of future generations are taken into account, the matter becomes, strictly speaking, unknowable. Even when objective outcomes are assumed to be known, how they are valued by people, much less how to aggregate those values, remains a serious problem.

The point is that the results of ecological understanding can inform decisions, but such understanding alone cannot dictate them. Ultimately, those decisions rest on the values of the persons making them. Ecology's job is to make sure that the full consequences of alternative decisions are illuminated. Some of the ways in which such knowledge could be used, and in fact is essential, have been outlined here. It is the premise of this essay that the task of making ecological insights available to decision makers is of service whatever the value system used for final decisions.

RELATED READINGS

Brundtland, G.H. 1987. Our common future. Report of the World Commission on Environment and Development. Oxford: Oxford University Press.

Dubos, René. 1980. The wooing of earth. New York: Charles Scribner's Sons.

Hardin, G., and J. Baden, eds. 1977. Managing the commons. San Francisco: W.H. Freeman and Co.

MacLean, D., ed. 1986. Values at risk. Totowa, New Jersey: Rowman and Allenheld.

Partridge, E., ed. 1981. Responsibilities to future generations: Environmental ethics. Buffalo, New York: Prometheus Books.

Sagoff, M. 1988. The economy of the earth: Philosophy, law, and the environment. Cambridge: Cambridge University Press.

Stokey, E., and R.Z. Larsen. 1978. A primer for policy analysis. New York: W.W. Norton.

The Relationship Between Strategies of Social Development and Environmental Protection

MICHAL J. MAREK
Institute of Public Administration and Management
Warsaw

ANDRZEJ T. KASSENBERG
Institute of Geography and Spatial Organization
Polish Academy of Sciences

Different concepts of social development will be described in this chapter to demonstrate how they influence approaches to problems of the natural environment. These concepts are described here as existing or potential dominant paradigms, i.e., basic perceptions, thoughts, and actions related to a given view of reality (Harman, 1975; Linstone et al., 1977; Wyka, 1987). In such considerations, alternative thinking proves useful (Suchodolski, 1981; Picht, 1981). Within this context, different aspects of environmental protection in Poland will also be discussed.

Such a discussion might allow us:

- to see the present form of development and environmental protection as relative, not as absolute;
- to understand not only how to improve the existing solutions but also to determine possible ways of identifying new approaches;
- to show more clearly the relationships between development and concepts of environmental protection;
- to enhance chances for planned evolutionary development instead of a series of crises;
- to accustom ourselves to thinking about the future and to encourage working for its sake;
- to discharge part of our obligations to future generations following the principle of due care (the French concept of *obligation de moyens*) and not of results (*obligation de resultat*) (Domanski, 1972; Grzybowski, 1981);
- to integrate partial solutions currently dispersed in various branches of science;
- to identify barriers to possible adaptations of future societies, the nature of these barriers, and the difficulties associated with them;

- to limit resistance even towards radical change, caused not necessarily by the negation of reasonability of the solution but by the uncertainty about the character of the costs associated with change;
- to create pilot solutions on a regional or local scale which may facilitate the development of spatial organization on a wider scale (i.e., a country or a group of countries);
- to introduce new, important modifications to solutions which already exist.

The ideas presented in this chapter pertain to various forms of progress. However, within the framework of this brief account we did not attempt:

- to outline alternative forms of progress, i.e., economic, technological, organizational, and scientific;
- to apply the holistic approach (Boulding, 1970; Sahal, 1977; Brown, 1981);
- to relate our discussion to the reality of those countries which greatly influence the ecological status of the world (e.g., Brazil, the United States, Canada, and the USSR);
- to present analysis concerning the feasibility of particular concepts and the scenarios leading to them;
- to present a vision of a world disaster as equally possible as a vision of progress.

The analysis which follows merely signals some issues of future development and environmental protection.

THE EXISITING PARADIGM

Authors dealing with the idea of social progress indicate that the time in which it serves as a dominant paradigm is short. Dawson (1938) emphasizes that each period in the history of civilization holds characteristic ideas as its particular property. These ideas were considered not merely as popular ideas of the time but as eternal truths and thus understandable, regardless of the method of reasoning applied.

In the 18th century, the idea of progress started to dominate the social awareness in many countries of Europe (Mounier, 1948). In spite of a widespread following, the idea of progress remains very equivocal; thus, efforts of researchers have concentrated on identifying common elements in diverse ideas of the meaning of progress. Most of all, it is anthropocentric—oriented exclusively towards humans, with its broadly understood goal the prosperity of *Homo sapiens*. Moreover, as pointed out by J. Dalvaille (1910), progress encompasses the concept of *improvement* (i.e., what was in the past is poorer than achievements of the future); the concept of *novelty* (i.e., all that is old will be replaced by something different and better); and the concept of *permanent growth and accumulation* (i.e., the constant gathering of material property). This view of economic growth has been very popular in modern societies since the 18th century. Studies by S. Kuznets (1971) clearly indicate that since several countries entered the stage of modern development, GNP has increased by an average of 3% per annum while the gross product per capita has increased by an average of 2% per annum. This means a five-fold increase in gross product per capita and at least a similar increment in GNP during one century. Such data support the classification of the last one and one-half to two centuries as a new economic era.

Economic progress is closely linked with developments in science, technology, and organization. Some scientists maintain that if there had been no development in technology and if progress had been based on multiplying existing solutions, the most developed countries of the world probably would have achieved only about 20% of the rate of growth observed so far (Beranek, 1978).

For several decades, human activity was based on the assumption of an unlimited supply of natural resources (Krader, 1970). There has, of course, always been an understanding in economics and in everyday life that resources are scarce in the short term and that allocation or rationing of those resources is often necessary. The difference lies in the perception of the longterm. Historically, the world was viewed as a "bundle of hay," i.e., once its resources were used, they could not be replenished (Jonston, 1960). Ultimately, the world's resources would diminish and progressive poverty would ensue. Approximately 50 years ago, the argument was developed that resources could expand with greater knowledge and technology and thus the world would become a "field of grass" that would be able to continuously replenish itself (Zimmerman, 1951; Barnet and Morse, 1963). More recently, beginning in the late 1960s and popularized by the Club of Rome, the opinion that there might actually be limits to the world's natural resources, even in the long term, has come full circle (Meadows et al., 1972; Russell, 1988).

The brief, general description presented above ignores the distinct differences between the specific patterns of development existing in particular countries and regions of the world (Kerr et al., 1964). These specifics are crucial to better understanding the relationship between social development and environmental protection. For example, this chapter will focus on the situation existing in Poland, where the domestic economy has proven to be very inefficient. This has been attributed to inherent characteristics within the Polish system, as is widely recognized now. Through the mid-1980s, a supreme role was granted to state-owned property. The creation

TABLE 1 Comparison of environmental situation in Poland and six COMECON countries (excluding the USSR) in the mid-1980s (expressed as percent of Europe).

Issues	Poland	COMECON (6)
Population	7	23
Territory	6	20
GNP	3	12
Consumption	8	27
Emission of SO_2 (a)	11	38
Emission of solid particulate (b)	26	53

(a) The smallest European countries and Romania were excluded because of the lack of reliable data. Therefore, emissions in Poland are slightly lower than reported here.

(b) Fifteen countries were included: Austria, Czechoslovakia, Finland, France, Greece, Spain, Holland, Hungary, Ireland, Norway, Poland, Portugal, FRG, Switzerland, and Sweden. Therefore, emissions in Poland are lower by several percent than reported here.

SOURCE: T. Zylicz (in press)

TABLE 2 Comparative data on anti-pollution investments expressed as percent of GNP.

Year	Poland	Holland	USA	FRG	Bulgaria	Hungary
1975	0.4	0.1	0.4	0.3	0.3	--
1976	0.4	--	0.4	0.2	0.3	0.4
1977	0.5	0.2	0.4	0.2	0.5	0.5
1978	0.4	0.1	0.3	0.2	0.6	--
1979	0.4	0.1	0.3	0.2	0.7	0.8
1980	0.3	0.1	0.3	0.2	0.7	0.6
1981	0.2	0.1	0.3	0.2	0.6	0.6
1982	0.2	0.2	0.2	--	0.6	0.5
1983	0.4	--	0.2	--	0.6	--

SOURCE: A. Budzikowski, 1988.

of an enterprise, its size, and specialization were the result of administrative decisions by the central government or a particular minister. In this system, bankruptcy could not exist. The country's economy was organizationally dominated by state-owned enterprises (particularly monopolistic enterprises); the state also maintained a monopoly in foreign trade, resulting in the isolation of the domestic market from competitive foreign products.

Poland also had a centrally planned economy, with the economic goals of the state-owned enterprises established by the central planner. The central planner also specified ways in which goals should be achieved (e.g., determining prices, wages, salaries, etc.). The primary concern of managers of state-owned enterprises was not profit but to achieve the goals established in the government's economic plan. Under these conditions, no managerial system (including environmental) could be very effective.

The resultant inefficiency of the Polish economic system has meant:

- *A long-standing central government policy to develop heavy industry.* Currently, 41.1% of all industrial capital is invested in industry (Statistical, 1986). Simultaneously, outdated technologies dominate, and coal is used as virtually the only source of energy. This results in a very high level of pollution, as is illustrated in Table 1.
- *An insufficient amount of financial resources allocated for environmental protection.* The share in GNP of all anti-pollution investments is equal or even higher in Poland to that of developed countries (Table 2). However, the efficacy of the investment process (measured by comparison of effects to outlays) is at least two times lower than that in France or West Germany. In addition, the anti-pollution equipment that is installed is often outdated and of poor quality. The negative impacts of this situation on the environment are presented in other chapters of this book.
- *An ever-increasing use of raw materials and energy in all sectors of the economy.* To measure the consumption of raw materials, Sitnicki (1986) compared the net output of material production in Poland to its material costs. The results produced are the following: 1.3% in 1960, 1.34% in 1978, and 1.41 in 1981%. During the period 1960-1978, consumption of raw materials increased by 3.1% and from 1978-1981 it increased even more, by 5.2%.

Other experts have estimated the losses of raw materials in all phases of the industrial process, from mining minerals to manufacturing final products. According to Ney (1983), these losses are substantial, equaling 38% for copper, 50% for zinc, 55% for lead, 46% for sulfur, 72% for pit-coal, and 78% for brown coal. These data suggest that the utilization of raw materials is extensive in Poland.

In the mid-1970s, the consumption of energy decreased in Poland, but in the period 1979-1982 it increased by 80 million tons of coal equivalent when compared with 1975. This amount of energy is equal to 100 million tons of pit-coal, which is worth 4.5-5 billion U.S. dollars (Albinowski, 1988). According to most estimates, present energy consumption in Poland is 2.5-3 times higher per unit of GNP than in developed countries of the European Economic Community (EEC) (Szpilewicz, 1977; Chandler, 1987).

Poland's inability to transform itself into an important supplier of industrial products to the world market in spite of the existence within its territory of rich supplies of raw materials has shaped Poland's export model to emphasize the export of raw materials (e.g., copper, pit- and brown coal, silver, sulfur, and timber) as well as intermediate products. This puts pressure on the natural environment and limits Poland's economic growth due to a deficit of energy, difficult geological conditions for mining, and the deterioration of forests which are exploited too intensively (Molenda, 1974).

- *The extensive use of land.* The average harvest in Poland of four crops (rye, wheat, barley, and oats) is equal to 0.3 metric tons per hectare; in West Germany it is equal to 0.5 metric tons, even though the agro-ecological conditions are comparable in both countries. This means that Poland is still unable to carry out its own conservation reserve program for crop production on idle cropland which is highly erodible or located adjacent to streams, lakes, or estuaries, as is done elsewhere (Thiede, 1975; Wolcott et al., 1988).

However, in the late 1980s, the Polish economy started to undergo dramatic and rapid changes. In the newly-created system:

- the central planners decide only general trends of development, and the market is recognized as an important regulator of economic life;
- enterprises are steered by economic tools (e.g., taxes, prices, and discount rates) to a much greater extent than before;
- profit is used as a criterion for enterprise evaluation;
- foreign investment is possible;
- bankruptcy of enterprises is possible;
- the first attempts to create a capital market have been observed;
- favorable conditions to create all kinds of small firms exist, thereby granting more economic freedom to individuals;
- regional and local authorities are less dependent on the central government.

Under these new reforms, each domestic managerial system can be much more effective than before. Despite these promising changes, however, the severity of the environmental crisis in Poland, compounded by the weakness of the economy, makes rapid progress in the field of environmental protection impossible. It is apparent that the deterioration of the natural environment in Poland is a limit to growth, but it is also clear that a lack of growth is a limit to environmental protection. It is very difficult to stop such a vicious circle.

Today, developed countries are able to effectively handle more environmental problems than can Poland. Still, substantial changes have been initiated due to the understanding that we live in a world of scarce resources, requiring a search for new approaches to social development and the environment.

A NEW ANTHROPOCENTRIC PARADIGM: HUMAN HEALTH AND QUALITY OF LIFE

One of the attempts to abandon (or rather modify) the dominant paradigm can be found in the concept of the "Social Indicator Movement" which has developed since the late 1960s. Researchers affiliated with this movement maintain that the indices of national income and GNP are not sufficient for analysis of human well-being. They suggest the introduction of systems of measures termed "the level and quality of life" (Rutkowski, 1984). In such considerations, scientists turn their attention to three components of social reality: the ability to satisfy social needs, the degree to which these needs are satisfied (in cases where this can be determined in an objective way), and the assessment by society of satisfying needs (Rutkowski, 1987).

Problems of quality of life in Poland will be presented with a focus on mortality issues, since they are both crucial and well documented. The rate of mortality in Poland is much higher than the rate in developed countries, particularly for the male population (Figure 1). In 1984, the average death rate for men was 1,069 cases per 100,000 people, exceeding the death rate for women by 17%. Substantial differences also exist among particular regions of Poland (Velrose, 1984). In this case the highest level of mortality surpasses the lowest by 32.6% (Szarski, 1985). Fifty percent of all deaths in Poland are caused by respiratory and heart disease, with an additional 25% caused by cancer, accidents, and intoxication (Dzienio, 1984).

Current concepts of quality of life are based on contemporary views of family, health, and public safety. Future concepts will probably be concerned with the same elements but understood in a different way. As pointed out by Rodenstein (1986), in ancient times the science of dietetics was developed to treat health issues in a highly integrated manner—addressing social and biological aspects simultaneously. This strongly echoes the contemporary philosophy of the unity of the world (macrocosm) and man (microcosm). Later, in the development of European culture, this philosophy ceased to exist, resulting in the consideration of human health on a much smaller scale, e.g., in terms of a patient/doctor relationship.

This attitude persists today, allowing effective control of infectious diseases but not chronic ones. The prevention of chronic disease is regarded by specialists as requiring increased control of the individual over his own health. Thus, modern medicine tries to influence human social behavior as well as the social and ecological environment. The environment will be considered healthy if it helps people to develop their various abilities to improve quality of their lives. This new approach demands a new philosophy that would combine the social and purely medical approaches to health into

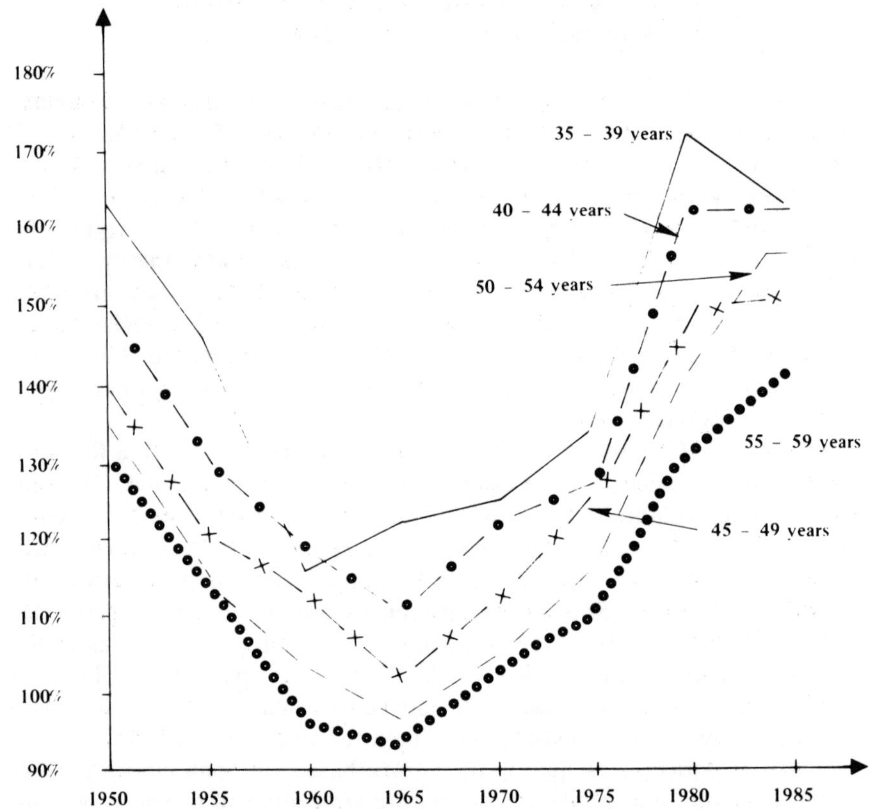

FIGURE 1 Mortality rates of males (age 35-59) in Poland as compared to the Federal Republic of Germany. Data on mortality rates in the Federal Republic of Germany were not available to the authors, therefore 1980 mortality rates were used for both 1980 and 1985 as a base for comparison with Poland.

one entity. Therefore, Rodenstein expects the ultimate development of a holistic theory similar to ancient dietetics.

The health of an individual usually depends strongly on his economic position. This holds true not only in poor countries but in wealthy ones, as well. It has been estimated, for example, that Canadian males from the highest income group could expect an additional 14 years of life free of activity restrictions when compared to males from the lowest income group. For the wealthiest females, the corresponding advantage amounted to an additional 8 years of life (Wilkins, 1983). Similar relationships were found in statistics regarding the frequency of lung and breast cancer (Davessa and Diamonds, 1983; Kelsey and Hildreth, 1983). Therefore, it may be presumed that in the future there will be a continued drive towards elevating the general standard of living, while solutions ensuring each individual good health regardless of his socioeconomic status will also be developed.

Recently, three approaches toward the realization of this goal have emerged. The first pertains to the change in human behavior in the areas of food consumption, use of technical facilities, and ways of spending free time. The second includes community activities like setting up self-help groups and enhancing cooperation within neighborhoods to facilitate common actions to further promote health. The third involves creating solutions in the areas of urban planning, agriculture, and public safety. This trend had a great effect upon the new understanding of environmental protection and management.

In 1975, European urban localities were inhabited by 67%—and in North America by 77%—of their respective total populations. These figures are expected to rise in the year 2025 to 88% and 93%, respectively (Hancock and Duhl, 1986; UN Population Division, 1982). These statistics alone are reason enough to assign fundamental importance to urban solutions in any consideration of quality of life. The previous conception of health as a purely medical issue entails its consideration far removed from many aspects of urban life. The emergence of new concepts, however, has led to changes. The traditional idea of life quality issues subordinated to economic development have also been questioned.

A brief characterization of the new shape of a city, although still not completed, conveys a sense of future. According to Hancock and Duhl (1986):

> A healthy city is one that is continually creating and improving those physical and social environments and expanding those community resources which enable people to mutually support each other in functions of life and in developing their maximum potential.

Within the context of this definition of a city's health, the following parameters merit consideration:

- quality of the physical environment;
- stability and sustainability of an ecosystem;
- strength of a mutually supportive community;
- degree of public participation in decision making;
- meeting public need;
- access to a wide variety of experiences and resources;
- a diverse, vital, and innovative city economy;
- an optimum level of appropriate public health care; and
- high health standards.

These considerations show how important to this new concept of health is the issue of urban populations living in a stable natural environment. However, it is difficult to determine future solutions of this kind, as there are currently only assumptions based on the most recent theoretical and practical achievements.

For example, during the early 1980s a building was erected in Vienna according to the principle of alternative accommodation (*die Alternative-wohnungen*) with the support of the municipal authorities. This design was created not by an architect but by a painter who is well known under the pseudonym Hundertwasser. This building, in addition to walls, rooms, and roofs of surprising shapes, contains playrooms for children and meeting rooms for adults. Thus, facilities provide for further integration of inhabitants, which is very important in the new concept of health. Grass grows on the steep parts of roofs, while flat roofs have summer gardens with trees and shrubs which can also be found in niches in walls. In addition, a winter garden is located inside the building (Otulak, personal communication). The vegetation mentioned here not only improves conditions for recreation but may also contribute to an improvement in urban climate after introducing such designs on a wider scale (Ryczywolska, 1987), as well as facilitate the stabilization and integration of intra-city systems of green areas (e.g., parks). The main element of such an environmental stabilization in urban areas would be provided by a few large protected areas of natural park, which would probably be situated on the edges of cities and be of high biological, aesthetic, and recreational value.

THE ECOLOGICAL PARADIGM

The discussion thus far has provided a brief outline of some features of alternative development and the spatial solutions resulting from it. However, some other very important features were omitted, e.g., a new view of agriculture and the relationship between current activities and the welfare of future generations. Nevertheless, the above considerations help to develop a potentially new concept of preservation and management of the environment.

In the alternative development presented above, only anthropocentric solutions are considered, i.e., those made solely with an intention to satisfy human needs. A different approach emerges based on vitacentric ideas. Both in Europe and the United States, a universal interpretation of holy writings states that human nature dominates over any other form of biological life (White, 1967; Zdziechowski, 1928; Nowosielski, 1982). The realization of vitacentric ideas calls for harmonizing human needs with those of nature rather than denying them. These ideas are supplemented by a new, equally important one: preserving full diversity of wildlife and living systems in large carefully selected areas (Woodwell, 1978; World, 1978; World, 1982). Moreover, the continuity of ecological systems and the adequacy of abiotic conditions should also be ensured (Andrzejewski, 1980).

It is also important to recognize the spatial conflicts that appear between ecological systems and remaining elements of spatial arrangements similar to those occurring among natural, nondegraded systems (Woodwell, 1985). This approach can be envisaged on a pan-European scale, encompassing many countries (e.g., EEC and COMECON), or for individual countries. For most European countries, the latter suggestion is the least attractive as activities carried out in a neighboring country may disturb domestic ecosystems. The system of environmental protection and management outlined here should maintain the following on a large scale:

- ability to produce biomass;
- ecological balance, which is manifested by ecological stability;
- pools of genetic and ecological information;
- a spatial arrangement of ecosystems which would inhibit excessive mobility of chemical elements, waters with dissolved chemicals, and rocks and humus in both air and water (i.e., erosion); and
- complete matter cycling within possibly small area (Andrzejewski, 1984).

In spatial terms, the above requirements may be satisfied by a type of ecological structure in a given area (Kassenberg and Marek, 1986). One of the theoretical proposals of this kind is called an "Ecological System of Preserved Areas," which would involve defining national, ecologically diversified, adjoining areas including various ecological systems (Gacka-Grzesikiewicz, 1977). Such a system would include:

- basic *elements* (e.g., national parks) where biological life has the best chance of survival;
- ecological *areas* that can link them together and serve as a buffer against anthropogenic pressure; and
- ecological *corridors* (at least one kilometer wide) to connect the elements of the system where links of the aforementioned type are not feasible.

Such a system could ensure proper spatial conditions for survival of nonhuman forms of life by allowing them unrestricted penetration contacts, exchange, and complement of genetic pools.

Before it is possible to define the shape of such a system, many parameters have to be assessed for the whole country, such as the degree of preserved abilities to regenerate the ecological systems (i.e., the level and rate for regeneration of biomass, structures, and processes); and the degree of mobility of chemical elements, waters, rock mass, and humus. To date, it has not been possible to cover all of Poland by this kind of study.

TABLE 3 Comparisons between various concepts of environmental protection.

CRITERIA	CONSERVATION CONCEPT	ECONOMIC-TECHNICAL CONCEPT	SOCIO-ECONOMIC CONCEPT	ECOLOGICAL AND SOCIAL CONCEPT
I. Man's attitude towards non-human forms of life and inanimate nature				
PERCEPTION OF NATURE	as an object (primary) or subject-object (primary-secondary)	as an object	as an object	as an subject-object
KEYWORDS OF THINKING	nature formations; specific ecosystems, species, specimens; selected forms of non-living nature (e.g., land forms and geological phenomena	natural resources; economic effectiveness	quality of life; interests of future generations	Ecological System of Protected Areas (ESPA) and its elements
FOCUS OF INTEREST	rarity of occurrence of biological forms; special significance for society (such as cultural, aesthetic, gene banks, etc.)	usefulness of natural resources to satisfy economic needs	satisfying present needs for life, health, security, aesthetic values and providing for development of future generations	ecological balance as a basis for proper conditions for survival of human and non-human forms of life
LIMITS OF PERMISSIBLE CHANGES IN CONDITION OF NATURE	inviolability of functioning unity of nature in chosen areas; preserving single forms of nature	defined by existing economic needs	defined by the needs of future and present generations	inviolability of ecological balance within ESPA

Also, many issues involved in creating an Ecological System of Preserved Areas have not yet been solved by science. Thus, such a system may be presented only theoretically at this time.

In conclusion, the principal differences between alternative anthropocentric and vitacentric concepts of development presented above should be emphasized. According to the former, all nonhuman forms of life can be protected if they are regarded as useful to humans, while in the latter concept these life forms are to be protected regardless of human needs. The anthropocentric concept does not assume creating proper conditions for survival of nonhuman forms of life, while the vitacentric concept makes it an imperative of development.

ENVIRONMENTAL PROTECTION AND MANAGEMENT

So far, we have distinguished two possible ways of thinking about nature: anthropocentric and vitacentric. Within the former approach, the

(TABLE 3, Continued)

CRITERIA	CONSERVATION CONCEPT	ECONOMIC-TECHNICAL CONCEPT	SOCIO-ECONOMIC CONCEPT	ECOLOGICAL AND SOCIAL CONCEPT
II. Satisfaction of human needs vs. development of natural environment				
PRIORITIES IN SATISFYING NEEDS	needs limited to sightseeing, research related to gene bank	satisfaction of needs does not destroy economic balance	satisfaction of needs does not destroy social balance	satisfaction of needs does not destroy ecological balance of ESPA
COSTS OF SATISFYING NEEDS	------	covering costs of present social development with non-human forms of life and future generations of humans	bearing all costs of present social development and some costs of future development	bearing all present social development costs and costs of future human and non-human development
GAINING NATURAL RESOURCES AND USING NATURAL AMENITIES	limited use of natural amenities	complete utilization of existing natural potential for economic and social development	partial utilization of existing natural potential for social development; remaining potential is reserved for the future	maintaining balance of ESPA along with utilization of remaining part of nature for social development
RECULTIVATION OF NATURAL ENVIRONMENT	protection of natural objects from disturbances	recultivation of natural environment when satisfaction of economic needs is threatened	restoration of ecological balance when required by life quality demands or interests of future generations	restoration of balance in ESPA when there are technical possibilities

following concepts are now working or are now being introduced: *conservation* (created by Humboldt in the 19th century), *economic-technological* (a part of the existing paradigm), or *socio-economic* (a part of the new anthropocentric paradigm). The latter approach (vitacentric) gives a basis for the development of a new *ecological and social* concept within the ecological paradigm, which also borrows from the conservationists' ideas. Thus, conservation actually appears in both trends and represents the necessary complement to any of the remaining concepts.

To know more about all of these concepts, we tried to identify both similarities and differences among them using a unified set of criteria. When creating these criteria we assumed that they have to consider two issues: attitudes of humans towards other forms of life and the abiotic environment, and ways of satisfying social needs. The description of all these concepts using these two groups of criteria is given in Table 3 (Kassenberg and

Marek, 1989). As can be seen from this table, a separate principle has been formulated for each concept and criterion.

The first arrow (pointing right) indicates the sequence of conceptual development, with newer ideas incorporating some existing principles. The arrow is drawn as a broken line to emphasize that it indicates only a perceived, but not quite explicit tendency. There is also an arrow pointing left, since future trends might also influence current ones.

Different forms of social development and environmental protection demand different interpretations of acceptable human pressure upon nature. Today, many newly invented substances are introduced into the market. This is, of course, indispensable to stimulate economic development and to better satisfy social needs. Also, environmental protection and resource conservation often benefit from new synthetic materials which, among other advantages, allow many enterprises to earn greater profits than their competitors.

In Poland the producers' market prevails over that of consumers', which means that there is no need to invent new synthetics to gain high profit. In spite of this fact, many new chemicals are introduced to the Polish market each year, because firms are often compelled to use new compounds due to shortages in supplies necessary for production. As a result, for instance, substances used for domestic dishwashing could legally be made of more than 335 chemicals (Kiss, 1986).

Use of new substances may bring advantages to society, but it also creates many hazards for humans and the environment. Today two strategies are generally applied in the field of social safety: to *not* test or regulate chemicals at all, or to test chemicals and regulate *only* if the test reveals serious adverse effects (Weinstein, 1979).

In the past, the first *laissez faire* approach dominated due to lack of proper knowledge. Currently, this strategy persisits in those countries where governments do not have sufficient resources to organize proper systems for societal protection, or where the domestic industry effectively blocks all governmental activities in this area. In all developed (including socialist) countries, the second strategy is obligatory, but its implementation cannot do much to overcome many difficulties. First of all, new man-made substances are very numerous. In the 1970s, specialists estimated that the number of chemicals produced in the world had reached nearly half a million. Moreover, thousands of new compounds have been added each year according to UN estimates (Witoszynski, 1977). Many of these chemicals can be mixed; and the mixtures that are created can be more dangerous than their components due to a synergistic effect. The abundance of these substances and their compositions limits application of the testing/regulation strategy, since studies of even one compound are expensive, time-consuming, demand sophisticated equipment and well-trained specialists, and still produce inconclusive results (Russell and Gruber, 1987; Oser, 1977). There is also evidence that tested chemicals, though not dangerous for animals exposed to them, still could cause adverse effects to humans. Moreover, even in the case of proven carcinogenicity of a synthetic, its use is often continued, e.g., asbestos and vinyl chloride in Poland, and chlordane in the United States (U.S. News, 1987).

Some experts maintain that existing safety strategies are very inefficient, and suggest establishing new ones based on the following principles:

Act toward maximal risk elimination, conveying a new concept of safety as a situation without risk. Safety has been defined as "a situation with negligible risk which is experienced by the population in the same way" (Eggink, 1980). The rationale for accepting a new approach to risk could be as follows:

> Since it is difficult to determine the "safe" level of exposure to any suspect chemical with any degree of confidence, the concept of "acceptable risk" and "risk-benefit analysis" are emerging in counterpoint to the issue of "risk free [life]. . ." (Schottenfeld and Haas, 1978).

Implement a new strategy of regulation of chemicals without testing, to complement current testing/regulation strategies (Weinstein, 1979). This new strategy could:

- minimize the number of newly invented chemicals in production and commercial use;
- minimize the number of chemicals used in social and economic activities;
- give priority to naturally occuring, as opposed to synthetic, chemicals (Commoner, 1971):

> The third law of ecology suggests that the artificial introduction of an organic compound that does not occur in nature, but is man-made and is nevertheless active in a living system, is very likely to be harmful (Dubos, 1973);

- give priority to the production of solid substances over liquids and liquids over gases; and
- minimize the spatial scope and time of their presence in the environment (Marek, 1987; Kassenberg and Marek, 1989).

Such actions are based on the implicit assumption that "it was a mistake to concede to chemicals the constitutional right to be judged innocent until proven guilty" (Evans, 1987).

This alternative strategy could be effective if new forms of social pressure upon industry and government were exerted. Industry would positively respond to the mass economic demand of the market for "new safety." The government would effectively act if, for instance, mass migrations of people from the territories of low ecological standards occurred. Applications of the new strategy should not limit fulfilling social needs, but rather help to find new agreement between humans and nature. The approach could encourage new technological innovations which would make it possible, for example, to wash dishes or clothes in mechanical devices that do not require the use of any chemicals. In fact, several pieces of such new equipment already have been invented (Eljari, 1985; Gornig, 1986; Majer, 1985).

This alternative approach does not necessarily deny the positive role of many new substances for greater economic development and better fulfillment of social needs. It tries rather to reduce significantly the number of chemicals now in use without prior testing. Of course, this alternative approach would be applicable only if many important questions could be satisfactorily answered. For instance: What are the new criteria to exclude use of a given synthetic? Who should pay for damages caused to the producer? What socially adverse effects could result?

CONCLUSION

Several possible long-term trends concerning development and environmental protection have been outlined in this chapter. It would appear that for next several decades Poland will be almost entirely preoccupied with strengthening its economy by lowering natural limits to growth and reversing high mortality rates. For this reason, healthy cities or an ecological grid in Poland will probably remain of low social priority, and therefore will be discussed within a time frame of many years rather than in the immediate future. In developed countries, however, social priorities could be quite different. Large agricultural areas are now abandoned, so that creation of an ecological grid could be a current issue in these countries. Strong economies could also afford creation of healthy cities.

In Poland, of course, scientific priorities will be shaped in accordance with those of the national economy. This does not necessarily mean that studies of the relationship between social development and natural environment, which are not for immediate application, will be neglected. Analyses of this kind can help Polish society to take into account the problems of tomorrow while resolving the problems of today. Also, Poland could be involved in international environmental programs based on European priorities different from its own. Finally, specific interests of concerned specialists will also stimulate other new studies in this field.

Acknowledgement

Through discussion, comments, and translation, a number of people contributed to the final version of this chapter. These include two Americans—Mary R. English and Milton Russell—and four Poles—Zofia

Gadaj, Zygmunt Gadaj, Stanislaw Sitnicki, and Roman Tertil. The authors thank all of these people for their time and help.

REFERENCES

Albinowski, S. 1988. Energy trap of Polish economy, Warsaw. p. 159 (in Polish).

Andrzejewski, R. 1980. Ecophysiography and ecological shaping of a biotic environment of urban areas. Czlowiek i srodowisko 4:15,17 (in Polish).

Andrzejewski, R. 1984. Ecological goals of spatial planning. In: Uwarunkowania srodowiska przyrodniczego w planowaniu przestrzennym. PZITB, Committee on Urbanisation TUP, Conference in Jadwisin, December 11-12. pp. 37,38 (in Polish).

Barnet, J.H., and C. Morse. 1963. Scarcity and growth: The economics of natural resources availability. Resources for the Future Inc. Baltimore: The John Hopkins Press (Polish translation, 1968). pp. 241, 243, 260, 298 (pages indicated are from the Polish translation).

Beranek, W. Jr., and G. Ranis, eds. 1978. Science, technology, and economic development. New York: Praeger Publishers, p. 3.

Boulding, K.E., 1980. Economics as a science. Ch. 2 in J. Semkow. 1980. Ekonomia a ekologia. Warszawa: PWN and McGraw Hill, p. 88.

Brown, L. 1981. Building a sustainable society. New York: W.W. Norton.

Budzikowski, A., M. Lubinski, and M. Rusinski. 1988. Poland and ecological problems of the world. Warsaw: Institite of CRZZ (in Polish).

Chandler, W. 1987. Designing systemable economies. In: State of the World, L. Brown, ed. Worldwatch Institute. New York and London: W.W. Nord and Company. p. 182.

Commoner, B. 1971. The closing circle. Nature, Man, and Technology. Bantam Books, p. 40 (Polish translation, 1974).

Dalvaille, P.J. 1910. Essai sur l'histoire de l'idee de progres jusqu'a la fin du XVIII-e siecle in Social Progress, B. Miszewski. 1977. Warsaw, p. 13,14 (in Polish).

Davessa, S.S., and E.L. Diamond. 1983. Socio-economic and racial differences in lung cancer incidence. American Journal of Epidemiology 118:818-831.

Dawson, C. 1938. Progress and religion: An historical enquiry. London, p. 3 (the Polish translation, 1958).

Domanski, G. 1972. Obligation of due care and of result. Studia Cywilistyczne 20 (in Polish).

Dubos, R. 1968. So human an animal. New York: C. Scribner's Sons, p. 219 (Polish translation, 1973).

Dzienio, K. 1984. Information about demographic situation of Poland in the period 1982-1983. Studia Demograficzne 4:87 (in Polish).

Eggink, E. 1980. Hazard management in Rijnmond. Paper Presented at the Seminar of the International Management Institute, Geneva, Switzerland, p. 6-7.

Eljari, H. 1985. Des ultrasons laveurs sans lessive ni detergent. Science et Vie 814:108,109.

Evans, H. 1987. The Missouri standard. U.S. News and World Report. October, 12:90.

Futurist. 1978. April, p. 127.

Gault, F.D., K.E. Hamilton, R.B. Hoffman, and B.C. McInnis. 1987. The design approach to socio-economic modelling. Futures 1:3-25.

Gornig, D., and J. Olender. 1986. Clotheswashing without use of chemicals (Patent Research, CPBP 04.10.11.01.01.01), Gliwice, msc. (in Polish).

Grzesikiewicz-Gacka, E.W., and Rozycka. 1977. Relationship between areas of high ecological status and urban aglomerations. Warsaw: IKS (in Polish). pp. 42-52.

Grzybowski. G., ed. 1981. System of civil law. vol. III, part 2. Warsaw (in Polish).

Hancock, T., and L.J. Duhl. 1986. Healthy cities: Promoting health in the urban context. Working paper for the Healthy Cities Symposium, Lisbon, Portugal. pp. 20,35,36,42.

Harman, W.W. 1977. The coming transformation. Futurist(February):5.

Johnston, J. 1960. About stability of nature. Warsaw: PWN (in Polish).

Kassenberg, A., and M. Marek. 1986. Ecological problems of spatial economy. Warsaw: PWN (in Polish). pp. 63-67,71.

Kassenberg, A., and M. Marek. In press. Perspectives of eco-development in Ekonomiczne i spoleczne problemy ochrony srodowiska, Ginsbert-Gebert ed. part 2, Ossolineum, Wroclaw (in Polish).

Kelsey, J.L., and N.G. Hildreth. 1983. Breast cancer and gynecologic cancer epidemiology in the Healthy Toronto 2000 Policy Conference, City of Toronto, October 13, 1984. Boca Raton, Fla.: CRC Press.

Kerr, C., J.T. Dunlop, F. Harbison, and C.A. Mayer. 1964. Industrialism and industrial man. New York: Oxford University Press.

Kiss, E., and M. Malanowska. 1986. The list of chemicals used for dish washing in Poland. Warsaw (CPBP 04.10.11.01.01.01.), msc., (in Polish).

Krader, I. 1970. Environmental threat and social organisation. The Annals of the American Academy of Political and Social Science(May) 298:11,12.

Kuznets, S. 1971. Economic growth of nations: Total output and production structure. Cambridge Mass.: The Belknap Press of Harvard, University Press, p. 300 (Polish translation, 1976).

Linstone, H.A., and W.H. Clive-Simonds, eds. 1977. Futures research. New Directions. London: Addison Wesley Publishing Company. p. xv.

Majer, H. 1985. Technology measurement: The functional approach. Technological Forecasting and Social Change 27:344 and footnotes.

Marek, M. 1987. Reality of eco-development. Paper Presented at the Conference of the Polish Academy of Sciences Committee "Czlowiek i srodowisko," Jablonna k. Warszawy, msc. (in Polish).

Meadows, D.H., D.L. Meadows, et. al. 1972. The limits to growth. London: A Potomac Associated Book.

Mlekodaj, S.T., Z. Piasecki, and P. Krasucki. 1986. Mortality of males in Poland in the period 1950-1985. Warsaw, msc. (in Polish).

Molenda, T. 1974. Polish forests in perspective of year 2000 in "Prognoza rozwoju lesnictwa w Polsce w okresie 1976-2030." Warszawa: IBL. pp. 91,92,182,183,185 (in Polish).

Mounier, E. 1948. La petite peur du XXe siecle in Le Christianisme et la notion de progres pp. 86, 87 (Polish translation, 1968).

Ney, R. 1983. Use of raw materials in Poland. Przeglad Techniczny. Polish Academy of Sciences 11:13 (in Polish).

Nowosielski, J. 1982. Symbol and icon. Znak 3:111-117 (in Polish).

Oser, B.L., and R.L. Hall. 1977. Criteria employed by the expert panel of FEMA for the GRAS evaluation of flavoring substances. Food Cosmetic Toxicology 15:464,465.

Picht, G. 1981. Boldness of utopia. Warszawa: PIW (Polish translation of selected papers).

Rodenstein, M. 1986. Health in the city: An historical view. Symposium on Healthy Cities, Lisbon, Portugal, pp. 2,3.

Russell, M. 1987. An opinion concerning this chapter, msc.

Russell, M., and M. Gruber. 1987. Risk assessment in environmental policy making. Science 4799:286-29O.

Rutkowski, J. 1987. Quality of life, concepts and research project. Warsaw: GUS and PAN (in Polish).

Rutkowski, J. 1984. Economic development and standard of life. Warsaw: GUS and PAN, p. 10 (in Polish).

Ryczywolska, E. 1987. New studies on urban climate problems 9:37-40 (in Polish).

Sahal, D. 1977. The conception of futures in a systems framework. Futures Research: New Directions, H.A. Linstone and W.H. Clive-Simmonds, eds. London: Addison Wesley Publishing Company, p. 158.

Schottenfeld, D., and J.F. Haas. 1978. The workplace as a cause of cancer. The Clinical Bulletin 3:117.

Schweitzer, A. 1976. Die ethic der ehrfurcht vor dem leben in Aus Meinem Leben und Denken:137. After: I. Lazari Pawlowska. 1976. Schweizer. Wiedza Powszechna, Warszawa, pp. 36-37.

Sitnicki, S. 1986. Ecological conditions of economic development. Warsaw: SGPiS (in Polish) pp. 49,178.

Statistical Yearbook of Industry, 1986. Warsaw: GUS (in Polish).

Suchodolski, B. 1981. Alternative thinking. Problemy 4:2 (in Polish).
Szarski, C. 1985. Mortality rates of human population in particular regions of Poland—Basic reasons of deaths in 1984, msc. (in Polish).
Szpilewicz, A. 1987. Use of energy by polish economy and Poland. Part 1, pp. 108-109, msc. (CPBP 04.10.01.01.01.) (in Polish).
Thiede, G. 1975. Europa grune zukunft. Dusseldorf-Wien: Econ Verlag.
United States Congress, Office of Technology Assessment. Identifying and regulating carcinogens. Background Paper.
United Nations Population Division, Urban Rural and City Projection (1982 assessment) in Urbanization and Urban Development. Biuletyn. 1984. Warszawa: Rzadowa Komisja Ludnosciowa, p. 35 (in Polish).
U.S. News and World Report. June 1987, p.65.
Velrose, E. 1984. Remarks on territorial distribution of reasons of deaths in Poland. Studia Demograficzne 3:123 (in Polish).
Weinstein, M.C. 1979. Decision making for toxic substances control: Cost-effective information development for the control of environmental carcinogens. Public Policy 3:345-357.
White, L. Jr. 1967. The Historical Roots of Our Ecological Crisis. Science 3767:1205.
Wilkins, R., and O.B. Adams. 1983. Health expectancy in Canada, late 1970s. American Journal of Public Health 9 in the Healthy Toronto 2000 Policy Conference. City of Toronto, October 13, 1984.
Witoszynski, S. 1977. The Report of the director of the U.N. program of environmental protection. Aura 11:17 (in Polish).
Wolcott, R.M., S.M. Johnson, and C.M. Long. 1988. Intensive crop production and the use of agricultural chemicals: An american case study. Paper prepared for the Organization for Economic Cooperation and Development, msc.
Woodwell, G. 1985. On the limits of nature in Global Possible, R. Repetto ed. New Haven: Yale University Press, p. 48.
World Charter for Nature (U.N.). October 1982, p. 28.
World Conservation Strategy. 1978. IUCN (Polish translation, 1985).
Wyka, A. 1987. Introduction to the polish translation of The Turning Point, F. Capra, 1982, in Sciences, Society, and the Rising Culture, pp. 5-27.
Zdziechowski, M. 1928. About cruelty. Krakow (in Polish) p. 47.
Zimmerman, E.W. 1951. World Resources and Industries. New York: Harper and Brothers.
Zylicz, T. In press. Economic sciences and environmental deterioration: Elements of theory and practical implications, Warsaw, msc., p. 161 (in Polish).

Environmental Protection as an Element of International Economic Cooperation in Poland

STANLEY J. KABALA
University of Pittsburgh

GROWTH: REDOUBLING EFFORT OR RETHINKING DIRECTION?

Economist Herman Daly pointed out a condition of far-reaching implications for world economic development when he wrote that as the world economy "grows beyond its present physical scale, it may increase costs faster than benefits and initiate an era of uneconomic growth which impoverishes rather than enriches" (Daly, 1986). While Daly's remarks refer to global levels of industrial activity, resource use, and economic momentum, they could also apply to Poland, a country where a massive thirty-year drive to create an industrial economy has left it with energy-intensive, resource-inefficient industry and severe water, air, and soil pollution.

In the 1980s, Poland reached the threshold where economic effort of the prior three decades no longer generated authentic welfare for its populace, and where the marginal gains of increased economic activity began to dwindle. The intensive use of energy and material resources that are characteristic of the Polish economy produce disproportionately small gains in material benefit and reduce opportunities for more benign development by consuming scarce economic resources (Kozlowski, 1986). Events since the economic crisis of 1981 indicate that for Poland to restore its economic vitality and avoid social, political, and ecologicial strains catastrophic in scale, it must modify the ponderous structure of its economy (Ocena, 1987). Poland is thus confronted by a dual crisis of economics and ecology: in short, a crisis of development.

THE ROLE OF INTERNATIONAL INVESTMENT

This chapter explores the premise that new international lending to Poland can be structured in such a way as to link economic gain with environmental protection. Poland's recent accession to membership in the World Bank and the increasingly likely prospect of the Bank's making substantial loans there constitute perhaps the most positive feature of its current economic situation. Accompanying this major move are other new private international economic and ecological initiatives, including the Polonia Ecological Foundation, the Fund for the Development of Polish Agriculture, and the Swedish-Polish Association for Environmental Protection (Polsko-Polonijna, 1988). Each of these have as their objective the dual goal of supporting environmental improvement while stimulating economic development in the agricultural, food processing, export, and pollution engineering sectors of the Polish economy.

Poland is a country in some ways outside the norm for World Bank lending. While the Bank invests in order to stimulate economic development, it does so most frequently in countries readily identifiable as "less developed" and considerably less often in such countries as Poland, which is of middle income in global terms and possesses a quite extensive industrial sector. Yet, as the Bank takes on an ever larger role among international organizations in managing loan repayment by the world's debtor nations, one element of its policies is to finance economic projects whose eventual profits will make possible the eventual retirement of debt. The opportunity presented by the Polish case to international financial institutions is that of supporting the reorientation of a developed industrial economy that has faltered in maintaining foreign trade, fulfilling domestic consumer demand, and efficiently using material resources.

In framing a lending program with Polish government, the Bank has the opportunity to support increases in economic productivity and productive output, activities which would address the immediate pragmatic goal of enabling Poland to redress its international debt condition. In Poland there is also the further possibility of such a lending program concomitantly making possible the relaxation of industrial pressure on the country's natural environment. International development assistance that takes into account the requirements of environmental protection holds the promise of improving economic performance as it enhances quality of life.

The issue of the compatibility of regulatory measures and economic systems affects prospects for economic restructuring in Poland and Eastern Europe. To a great degree, command economies rely on administrative measures to gain compliance throughout the system, whereas in market economies—where opportunities for effective administrative control are comparatively few—financial instruments prevail. Thus, there is some question whether the economic mechanisms of environmental protection (i.e., use charges and penalties) which Polish regulators currently value so highly can really work without first making fundamental changes in the country's economic system. Economic reform in Eastern Europe regularly takes the form of increasing the role of the market in the region's economies. Once market forces are in place, then financial incentives for pollution control and production efficiency will operate as effective checks on economic behavior. Because Poland's environment is so greatly affected by its pattern of economic development, the question of how the reforms will affect environmental quality deserves brief mention here.

Some Polish analysts have observed that the strong emphasis on profitability in the recently enacted "Second Phase of Economic Reform" bodes poorly for the environment, because firms concerned with profit will divert even less of their scarce financial resources from investment in production to pollution control, and will use whatever means at their disposal to avoid compliance with pollution regulations (Chapter 22, this volume; Kozlowski, 1986). Others point out that two fundamental factors have hindered effective regulation of resource use and pollution control. The policy of pricing coal and other forms of energy below not only world prices but also domestic production costs has led to excessive use of this input by producers (Gomulka, 1988). Simultaneously, the generally inadequate levels of pollution fines has made penalties little more than a symbolic gesture in regulatory policy. This school places great emphasis on new policies that will price raw materials and energy high enough so that firms will be encouraged to use these inputs more carefully, creating a new regulatory context that backs pollution standards with an increasingly costly system of fines for violation of pollution norms (Gorka and Poskrobko, 1987; Broda, 1988; Peszko, 1988).

A third position on reform efforts in Poland has been taken by those economists who argue that no pressure can be exerted on firms to act in favor of the environment in the conventional financial context of central planning without the introduction of two additional crucial elements of economic restructuring. Appropriate pricing of resources and stiffer fines for pollution violations are necessary but not sufficient to assure compliance with environmental regulations. Of these additional elements, first is elimination of the "soft budget constraint" that allows firms to cover increased production costs by requesting higher budgetary allocations in the next year (Rostowski and Gomulka, 1984). Next is the establishment of authentic market conditions in the national economy as the only means of forcing firms to reduce wasteful and costly practices as they seek to bring production costs into line with revenues and enable them to turn a profit (Zylicz, 1987a, 1987b).

INVESTMENT AND THE ENVIRONMENT

Investing in environmental protection does not always appear to make direct, immediate economic sense, as it often generates benefits not readily quantifiable in narrow financial or short-term calculations. This does not mean, of course, that there are not substantial, even at times massive, gains in social or cultural terms; they are merely hard to measure and relate to specific investments. Increasingly, this has become the case as environmental protection has moved from cleaning up offending point sources to managing very broad non-point source ambient quality questions, a practice in which most developed industrial countries have only limited experience.

Accurately identifying and quantifying payback can be difficult for a number of reasons. Gains may be too diffuse, as in the case of improved health and longevity for large numbers of people as a result of breathing cleaner air. They also may be too indirect, as in the case of rivers and lakes being restored to use for recreation, fishing, or municipal water supply; or they may be non-quantifiable, as in the case of the preservation of species (Chapter 3, this volume). In many instances where quantification is possible, the economic problem arises from the difficulty of creating an institutional framework capable of gathering the return from the many people who benefit from a particular action.

In the context of borrowing for environmental protection, the seemingly esoteric questions of future value and discount rates and gathering returns discussed in Chapter 3 take on a more substantial cast when illuminated by the harsh light of repayment schedules. Investments that stand to yield even the most fortuitous environmental benefits do not necessarily produce the sort of clearly defined financial paybacks that justify hard lending for that purpose. An illustration of this point in the Polish context is borrowing for the construction of wastewater treatment plants at key municipal and industrial locations. The outcome would be a marked increase in the quality of the country's rivers, especially the Vistula, but the financial benefits would be extremely difficult to assess and tie back into repayment by borrowing institutions. In such a case, the collective good to be gained by the investment might be great enough to warrant financing it from general public revenues, but it could not be termed a profit-making investment. The World Bank regularly finances public sector infrastructure projects of this type whose purpose is the public good.

This chapter poses a different issue by adding the variable of ecological gain to the decision framework of investing in economic restructuring. It suggests that new profit-making investments in Poland supported by international financial institutions should be not only environmentally neutral, but environmentally beneficial as well. That is, it asserts that the possibility exists for investment in new productive facilities in Poland that serve the government's dual goals of restructuring the economy and stimulating economic growth while lessening the overall negative impact of the country's economy on its environment.

THE COSTS OF ENVIRONMENTAL DEGRADATION

Ecological degradation in Poland affects a significant segment of the country and places parts of its people in danger. In 1982, areas in 27 of the country's 49 *voivodships* (provinces) were identified as *Areas of Ecological Hazard*. Comprising 11.3% of the country's land area and 35.5% of its population, these areas account for 65% of the country's water consumption, generate 67% of its sewage and industrial discharge, and generate 82% of total particluate and 85% of total gaseous pollution (Chapter 12 and 22, this volume). Five of these areas—Upper Silesia, Krakow, Rybnik in the southwest, Legnica-Glogow in central Poland, and Gdansk in the north—have been designated areas of "ecological disaster" because of the severity of pollution they experience (Kassenberg and Rolewicz, 1987).

In the heavily industrialized and polluted region of Upper Silesia, coal mines generate a vast volume of saline wastewater that begins the degradation of the Vistula River near its headwaters. This mine water combines with municipal sewage and industrial wastewater which are discharged directly into the river and render it so polluted that over half of its 1,047 kilometers of length are unfit for any use. The tremendous outpouring of air pollutants in the Upper Silesian region is carried in high concentrations northeastward by prevailing winds, endangering large numbers of people in other voivodships as well as large areas of forest and national parks and the ancient city of Krakow (Grodzinski et al, 1984; Kabala, 1985).

The entire, physically intact Old Town of Krakow—designated by UNESCO as part of the World Cultural Heritage—lies at the center of another of the country's zones of ecological disaster. Pollutants from the nearby massive Nowa Huta steelworks and other industrial plants combine with coal-fired home and commercial heating emissions to corrode the historical, cultural, and architectural treasure of the buildings of Krakow and jeopardize the health of its citizens. The Nowa Huta steel mill employs 35,000 people, produces 6 million tons of steel per year, and is an important part of the national and regional economies. However, Polish environmentalists and Krakow city officials alike are pessimistic about the fate of the city unless emissions from the mill and other sources are significantly curbed.

Research from a number of sources, including the Polish Ministry of Environmental Protection and Natural Resources and the National Planning Commission, places the value of economically induced environmental losses at between 800 billion and 1 trillion zloties per year, roughly equivalent to 10% of the country's Net Material Product (NMP) (Chapter 22, this volume). This level of loss has come to be termed by Polish analysts nothing less than the "ecological barrier to the development of the country."

THE ECONOMIC BIND

Poland's acute problems of environmental degradation are paralleled and aggravated by the problems of its economy. With exports unable to generate sufficient hard currency to make full interest payments on its $42 billion foreign debt, during the years 1987-1989 Poland paid an average of only $2 billion per year of interest; that interest now exceeds $3 billion annually. Despite its standing as the world's eleventh-ranked industrial nation, Poland is a country in need of both economic growth and economic reorientation (Fallenbuchl, 1987), and is constrained by a history of four decades of intense economic transformation. Although it was primarily an agricultural nation before World War II, Poland transformed itself in the two decades after the war into one of Eastern Europe's industrial powers, along with the GDR and Czechoslovakia.

When Poland undertook its development program in the early 1950s, it followed the model formulated in the USSR twenty years earlier. The so-called "extensive" pattern of development emphasized the creation of heavy industry and import substitution as key elements of the road to socialism. Under this model, growth depended on the quantity of inputs into the productive process, i.e., increases in the labor force and heavy direct investment at the expense of both consumption and social infrastructure (Fallenbuchl, 1965). The resultant massive and concentrated investment in the steel, coal, metallurgical, chemical, and shipbuilding industries began a pattern of heavy emphasis on capital goods that remained a feature of Poland's economy for over twenty years. Consequently, through the 1970s the industrial and construction sectors continued to absorb 48% of gross investment—twice as much as in most developed countries (Gorka and Poskrobko, 1987). The high demand for energy required by this effort left Poland with one of the most intensive ratios of per capita energy use to per capital GNP in the world (Budnikowski et al., 1987; Haberstroh, 1977).

One object of Poland's investment strategy of the 1970s, which entailed borrowing from the West to purchase modern licenses and production technologies, was to reduce the extreme bias toward heavy industry in its economy and increase the share of other types of manufacturing. This strategy was intended to increase the efficiency of Polish industry and engage Poland in the vigorous export trade of high quality goods, thereby raising the country's standard of living and, at the same time, making the successful repayment of foreign debt possible.

This policy failed for a number of reasons, both internal and external.

The oil price shock and the worldwide recession of the 1970s were accompanied by the inability of Poland's centrally planned economy to effectively and flexibly utilize Western credits to boost output in priority sectors, and by the outright misdirection in the use of borrowed capital. A 1986 study by the Polish Ministry of Foreign Trade pointed out that only one-third of the $44.4 billion in loans obtained from 1971 to 1981 was spent on modernizing industry, while two-thirds were spent on importing raw materials and consumer goods (U.S. House of Representatives, 1987). The virtual collapse of output at the turn of the decade reduced the Net Material Product by 25% in four years, and left the country with production levels in 1986 that were still 8.5% lower than those in 1978 (Fallenbuchl, 1987). In this context, the needs of environmental protection yielded to economic pressure to maintain production by whatever means possible, including retaining obsolescent production lines inefficient in their use of energy and raw materials.

What is the structure of the economy that generated Poland's dual crisis of ecological destruction and declining real wealth? Poland's economic pattern—which is common to the centrally planned economies of Eastern Europe and acutely inefficient in its use of raw materials and energy per unit of economic product—lies at the root of the region's environmental problems. This pattern is manifested in the overall sectoral mix of the economy, the structure of the industrial sector within the economy, and the operating efficiency of processes within industry.

In the process of industrialization, market economies in a certain range of GNP per capita reach a point at which industry's share of the economy ceases to grow. Change in industry then takes place within the slowly shrinking industrial sector of the economy as a progressively greater share is taken up by the service sector. Consistently in centrally planned economies, the share of the industrial sector continues to grow through and beyond this phase, leading observers to postulate that distortionary systemic factors are at work to cause this phenomenon (Winiecki, 1988).

Figures comparing energy and material intensity of the national products of industrialized countries illuminate the Poland's situation in this context (Table 1). Compared to those of West European countries, the economies of the European members of the Council for Mutual Economic Assistance (CMEA) use on average 18% more energy and 36% more steel per capita, while national income per capita in dollars is lower. Consumption of steel per dollar of national income hovers at three times the West European average. As a result, the energy intensity per dollar of national income of the CMEA economies is on the whole more than double that of even the more industrialized countries of Western Europe: 1.29 kg. coal equivalent versus .54 kg. Among European countries overall, Poland has a slightly higher than average primary energy consumption (5,590 kg. coal

TABLE 1 Energy and material intensity of selected industrialized countries, 1980 (author's calculations).

Country	Gross energy consump. kg CE/cap	Steel consump. kg/cap	National income $/cap	Energy use kg. CE/$	Steel use kg/$
CMEA AVG.	5,600	509	4,590	1.29	0.12
Bulgaria	5,678	320	4,150	1.37	0.08
Czecho.	6,482	706	5,820	1.11	0.12
Hungary	3,850	340	4,180	0.92	0.08
GDR	4,708	566	7,180	1.03	0.08
Poland	5,590	516	3,900	1.43	0.13
Romania	4,593	562	2,340	1.96	0.24
USSR	5,595	554	4,550	1.22	0.12
DEVELOPED MARKET AVG.	4,770	373	9,760	0.54	0.04
Austria	4,160	355	10,230	0.41	0.03
Denmark	5,225	356	12,950	0.40	0.03
France	4,351	422	11,730	0.37	0.04
FRG	5,727	609	13,590	0.42	0.04
Italy	3,318	412	6,480	0.51	0.06
Japan	3,690	512	9,890	0.37	0.04
Spain	5,727	244	5,400	1.06	0.05
U.K.	4,835	237	7,920	0.61	0.03
U.S.A.	10,410	490	11,360	0.92	0.04

SOURCE: Budnikowski et al., 1987;

equivalent per capita versus 5,000 kg. for Europe) and a notably higher relative consumption of new steel (516 kg. per capita versus 441 kg. for Europe). Consumption of steel and energy per dollar of national income are also quite high in Poland, at .13 kg. of steel and 1.43 kg coal equivalent per dollar.

The disparity in economic structure between market and planned economies exists not only at the sectoral level but also within the industrial sector. In centrally planned economies, the extractive industries continue to display a much larger share of industrial output and employment in comparison with mature market economies at the same level of GNP per capita. The share of extractive industries in centrally planned economies is on average three times greater than in mature market economies, and almost twice that in semi-developed market economies. Poland exhibits one of the highest such ratios, with extractive industries constituting 10% of industrial employment (Table 2). This pattern of concentration in certain forms of heavy industry in centrally planned economies also appears in data on the engineering industries, which shows this subsector to be 50% greater

TABLE 2 Share of extractive industries in industrial employment, 1963-80 (% of total)

	1963	1970	1980
Centrally Planned Economies (CPE)			
Unweighted average	10.5	8.8	7.8
Bulgaria	10.5	10.6	8.0
Czechoslovakia	9.0	6.7	6.5
Hungary	11.1	8.7	7.8
Poland	13.1	11.2	10.2
USSR	8.7	6.8	6.2
Developed Market Economies (DME)			
Unweighted average	3.9	2.6	2.6
Belgium	7.5	4.0	3.0
France	--	--	2.6
FRG	6.7	4.0	3.0
Italy	2.3	3.3	3.3
Sweden	1.9	1.5	1.5
U.K.	7.0	4.3	4.4
USA	2.9	2.3	2.8
CPEs/DMEs ratio	2.69	3.38	3.00

SOURCE: Winiecki, 1988 (from United Nations statistical yearbooks).

TABLE 3 Energy consumption in selected industries in MJ per ton of product.

Product	Poland	International indicative level	% difference
Raw steel	15,632	13,400	17
Open hearth steel	6,231	4,600	35
Electric arc steel	3,192	5,900	-45
Steel rolling	5,850	4,200	40
Aluminum smelting (in kilowatthours)	16,237	13,500	21
Cement manuf.	5,010	3,700	35
Ammonia prod.	41,120	40,000	0

SOURCE: Rocznik Statystyczny, 1987; World Bank, 1987.

than would have been expected compared to other countries with different economic systems but comparable levels of GNP per capita (Winiecki, 1988).

In Poland, the predominance of industry in the economy and the emphasis within industry on heavy industries is accompanied by generally low energy efficiency in industrial processes. This parallels general patterns of low labor and materials efficiency in the country's economy (Budnikowski et al., 1987). Table 3 characterizes the efficiency of Poland's industrial sector according to energy use in selected heavy industries. Except for ammonia production—where Poland's efficiency level is on a par with the international figure—and electric arc steelmaking—where its level is superior to the international figure—Poland's heavy industries trail attainable levels of energy efficiency by anywhere from 17% to 40%. These relatively low levels of energy efficiency mean greater consumption of coal-fired electricity to fuel Polish industry. Poland's steel industry is energy intensive because of the nature of the technology currently in use. Of the 17 million tons of steel produced in 1986, roughly 37% was produced in open hearth furnaces, 16% in electric arc furnaces, and 47% in basic oxygen furnaces (Rocznik Statystyczny, 1987). While the open hearth furnace accounts for some 55-60% of steel production in Eastern Europe and the Soviet Union, it is virtually out of use in Western Europe and accounts for only 8% of U.S. output (World Bank, 1984).

The overall effect of this pattern is that, compared to other economies of roughly similar size and level of development, Poland's economy is two to three times as intensive in its use of energy (Table 4). This pattern has concomitant environmental outcomes when linked with inadequate pollution control. Poland is a high emitter of sulfur dioxide and the only country in Europe whose emissions of this pollutant are projected to rise over the next decade (UNECE, 1987). Comparisons of gross energy consumption to national emissions of sulfur dioxide show dramatic differences in overall levels of pollution control efficiency among European countries (Tables 5 and 6).

POLICY EXPLORATIONS

At the end of the 1960s, the sectoral bias of Poland's economy toward heavy industy was recognized as offering limited possibilities for economic growth. Changing this bias was the aim of the development strategy undertaken by the Gierek government at the beginning of the 1970s. In the subsequent two decades, the situation has become more acute as this pattern has, in addition, come to place great pressure on the country's environment. References in Poland to environmental degradation as a barrier to the country's development allude to the fact that only the types of economic reform that recognize the country's ecological limits can build sustainable economic growth. The following survey of options indicates that even though Poland finds itself in difficult economic and ecological straits, there are aspects of this situation that have positive elements within them.

TABLE 4 Energy intensity in selected industrialized countries.

MARKET ECONOMIES:	Belgium	Spain	FRG	Units
GDP	79,080	164,250	624,970	mil. U.S.$
% industry	33	N.A.	40	%
Energy consumption				Mtce
--gross	60.82	105.53	381.28	
--final	41.71	63.12	254.73	
Energy intensity U.S.$				Mtce/bil.
--gross	0.77	0.64	0.61	
--final	0.53	0.38	0.41	

PLANNED ECONOMIES:	Yugoslavia	Hungary	Poland	Units
GDP	44,370	20,560	70,439	mil. U.S.$
% industry	33	41	49	%
Energy consumption				Mtce
--gross	62.62	42.42	180.67	
--final	31.86	28.97	109.44	
Energy intensity				Mtce/bil. U.S.$
--gross	1.41	2.06	2.56	
--final	0.72	1.14	1.55	

SOURCE: World Bank, 1988; from World Development Report, 1987 and UN ECE, 1986.

TABLE 5 National emissions of sulfur dioxide (1,000 tons per year)

Country	1980	1985	1990	1995
Belgium	799	467	506	544
Spain	3,250	3,250	3,053	---
FRG	3,200	2,400	---	1,100
Hungary	1,633	1,420	1,400	1,140
Poland	4,100	4,300	4,900	---
Yugoslavia	1,175	1,800	---	---

SOURCE: UN ECE, 1987.

Energy

Poland's economy, particularly the industrial sector, is adapted to domestic fuel sources. Thus, it is to a great extent free from reliance on imported energy and foreign exchange considerations. This is a fortunate circumstance given Poland's debt situation, and it can only take on greater significance as world energy prices start their temporarily deferred rise and

TABLE 6 Emissions of sulfur dioxide in realtion to gross energy (tons per year SO_2 per millions tons coal equivalent)

Belgium	7,656
FRG	6,299
Spain	30,660
Hungary	33,800
Poland	23,757
Yugoslavia	28,571

SOURCE: UN ECE, 1987.

the exportable share of Poland's coal production becomes more valuable on world energy markets.

As noted previously, Poland's industrial sector is characterized by generally low levels of efficiency in the use of fuel and energy, which means, ironically, that relatively modest improvements in process efficiencies will generate notable marginal gains in fuel conservation and cost savings. Reduced energy use has the potential to simultaneously improve operating costs for firms, maintain the nation's balance of payments, and lessen the strain placed on the country's environment by reducing the quantities of air pollutants emitted from burning coal. The implementation of aggressive energy-efficient measures could allow for increased use of coal for domestic needs while maintaining coal exports at just half of the 1986 volume in the year 2000.

Transportation

Poland's transportation sector is significantly oriented toward relatively efficient modes (e.g., railroads and buses) as opposed to less efficient ones (e.g., private autos and airplanes). To the extent that a transportation system can be operated on electricity produced from domestic coal, this fundamental orientation is an inherent strength for a country whose import of oil is constrained by its balance of payments situation.

However, while it is a boon for its environment that cars in Poland are fewer and smaller than in Western countries, the potential decrease of pollution this causes is offset by the fact that the general fuel efficiency of Polish-produced automobiles is poor by European standards. Therefore, improving the overall efficiency of motorized vehicles will both conserve imported fuel and limit emissions of air pollutants. Specifically, investing in upgrading the combustion efficiency of fleet vehicles such as urban diesel-fueled buses and trucks would be a step toward improving air quality in Poland's cities (Wharton, 1986).

Some 600,000 tons of liquid fuel could be saved annually by improving the efficiency of the country's delivery transport system, which would virtually maintain the balance between current liquid fuel supplies and projected growth of motorization (Skorkowska, 1986). Capital might be made available to assist the simultaneous conversion of Poland's automotive industry to lead-free and more efficient engines and its refining industry to lead-free gasoline. This would speed the removal of a serious environmental health threat as it abetted fuel conservation.

Housing

Poland's housing sector has not responded to the potential posed by energy-saving housing construction and heating regulation methods. Poland's buildings, new as well as old, are in desperate need of weatherizing and insulation due to prior efforts over decades to ease the country's housing shortage by the rapid construction of a huge number of units with negligible attention to energy efficiency (Dienes and Merkin, 1985). Energy use in heating is estimated to be twice the technically required level (Skorkowska, 1986). Automatic regulation of district heating systems combined with weatherized buildings constitute a large and untapped "source" of energy which Poland's energy planners can utilize.

Agriculture

The country's agricultural system, though in some ways antiquated, is by West European and North American standards relatively free of dependence on imported fuels, fertilizers, and pesticides. Because 75% of the country's farms are in private hands and operate on a rather lean budget of purchased inputs, agriculture in Poland holds the potential to generate greater marginal gains in output in response to relatively modest inputs of fertilizer and technology than do more developed agricultural sectors of other industrialized countries. Poland's "low-tech" private farming should be examined as a case study of economically viable, labor-intensive, ecologically sustainable, low-energy agriculture in an advanced economy (Cook, 1984; Chapter 15, this volume).

At present, Poland's 2.8 million small farms rely on purchased energy from conventional sources, including coal, wood, and electricity, which contribute to the pollution of water in the form of farm runoff. An estimate prepared for the World Bank states that 20,000 of these farms under 50 hectares are suited to the production of biogas energy from farm wastes, a practice that would satisfy farmstead energy needs, reduce effluent runoff, and improve farm financial performance. Experimental activities along these lines are planned as part of the program sponsored by the Fund for the Development of Polish Agriculture.

Food Processing

Poland's food processing industry, though technologically obsolescent, produces products of good quality. Improvements in processing and marketing should be sufficient to facilitate the expanded export of the country's unique food products to the international market. Food production and processing is not an environmental measure per se, but, if properly carried out, can be ecologically benign, function on low energy inputs, and produce quite favorable value-added ratios—all elements of the type of reorientation necessary for Poland's economy overall. As a component of national plans for investment in export products, it could constitute a tonic for the country's heavily industrial economic mix. Poland ought to be internationally known, for example, for its fruit juices and fruit products. Both fruit production and processing are well established in Poland, and the technology and inputs this industry needs for modernization are only modestly sophisticated and relatively non-capital intensive.

Pollution Control Technology

Poland has a sophisticated engineering profession and a highly skilled manufacturing work force. However, in terms of environmental protection, it has an undeveloped sector for the manufacture of pollution control and waste treatment equipment which currently cannot supply the country's needs. This failing has been pointed out by Polish observers not only as a hindrance to pollution control policies, but also as a missed opporunity for both domestic industrial restructuring and export of manufactured goods. This prospect has recently received attention in Poland as well as abroad. Because of the acute load of transboundary air pollution it receives from Poland and other countries of Central Europe, Sweden has indicated a willingness to make available certain advanced pollution control technologies and credits for their manufacture and installation in Poland in return for the long-term gains that will ensue. In a similar vein, two foreign firms—one British and one Swedish—have joined with the Polonia Association to create the Polonia Ecological Foundation for the purpose of gathering foreign capital for investment in Polish firms engaged in the manufacture of pollution control and wastewater treatment devices.

PROSPECTS

To free itself from its economic impasse and reduce the threat of ecological crisis, Poland must exploit the potential of investment with an eye not merely for profit but profit with environmental protection. What are the prospects for a policy of investment that takes into account the demands and constraints of environmental protection? Polish authorities appear to be aware that the country is approaching a day of reckoning with regard to its environment which could be as politically disruptive as the country's economic problems have been. Continuing decline in Poland's standard of living and quality of life, due to acute ecological stress and leading to increased social discontent, is a prospect that was pointed out by the Committee on Man and the Environment of the Polish Academy of Sciences in a report on the effectiveness of environmental protection measures (Polish Academy of Sciences, 1987).

Events of recent years indicate that Poland's leaders are now willing to take a stance that places environmental issues on a par with the country's other serious problems. One of the first actions of the government of Prime Minister Mieczyslaw Rakowski after it assumed power in mid-1988 was to announce that environmental protection was to be one of the three elements of its program of economic recovery, along with the development of agriculture and the expansion of the housing construction industry. The Solidarity labor union's proposal for the 1989 government-opposition "roundtable" talks placed environmental protection in a central position among its issues on the agenda (Protokol, 1989).

The non-communist Mazowiecki government which assumed power in mid-1989 retains concern for environmental protection, despite the fact that it is only one of a series of major problems the country must address. That environmental protection has figured prominently in the programs of two successive governments from opposite ends of the political spectrum is significant in Poland's current political context; Poland's leaders are now pressed to redeem the country's economy in the eyes of both the international financial system and their own people. By addressing the interwoven problems of economic stagnation, environmental havoc, and economic reorientation, Poland may be able to chart a course that progressively satisfies all of its real creditors—its people, the banks, and the environment.

REFERENCES

Broda, Z. 1988. Dyskusyjne problemy gospodarki funduszami kologicznymi (Current issues in the management of the ecological funds) in Zastosowanie Instrumentow Ekonomicznych i Prawnych w Ochronie Srodowiska Naturalnego (The Application of Economic and Legal Instruments in the Protection of the Natural Environment), Gorka, K., ed. Krakow: Polish Economic Association.

Budnikowski, A., M.J. Welfens, and S. Sitnicki. 1987. Rozwoj gospodarczy a ochrona srodowiska w krajach RWPG (Economic development and environmental protection in the countries of the CMEA). Warsaw: National Economic Publishers.

Cook, E. 1984. Agricultural reform in Poland: Background and prospects. Soviet Studies 36(3). [For examples of U.S. research on low-energy, biologically sustainable and ecologically sound agriculture, see the work of the Land Institute of Salina, Ks., the Rodale Institute of Emmaus, Pa., or the Institute for Alternative Agriculture, of Greenbelt, Md.]

Daly, H.E. 1986. Toward a new economic model. Bulletin of the Atomic Scientists 4.

Dienes, L., and V. Merkin. 1985. Energy policy and conservation in Eastern Europe in East European Economies: Slow Growth in the 1980s. vol. 1. U.S. Congress Joint Economic Committee. Washington, D.C.: U.S. Government Printing Office.

Fallenbuchl, Z.M. 1965. Some structural aspects of the soviet-type investment policy. Soviet Studies 16(4).

Fallenbuchl, Z.M. 1987. The Polish economy in the year 2000: Need and outlook for systematic reforms, recovery, and growth strategies (the Western view). Paper presented at the conference "The Polish Economy in the Year 2000," February 5-6, 1987. University of Pittsburgh, Pittsburgh, Pa.

Gomulka, S. 1988. Reforms the IMF and the World Bank should promote in Poland in Creditworthiness and Reform in Poland: Western and Polish Perspectives, Marer, P. and W. Siwinski, eds. Bloomington, Ind.: Indiana University Press.

Gorka, K., and B. Poskrobko. 1987. Ekonomika ochrony srodowiska (The economics of environmental protection). Warsaw: National Economic Publishers.

Grodzinski, W., J. Weiner, and P.F. Maycock. 1984. Forest ecosystems in industrial regions: Studies in the cycling of energy, nutrients, and pollutants in the Niepolomice Forest, southern Poland. Berlin: Springer-Verlag.

Haberstroh, J.R. 1977. Eastern Europe: Growing energy problems. In East European Economies Post-Helsinki. U.S. Congress Joint Economic Committee,. Washington, D.C.: U.S. Government Printing Office.

Kabala, S.J. 1985. Poland: Facing the hidden costs of development. Environment 11.

Kassenberg, A., and C. Rolewicz. 1985. Przestrzenna diagnoza ochrony srodowiska w Polsce (The spatial diagnosis of environmental protection in Poland). Warsaw: National Scientific Publishers.

Kozlowski, S. 1986. Kierunki zmian w srodowisku przyrodnicznym (Directions of change in the natural environment). Kosmos 1(190). Warsaw: National Scientific Publishers.

Kramer, J.M. 1986. The environmental crisis in Poland. International Institute for Environment and Society, Berlin.

Peszko, Z. 1988. Funkcjonowanie systemu instrumentow ekonomicznych ochrony srodowiska (The functioning of the system of economic instruments of environmental protection) in Gorka, K. 1988.

Polish Academy of Sciences. 1987. Ocena skutecznosci i kierunkow zialania w dziedzinie ksztaltowania i ochrony srodowiska (An evaluation of the effectiveness of measures in the field of the shaping and protection of the environment). Committee on Man and the Environment. Warsaw.

Polsko-Polonijna Fundacja Ekologiczna. 1988. Statut Fundacja (Articles of Incorporation). Polonia Ecological Foundation, Warsaw. Unpublished.

Protokol Podzespolu d/s Ekologii Okraglego Stolu (Report of the Roundtable Subcommittee on Ecology). 1989. Warsaw. Unpublished.

Rocznik Statystyczny. (Statistical Yearbook). 1987. Main Statistical Office, Warsaw.

Rostowski, J., and S. Gomulka. 1984. The Polish reformed economic system, 1982-83. Soviet Studies 36(3).

Skorkowska, B. 1986. Energy resources and the restructuring of the Polish economy in Energetyka, Srodowisko, i Planowania Rozwoju (Energy Policy, the Environment, and Development Planning). Committee on Environmental Engineering and Institute for Basic Technical Problems. Warsaw: Polish Academy of Sciences.

United Nations Economic Commission for Europe. 1987. National strategies and policies for air pollution abatement. New York: United Nations.

U.S. House of Representatives. 1987. Poland's renewal and U.S. options: A policy reconnaissance. Committee on Foreign Affairs. Washington, D.C.: U.S. Government Printing Office.

Wharton Centrally Planned Economies Series. 1986. Analysis of Current Issues. 6(42).
Winiecki, J. 1987. The distorted world of Soviet-type economies. Pittsburgh, Pa.: University of Pittsburgh Press.
World Bank. 1984. Energy Efficiency in the Steel Industry. Washington, D.C.: World Bank,
World Bank. 1987. Poland: Reform, adjustment, and growth. Washington, D.C.: World Bank.
Zylicz, T. 1987a. Europejskie transfery zanieczyszczen a problemy modernizacji gospodarki polskiej (European transfers of air pollution and problems of modernizing the Polish economy) EKO Series Paper #8713. Warsaw: University of Warsaw Department of Economics.
_____1987b. Will Poland join the thirty percent club? EKO2 Series Paper #8707. Department of Economics. Warsaw: University of Warsaw.

The Role of Ecological Risk Assessment in Environmental Decision Making

ALAN W. MAKI
Exxon Company, USA

MICHAEL W. SLIMAK
U.S. Environmental Protection Agency

PREDICTIVE ASSESSMENT OF ECOLOGICAL RISKS

Potential impacts and effects of pollutants on ecosystems are often complex and far-reaching. The simplest counter to a chemical use that produces immediate and obviously deleterious environmental effects is to impose a ban on its use before the effects become irreversible. However, many of the less obvious and subtle environmental effects of pollutants are not readily recognized and therefore are considerably more difficult to handle. Thus, the effect of pollutants on ecosystems is an issue which requires the development and enhancement of methodology to predict the fate and effects of substances in the environment prior to their manufacture, use, or distribution.

Procedures for ecosystem risk assessment are currently evolving. An *ecosystem risk assessment* is defined as a set of procedures for measuring risk to the environment associated with the use of substances through an objective and probabilistic exercise based on empirical data and scientific judgment. The results of such a risk assessment can then be used to provide a consistent means of estimating the limiting concentrations of substances that will produce no unacceptably negative effects on ecosystems which are potentially exposed to the substance.

Thus, ecological risk assessments serve as the scientific basis for deciding whether the risks are acceptable or unacceptable for the environment, i.e., a risk management decision. Basically, the risk assessment process consists of two parallel lines of investigation and relates observed biological effects to expected exposure concentrations. Figure 1 represents the two

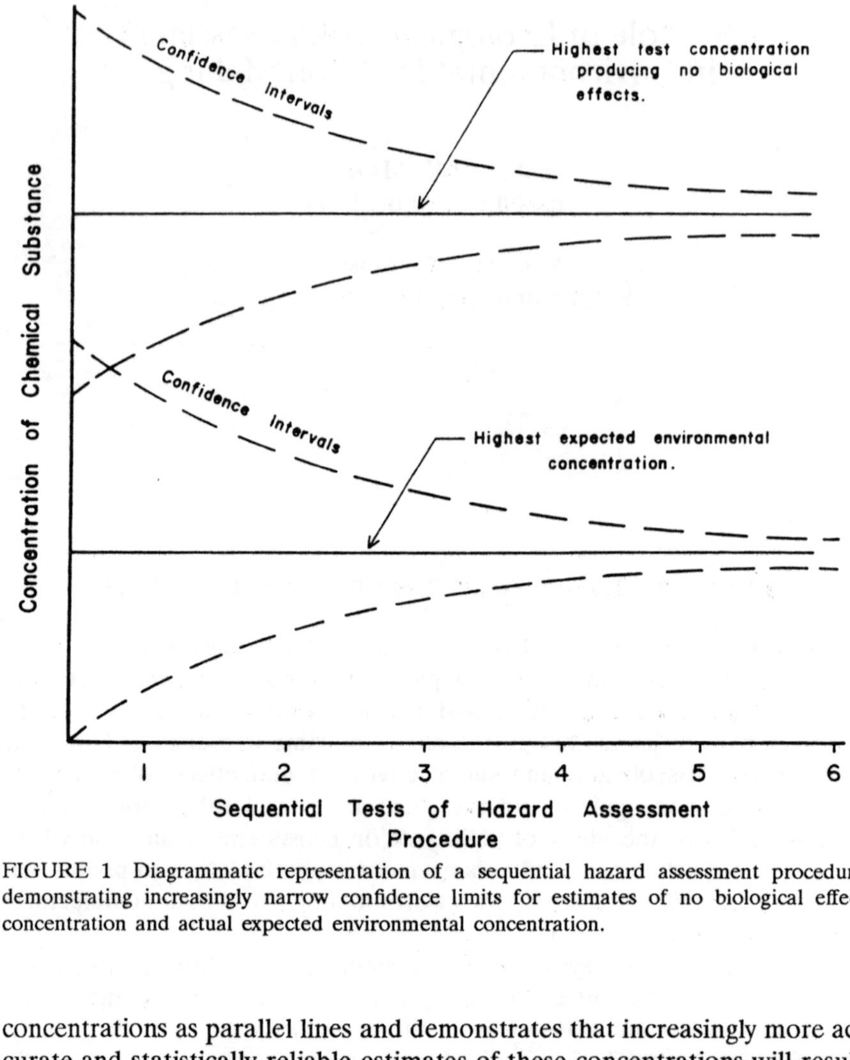

FIGURE 1 Diagrammatic representation of a sequential hazard assessment procedure demonstrating increasingly narrow confidence limits for estimates of no biological effect concentration and actual expected environmental concentration.

concentrations as parallel lines and demonstrates that increasingly more accurate and statistically reliable estimates of these concentrations will result from a sequential series of tests completed along the *x* axis of the graph. Using the risk assessment process, increasingly more accurate estimates of fate and effects can be made to the point where it becomes possible to state with a high degree of confidence that environmental concentrations and biological effects will likely result in negative environmental consequences. It is a matter of judgment to determine just how far into the hazard-evaluation process investigation must proceed to establish acceptable confidence regarding fate and effect concentration predictions (Cairns et al., 1978).

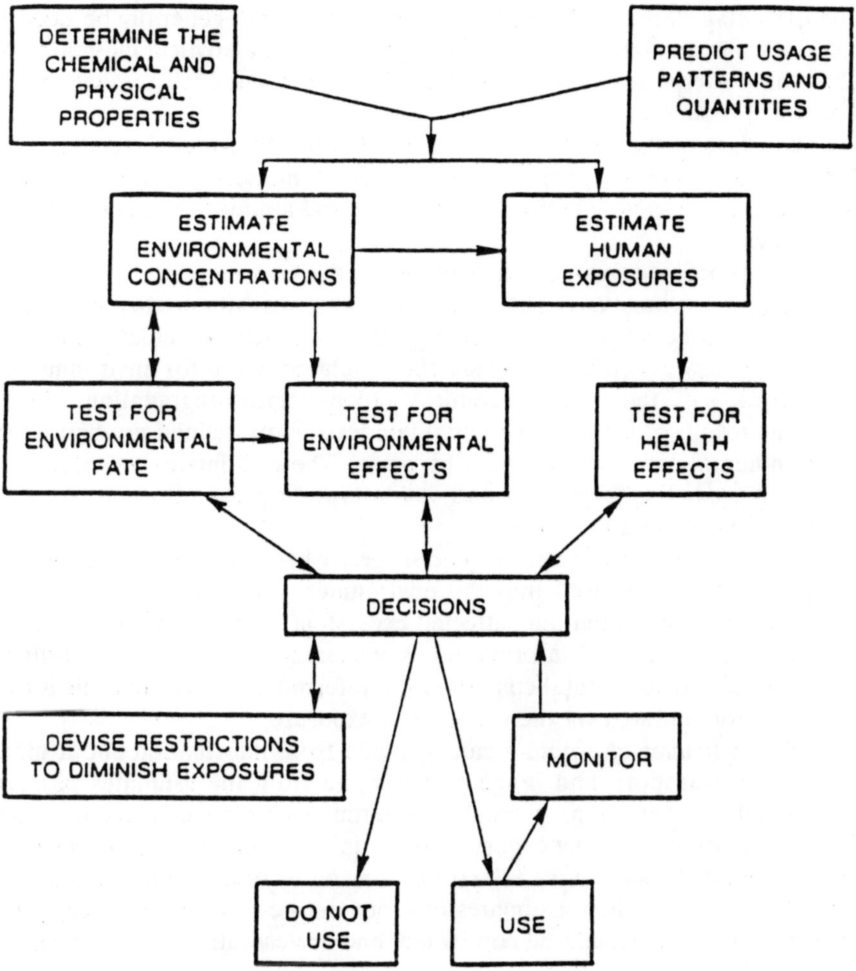

FIGURE 2 Flow chart for the 10 phases of the safety evaluation program.

The process of making and reviewing an environmental risk assessment can be broken down into ten stages (Beck et al., 1981). These stages and how they are related to each other are shown in Figure 2. The sequence of stages splits into two paths, one for environmental safety and one for human safety, which come together again at the point of decision making. Each stage is described briefly below:

1. The evaluation starts with a consideration of the *properties of the substance* (e.g., chemicals, materials) ascertained from the literature or determined in the laboratory, if necessary. Estimates may suffice in the early stages, but more precise determinations will generally be needed later. This is the time for initial development of the analytical methodology that will be required to measure the substance in the environment and in biological tissues.

2. Use of the substance, and quantities involved must be considered, since *usage patterns* (along with such related matters as manufacturing, shipping, and disposal) influence the routes and amounts of environmental exposure.

3. From anticipated usage patterns, the *concentrations* that can be expected in various environmental compartments are predicted. These estimates can be helpful in projecting *exposures*, but their chief value is in suggesting how extensively the material should be tested for environmental properties, e.g., the rate of chemical and biological degradation. Then, from the results of the environmental fate tests, more refined predictions of environmental concentrations can be made. These estimates are necessary both for predicting exposures and for interpreting the results of tests to evaluate environmental effects.

4. Tests for *environmental fate* suggest what may happen to the substance after it is released into the environment. The nature of both the substance and the potentially affected ecosystem is important in determining fate. This kind of information is necessary for making the refined estimates of environmental concentration referred to above, and this is one of the factors influencing the estimate of exposure.

5. *Estimates of exposure* can be made from information about manufacturing, transport, and usage patterns and from the estimates of environmental concentration. A number of estimates are needed to cover such diverse situations as direct exposure to pesticides or indirect exposures from a contaminated food source. Exposure may be by oral, respiratory, ocular, or dermal routes. These estimates of exposure are of value in determining which effects tests should be conducted, and are essential in evaluating the results of those tests.

6. Tests for hazard are concerned with the possible *effects* of the material on *non-target health*, but most of the tests are conducted with laboratory animals. From information about the kinds of effects produced in these animals, and the concentrations of chemicals necessary to produce them, it is possible to determine the exposure of chemicals that would be acceptably low risk to non-target organisms.

7. Tests for *environmental effects* indicates whether the concentrations expected may cause harm to the environment, particularly to the living organisms in it, the kinds of injury that may occur, and the species or processes most likely to be affected. Further discussion of ecological responses to stress can be found in subsequent chapters by Harwell et al. and Breymeyer (Chapters 7 and 8, this volume).

8. *Decision making* occurs repeatedly throughout the process. If the decision is to use the material, field monitoring may be needed to determine whether the resulting environmental concentrations correspond to those that were predicted from laboratory testing and modelling.

9. The first step in decision making is to compare the concentrations of a chemical that are predicted to cause unacceptable harm to the environment with the concentrations that will likely result from using the chemical. From this comparison, it is possible to *assess the risk* of causing adverse effects. Finally, the decision is made whether the risks are societally acceptable or not. It is not necessary to do all the tests that are listed before making this decision; there are provisions at many points in the process for deciding that no further testing is necessary and that regulation is or is not required.

10. If any of the risks associated with using the chemical as originally planned are judged unacceptable, it may be possible to *devise restrictions* on the use of the chemical that would diminish the anticipated exposure and therefore lower the risk. The restrictions might be of many sorts, ranging from warning labels to the construction of containment dikes around storage tanks. Once such restrictions have been devised, it will then be necessary to go through parts of the decision-making process again to see whether the risk becomes acceptable.

ECOTOXICOLOGY: THE PRACTICE OF RISK ASSESSMENTS

Ecotoxicology can be defined as the study of the fate and effects of toxic agents in ecosystems. Ecotoxicology is the study of toxic effects on biota—particularly on populations and communities—and their interactions with processes controlling the functioning of defined ecosystems.

The U.S. Environmental Protection Agency (EPA) has primary responsibility for determining the ecological risks associated with the use of pesticides and industrial chemicals (xenobiotics) in the United States. This responsibility comes from two major pieces of legislation: the Toxic Substances Control Act (TSCA) and the Federal Insecticide, Fungicide, and Rodenticide Act (FIFRA).

In practice, both programs assess risks to ecological resources using an ecotoxicological approach: laboratory toxicity bioassays to determine hazard; determination of exposure either from monitoring data or predicted from models; and a comparison of exposure to hazard using the quotient method. In the quotient method, the exposure value is directly compared with a toxicity endpoint (e.g., concentration in water to an LC_{50} value; 10 ppm/100 ppm). The closer the quotient is to 1 (or greater), the higher the probability that an adverse effect will occur. Interpreting this adverse effect (i.e., the likelihood that what is observed in the lab will actually occur in the field) is one of the biggest uncertainties in both programs. Although each program derives it differently, the final result is the application of either a safety factor or an assessment factor to account for uncertainty.

Although the two programs are similar in their approach to assessing ecological risk, the quantity of data used to make assessments is strikingly different: TSCA assessments tend to be data-poor while FIFRA assessments are usually data-rich. This difference is due to what is being regulated. TSCA regulates new and existing industrial chemicals, many of which are site-limited intermediates, are produced in low volumes with limited potential for environmental release, and are not manufactured primarily for their toxicological (biological) properties. Pesticides, on the other hand, are generally very active biologically, are designed to kill pests and other organisms, and are broadly released into the environment on a regional scale.

The Toxic Substances Control Act

The Office of Toxic Substances (OTS) at EPA is responsible for implementing provisions of TSCA. TSCA was enacted in 1976 to protect humans and the environment from unreasonable risks caused by industrial chemicals and mixtures. Section 4 of TSCA is primarily concerned with existing chemicals which were in production before the law was passed. Because of the thousands of chemicals on the inventory, a prioritization scheme had to be established to effectively assess potential risks. Thus, an Inter-Agency Testing Committee recommends to EPA chemicals it believes should be given priority testing for hazard determination. EPA evaluates these recommendations and can either refute them or proceed with hazard testing. Chloroparaffins are an example of a group of chemicals for which ecological testing has been required pursuant to this section of TSCA. OTS used and refined existing ecosystem-level models to address the risks to natural populations posed by indirect and direct toxic effects of these chemicals. The use of these higher level models was necessary since the quotient method of assessment indicated a likelihood of adverse effects.

Section 5 of TSCA covers chemicals which are new, i.e., those which were produced since the law was passed. Industrial chemicals regulated in this section are referred to as *Pre-Manufacturing Notice* chemicals or "PMNs." This section of the law requires a manufacturer to submit a premanufacture notice to EPA before it manufactures a particular chemical. EPA then has 90 days (or, with good cause, 180 days) to review the notice for potential risks to the environment. There were over 2,000 PMNs submitted to EPA in 1988. Generally, however, ecotoxicological data is not submitted by the manufacturer and the potential hazard must therefore be determined using predictive toxicological techniques (e.g., structure/activity relationships) techniques. If an unacceptable risk is judged to be a high probability (*risk = hazard + exposure*), then the manufacturer may be asked to conduct suitable tests.

Because of the large numbers of industrial chemicals that must be assessed in this program, a method was devised to insure uniformity and consistency in identifying chemicals for testing to determine ecological hazard. *Assessment factors* are used in conjunction with the hazard assessment to derive concentrations of concern in aquatic media which, if equaled or exceeded, provide a basis for further testing. Assessment factors are numbers which are used to adjust standard toxicological measurements (e.g., LC_{50}, EC_{50}, etc.) to derive a "concern level." An environmental concentration of concern is that concentration at which populations of organisms are adversely affected as found in a field study conducted under simulated or actual conditions of production, use, and disposal. Assessment factors take into account the uncertainties due to such variables as test species sensitivity to acute and chronic exposures, laboratory test conditions, and age-group susceptibility. There are four assessment factors currently being used: 1, 10, 100, and 1000.

Table 1, taken from EPA (1984), summarizes the application of assessment factors. OTS does not consider assessment factors to be equivalent to safety factors. Safety factors are usually interpreted as being a margin of safety applied to a no-observed-effect level to produce a value below which exposures are presumed to be safe. Assessment factors are used with acute or chronic toxicity values to arrive at a concentration which if equaled or exceeded could cause adverse effects. Assessment factors have been developed solely for the PMN process to identify those chemicals which require ecological testing to fully assess ecological risks.

In assessing risks to PMN chemicals, OTS uses the quotient or ratio method. The specific equation used is:

Environmental Concentration/Concern Level = Risk

If the quotient is equal to or greater than 1, the conclusion is that adverse effects are likely to occur to the population of organisms represented by the toxicity data. The quotient method is only used, however, to determine if actual testing is necessary. If actual hazard data is obtained, the quotient method is still used; however, more analysis is conducted using dose/response curves in conjunction with the measured or predicted environmental concentration. In addition, consideration is given to understanding the acute to chronic ratios, and inter- and intra-taxa dose relationships. In some instances, simulation models, such as the Standard Water Column Model (SWACOM) developed by Bartell et al. (1988), have been employed where the chemical impact on one trophic level is analyzed relative to the other trophic levels.

TABLE 1 Application of assessment factors to evaluate need for testing.

DATA AVAILABLE	ASSESSMENT FACTOR TO BE APPLIED
Structure-Activity Derived LC_{50}	1000
Single LC_{50} From Chemical Analog	1000
Single Test LC_{50} for PMN	1000
Two LC_{50}s for Same Analog (e.g., 1 Fish, 1 Algal test)	1000
Two LC_{50}s for PMN (e.g., 1 Fish test, 1 Invertebrate)	1000
Three LC_{50}s for Same Analog (Fish, Algae, Invertebrate)	100
Five LC_{50}s for Same Analog (3 Invertebrates, 2 Fish)	100
Five LC_{50}s for the PMN (e.g., 3 Algae, 2 Fish)	100
Maximum Acceptable Toxic Concentration for Analog	10
Field Study	1

SOURCE: U.S. EPA, 1984.

Federal Insecticide, Fungicide, and Rodenticide Act (FIFRA)

Under this law, EPA must determine whether a pesticide can be registered for a particular use. FIFRA states that the EPA Administrator shall register a pesticide if he determines that "when used in accordance with widespread and commonly recognized practice it will not generally cause unreasonable adverse effects on the environment."

The term "unreasonable adverse effects on the environment" means any unreasonable risk to humans or the environment, taking into account the economic, social, and environmental costs and benefits of the use of any pesticide. Under FIFRA, the process of determining whether or not a risk is unreasonable (i.e., factoring in benefits along with risks) is a risk management function. For this discussion, the important term used in FIFRA is "risk to the environment." In order for the EPA Administrator to determine if there will be an unreasonable risk to the environment from the use of a pesticide, an ecological risk assessment from a pesticides perspective involves estimating the likelihood or probability that adverse effects (e.g., mortality to single species of organisms; reductions in populations of non-target organisms due to acute, chronic, and reproductive effects; or disruption in community and ecosystem level functions) will occur, are occurring, or have occurred.

Ecological risk is a function of toxicological hazard and environmental exposure. *Toxicological hazard* is the intrinsic quality of a pesticide to cause an adverse effect under a particular set of circumstances. Toxicological

TABLE 2 Regulatory risk assessment criteria for pesticides.

PRESUMPTION OF NO RISK	PRESUMPTION OF RISK THAT MAY BE MITIGATED BY RESTRICTED USE	PRESUMPTION OF UNACCEPTABLE RISK
I. *Acute Toxicity*		
1) *Mammals*		
EEC* < 1/5 LC_{50}	EEC ≥ 1/5 LC_{50}	EEC ≥ LC50
mg/kg/day < 1/5 LC_{50}	mg/kg/day > 1/5 LC_{50}	
2) *Birds*		
EEC < 1/5 LC_{50}	1/5 LC_{50} ≤ EEC < LC_{50}	EEC ≥ LC_{50}
3) *Aquatic Organisms*		
EEC < 1/10 LC_{50} EEC ≥ 1/10 LC_{50}	1/10 LC_{50} ≤ EEC < 1/2 LC_{50}	EEC ≥ 1/2 LC_{50}
II. *Chronic Toxicity*		
EEC < Chronic No Effect Level	N/A	EEC ≥ Chronic Effect Levels Including Reproductive Effects

*EEC = Expected Environmental Concentration. This is typically calculated using a series of simple nomographs to complex exposure models.

SOURCE: Adapted from Urban and Cook, 1986.

hazard data includes, for example, laboratory fish, aquatic invertebrate, or bird LC_{50} values, and effect levels for fish and avian reproduction tests. *Environmental exposure* is a function of two data components. The first is the estimated amount of the pesticide residue that will be in the environment and available to non-target organisms. The second consists of the numbers, types, distribution, abundance, dynamics, and natural history of non-target organisms which will be used in contact with these residues. Information on the proposed label use of the pesticide is essential for such exposure estimates. Toxicological hazard is estimated first and environmental exposure separately; then they are compared to each other.

In the EPA Pesticides Program, the comparison of exposure with effects data is based on regulatory risk criteria. These criteria are summarized and presented in Table 2. Within the table are risk criteria which contain specific safety factors that were derived from a toxicological model developed by the Program in 1975. The model was designed to provide a safety factor that would allow for differential variability and sensitivity among fish and wildlife species. A detailed explanation of the derivation of these safety factors is found in Urban and Cook (1986).

Many theoretical questions can be raised about the use of risk criteria and safety factors in general. Currently, the Program does not use the model to predict the probability of the pesticide causing significant acute adverse effects to non-target organisms, since the model does not provide a mechanism for estimating model uncertainty. Thus, the risk criteria with their safety factors are used as "rough" estimates of potential risk to non-target organisms.

Specific information and testing data are necessary in order to conduct an ecological risk assessment for a pesticide. Under FIFRA, EPA is not responsible for producing the data needed to make an ecological risk assessment. That burden is placed upon the applicants for registration. The Office of Pesticides Programs has published regulations which specify the data that are required for registration (40 CFR, Part 158 as cited in Urban and Cook, 1986), and guidelines which provide recommended testing methods that are needed to produce the required data.

The EPA Pesticides Program follows four procedural steps in assessing ecological risk:

1. review and evaluation of hazard data to identify the nature of the hazards;
2. identification and evaluation of the observed quantitative relationship between dose and response;
3. identification of the conditions of exposure (e.g., intensity, frequency, and duration of exposure); and
4. combination of dose/response and exposure information to derive estimates of the probability that hazards associated with the use of the chemical will be realized under conditions of exposure experienced by the non-target population(s) under consideration.

These steps result in the comparison of toxicological hazard data and exposure data using regulatory risk criteria. Typically, toxicological hazard data may consist of acute LD_{50} and LC_{50} values, or chronic no-effect levels for the most sensitive indicator species. Exposure data normally consist of model-based estimated environmental concentrations in important media of concern (i.e., water, soil, and non-target organism food items). As the ratio of these input data equals or exceeds the regulatory criteria, a risk is inferred.

CONCLUSION

Both the toxic substances and pesticides programs at EPA recognize that the ratio method for assessing risk has numerous weaknesses. For example:

- it does not adequately account for effects of incremental dosages;
- it does not compensate for differences between laboratory tests and field populations;
- it cannot be used for estimating indirect effects of toxicants (e.g., food chain interactions);

- it has an unknown reliability;
- it does not quantify uncertainties; and
- it does not adequately account for other ecosystem effects (e.g., predator/prey relationships, community metabolism, structural shifts, etc.).

Therefore, at the present time, the state-of-the-art does not provide a complete characterization of the magnitude of risk or the degree of confidence associated with the characterization.

The development of the field of ecotoxicity, like that of risk assessment, has paralleled the increased awareness of the environment during the past two decades. This awareness is especially evident now in Eastern Europe. The science is complex and addresses a broad range of issues that are frequently the focus of public concern and international policy. Today, many ecological risk assessment protocols are modifications of methods used to characterize risk to public health. Unfortunately, these methods often lack environmental validity and may not effectively measure ecosystem integrity. New directions in this field reflect an increased emphasis on the role of sediments, biomarkers, and ecosystem assessments in controlling environmental contaminants. As these methods and protocols become more refined and are practiced by a larger body of scientists, their role and importance in environmental decision making will become much more important.

REFERENCES

Bartell, S.M. 1988. Community and ecosystem models for ecological risk assessment. Oak Ridge National Laboratory, Environmental Sciences Division, Oak Ridge, Tennessee.

Beck, L.W., A.W. Maki, N.R. Artman, and E.R. Wilson. 1981. Outline and criteria for evaluating the safety of new chemicals. Reg. Tox. and Pharmacology (1):19-58.

Cairns, J. Jr., K.L. Dickson, and A.W. Maki, eds. 1978. Estimating the hazard of chemical substances to aquatic life. ASTM STP 667. American Society for Testing and Materials, Philadelphia, Pennsylvania.

Urban, D.J., and N.J. Cook. 1986. Hazard Evaluation Division, Standard Evaluation Procedure, Ecological Risk Assessment 504/9-85-001; NTIS PD86-247-657. U.S. Environmental Protection Agency, Washington, D.C.

U.S. EPA. 1984. Estimating concern levels for concentrations of chemical substances in the environment. U.S. Environmental Protection Agency, Washington, D.C.

Human Effects on the Terrestrial Environment

Characterizing Ecosystem Responses to Stress

MARK A. HARWELL
CHRISTINE C. HARWELL
DAVID A. WEINSTEIN
JOHN R. KELLY

Cornell University

Environmental risk assessment and management involve the use of methodologies to assess risks to the health of biological systems, especially the stresses from human activities. The use of appropriate ecological indicators to measure environmental effects of these stresses can allow a realistic evaluation of risks.

ECOLOGICAL RISK ASSESSMENT

Effective protection and management of environmental systems requires an adequate understanding of stress ecology. Three facets are central to the science of stress ecology: 1) how various components of ecosystems are exposed to stress; 2) how ecosystems respond to these stresses; and 3) how ecosystems recover from or adapt to stress. When there is a solid understanding of these relationships, as well as the inherent uncertainties in predicting stress response, then risks to ecological systems can be properly balanced with risks and benefits to other systems of human concern, such as economic or societal systems. Instances of unprotected or unexpected adverse effects on the environment from a particular human activity will continue to occur, along with instances of expensive over protection from effects of other human activities. Risk assessment based on ecological science is essential to minimize these problems.

In principle, ecological risk assessment is intended to illustrate and accommodate differences in stress/response relationships and to provide a basis for balancing environmental concerns with other economic and societal issues through the associated process of risk management. However, the current reality of risk assessment and risk management is quite remote from this ideal, for a number of important reasons:

1. A satisfactory basis does not exist for making cross-comparisons among alternative risks and values; for example, what value is placed on a single human life or on an endangered species? How do you compare the loss of an endangered whale species versus an endangered liverwort species? Ecological and societal values typically cannot be reduced solely to monetary units; consequently, those factors in an environmental management decision that are easier to quantify, such as economic costs of compliance with regulations, often dominate in debates about the effects of human activities on the environment.

2. Even if there were a common method of valuation, there would be great difficulty in establishing a common level of acceptability of risk across different problems. Emotional and subjective factors play a major role in defining social acceptability. Many factors contribute to this disparity, including very different perceptions about voluntary versus involuntary risks, and society's inability to handle infrequent but catastrophic events on the same basis as frequent but relatively low-consequence events.

For example, the routine emissions from a nuclear power plant have few, if any, demonstrated adverse impacts on the health of the environment or the general public; nearby residents could receive a much greater dose of radiation from the thorium-daughter products in soils and rocks of the neighborhood, including radon in their homes, or from flying a few times across the country at high altitudes, than from a properly operated nuclear facility. In contrast, a fossil-fuel plant continuously emits a host of compounds that are known to cause adverse ecological and human health effects, including acid precipitation, greenhouse-induced global climate changes, and long-lived alpha-emitting radionuclides.

On the other hand, an accident involving the fossil-fuel plant could at most lead to local-scale fires and injuries, whereas an accident at a nuclear facility might cause extensive ecological and human health consequences, as witnessed by increased cancer risks for tens of thousands of inhabitants in Eastern Europe, and has the potential to eradicate a whole lifestyle for cultures in Northern Scandinavia based on herding reindeer populations, all from the single event at Chernobyl.

3. The effectiveness of ecological risk assessment is seriously limited by the inability of scientists accurately to predict ecological responses to stress. There are many reasons for this limitation, including:

- the considerable variety of ecosystems and potential types of human disturbances to those ecosystems;
- the wide range of spatial, temporal, and organizational scales inherent in any ecosystem;
- the lack of an adequate baseline database for comparison of disturbed and undisturbed ecosystems;

- fundamental limitations in ecological theory and understanding; and
- environmental variability and other irreducible forms of uncertainty associated with stress/response/recovery predictions.

ECOLOGICAL RISK MANAGEMENT: REGULATORY ENDPOINTS

Each state or nation has its own traditions and systems of environmental values and regulations. In the United States, the U.S. Environmental Protection Agency (EPA) is charged with the responsibility of protecting and managing many aspects of the environment. To date, the predominant focus at EPA has been on environmental threats to human health, rather than on human actions affecting the health of the environment. This reality may be shifting however, as EPA's interest grows in using ecological risk assessment as a tool for objective environmental decision making (Chapter 6, this volume).

A variety of legislative actions provides the framework for the role of EPA. The language of the laws written by the U.S. Congress typically contains both broad statements as to the law's general purpose and narrower statements or sections in the law which detail the particular activities to be regulated by a government agency such as EPA, sometimes including detailed instructions as to how that regulation is to occur. The exact wording of the legislation, often clarified by a study of the legislation's history, indicates to EPA the directions to follow in formulating regulations and mechanisms to enforce the law. Thus, regulations developed by EPA are one way to translate both general and specific legislative directions provided by Congress (or directions provided by the President's executive orders or by various courts' judicial interpretations) into regulatory actions and requirements. There are sometimes certain issues or phrases in a law that are key to deciding which regulatory actions are to be taken by the government agency; these are often called *regulatory endpoints*, defined as those regulatory norms that translate fundamental legislative purposes into regulatory decisions or actions (C. Harwell, 1989; Limburg et al., 1986) (Figure 1).

The regulatory endpoints for environmental protection can be very *specific*, such as the requirement that sulfur dioxide (SO_2) and ozone (O_3) concentrations in urban areas should not exceed particular numerical amounts; that lead (Pb) levels in gasoline should be below specified concentrations; or that fecal coliform counts in effluents from sewage facilities should remain under a certain value. In part, the degree of specificity in regulations often reflects the level of certainty about causal relationships or the simplicity and consensus within society about the endpoint of concern.

Other environmental regulatory endpoints, however, are quite *generic*;

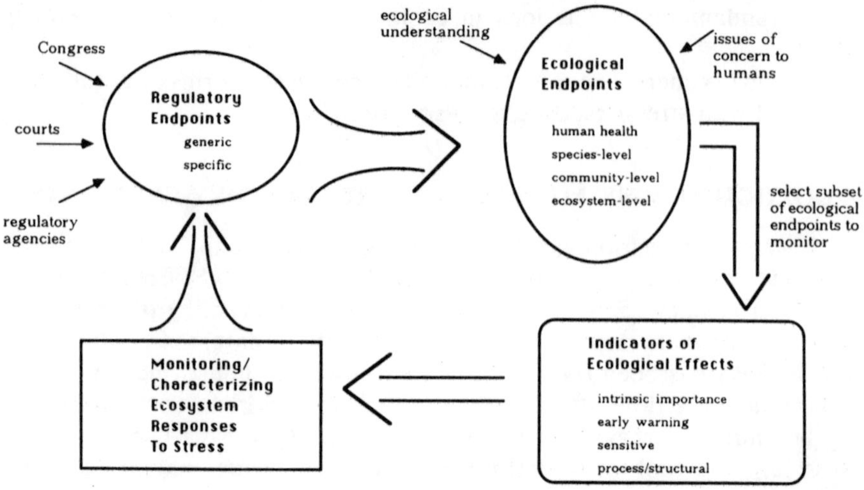

FIGURE 1 *Environmental decision making process.* Regulatory endpoints are specified by legislation, courts, or regulations written by agencies. These must be translated into ecologically meaningful endpoints, focusing on those properties of ecosystems of concern to humans. For each selected endpoint, one or more specific indicator is appropriate to measure or monitor for changes in that endpoint. The suite of selected indicators provides the basis for evaluating ecological responses to stress and characterizing ecosystem health.

they require EPA to take action on an area of legislative concern, though the specific action to be taken is not designated. For example, EPA is required by several laws to develop regulations that will accomplish the general, overall purpose of environmental and human health protection by:

- maintaining and propagating a "balanced indigenous population" in estuarine ecosystems exposed to less-than-secondarily treated municipal effluents (Harwell, 1984a);
- preventing point-source discharges to marine ecosystems that cause "unreasonable degradation of the marine environment" (Harwell, 1984b);
- minimizing "significant adverse impacts";
- protecting "areas of biological concern";
- preventing "irreparable harm" to the environment;
- maintaining "biological integrity"; or
- not allowing actions that result in "accumulation of toxic materials in the human food chain."

The remainder of this chapter focuses on such generic regulatory endpoints, because these tend to be associated with regulation based on ecological responses to human activity, as opposed to the highly specific regulatory endpoints, which tend to focus on technological capabilities or on chemical concentrations.

Generic environmental regulatory endpoints typically do not incorporate language that has clear, intrinsic ecological meaning. For instance, maintaining a "balanced indigenous population" as required by Section 301(h) of the Federal Water Pollution Control Act is actually interpreted in EPA's regulations to relate to maintaining a biological community that is similar to other communities in the region surrounding a local area of disturbance. Thus, the regulatory phrasing is not what was literally specified by the legislation, though it may be what was intended: EPA's regulatory interpretation of the law refers to a community, rather than a single "population"; the biota of concern are not necessarily "indigenous" to the area; and since natural populations continuously fluctuate and experience dynamic interactions, it is not clear that "balanced" has any biological meaning at all. Likewise, it is not immediately evident what constitutes "degradation" of an ecosystem, or what "biological integrity" means. Terms like these were often selected by legislators precisely because the words are subject to alternate interpretations or because they allow flexibility and discretion on the part of the regulatory agency. Nevertheless, there is a common theme in generic regulatory endpoints, i.e., they call for some measure of maintenance of the *health* of the ecosystem. The intent is not to preserve all ecosystems in their pristine state, uninfluenced by human activities; such an endpoint is simply not possible for an industrial nation with a quarter-billion human inhabitants. But, on the other hand, the intent is to prevent serious adverse impacts on ecosystems from human activities that are so extensive that the environment is perceived to be excessively degraded or irreversibly disturbed—that is to say, *unhealthy*.

Unfortunately, unlike measures of adverse human health effects (such as mortality, induced cancers, respiratory illnesses, or chromosomal aberrations), there are no readily comparable, integrative, simple measures or indices of adverse effects on ecosystem health caused by stress. Attempts at drawing an analogy between ecological health and human health (e.g., Rapport et al., 1985) have been unsatisfactory, in part because exposure of ecosystems to stress is very complex, with differential exposure to different parts of the ecosystem; in part because ecosystems are both more diverse and more complex than the human metabolic system; and in part because ecosystems are much less internally integrated, i.e., they have a far less coordinated and controlled response to stress, and fewer mechanisms for compensation and homeostasis. If ecosystems truly were superorganisms, then ecological stress/response/recovery predictions in principle could be as reliable as human health predictions; but the reality is that the science of ecology is not in a position now to meet the needs of ecological risk assessment. Nevertheless, we believe that reasonable environmental decision making can be accomplished in the presence of uncertainties (Harwell et al., 1986; Harwell and Harwell, 1989).

ECOLOGICAL ENDPOINTS

The single most useful criterion to apply to measure ecosystem health is the requirement of relevance to issues of concern to humans. That is, a change in an ecosystem is only considered relevant if it relates directly or indirectly to something affecting humans. By focusing on such human-centered *ecological endpoints*, a structured way of evaluating ecological effects can be developed, and a framework can be created for incorporating non-ecological issues into environmental decision making (Figure 1). While regulatory endpoints are specific goals or standards stated in laws or regulations, ecological endpoints are selected characteristics of ecosystems at various levels, the examination of which can allow evaluation of societally important environmental issues. These issues may have been covered by existing regulatory endpoints or may yet remain to be regulated.

Ecological endpoints are categorized vis-à-vis issues of human concern (Table 1). The first item, human health effects, actually dominates environmental regulation/protection in the United States. Our concern here, however, is limited to ecosystems as vectors for human exposure to potentially harmful substances; we do not consider human effects themselves as part of ecological endpoints. For example, a bathing beach contaminated by high fecal coliform counts involves a serious risk to human health. Similarly, radiocesium deposited in fallout on tundra ecosystems and subsequently biomagnified to dangerous levels for human consumption of reindeer is a serious concern. These examples are demonstrably ecological endpoints, even if ecologically no adverse reactions occur. Even though the radiocesium is unlikely to affect the biota population levels, or productivity, or nutrient cycling rates, its presence in potential human food-chain pathways is *prima facie* an ecological endpoint.

All of the other categories of ecological endpoints in Table 1 are not directly related to human health issues, but are associated with ecological responses and recovery. These can be separated into species-, community-, or ecosystem-level endpoints.

Species-level Endpoints: Primary

The simplest, *primary* ecological endpoints concern direct effects on particular species that have a direct interest to humans. Such interest can involve the economic value of the species, such as Douglas fir, salmon, or oysters. Similarly, direct importance can accrue to recreational species, such as the striped bass of the eastern United States, or the variety of deer and gamebird populations throughout North America. Other species do not have a specific, direct economic value, but are of special concern

TABLE 1 Ecological endpoints: issues of concern to humans.

- HUMAN HEALTH EFFECTS
  - *vector for exposure to humans*

- SPECIES-LEVEL ENDPOINTS
  - *direct interest*
    - economic, aesthetic, recreational, nuisance, endangered species
  - indirect interest (secondary endpoints)
    - bi-species effects (predation, competition, pollination)
    - habitat role
  - *ecological role*
    - trophic relationship
    - functional relationship
    - critical species

- COMMUNITY-LEVEL ENDPOINTS
  - *food-web structure*
  - *species diversity*
  - *biotic diversity*

- ECOSYSTEM-LEVEL ENDPOINTS
  - *ecologically important process*
  - *economically important process*
  - *water quality*
  - *habitat quality*

because of aesthetics or other human values, e.g., dolphins, eagles, wild horses, and grizzly bears.

Many species on earth are endangered or threatened with extinction; unprecedented losses of species are underway, especially in tropical biomes, from massive deforestation. A select few endangered species also hold particular recognition, usually because of aesthetic values rather than because they have a particular ecological value; after all, an endangered species typically is too rare to play a major ecological role. Examples are many species of birds (e.g., least tern, California condor); large predators (e.g., Florida panther, Peregrine falcon); or large herbivores (e.g., white rhinoceros, American bison).

Finally, many species are of direct concern to humans because of a negative role. Such nuisance species include disease-vectors (e.g., certain species of mosquitoes); exotic plants that outcompete native vegetation (e.g., kudzu, *Casurina*); and noxious species, such as blue-green algal blooms.

Species-level Endpoints: Secondary

Indirect effects on species, mediated by effects on other components of the ecosystem, must also be considered. If the ecosystem component is another species which is not directly important to humans, it then becomes a *secondary* ecological endpoint. Its relationship to the species of primary concern can involve several different mechanisms. For instance, bi-specific interactions can be important, i.e., where the primary and secondary species are closely linked through some interrelationships. One example is the predator/prey relationship: if society is concerned with the population levels of blue whales near Antarctica, then there must be a concern for the dynamics of the krill population. In this case, adverse impacts on the density of krill would constitute an ecological endpoint, even though the krill population itself might have little direct concern for humans.

Other bi-specific relationships include pollination (e.g., concern for a species of fruit tree might translate into concern for the insects necessary for fertilization); mutualism (where two species mutually benefit from the presence of each other); and competition. The latter has been shown through theoretical studies to be potentially quite important; for example, studies using computer simulation models of forested ecosystems have shown that direct effects of air pollutants on one species of trees may allow another species to become dominant, as it outcompetes the other by being less sensitive to the pollution (Weinstein et al., in prep.). Also, host/parasite interactions may prove important, such as when air pollutants impact forest tree species indirectly through enhancing the prospects for pest or disease outbreaks (Bedford, 1987).

Other secondary endpoints may be found in effects mediated by habitat alterations, such as changes in the physical structure of the environment that alter the vertical or horizontal heterogeneity of the ecosystem. This is particularly important when a single species dominates the environmental structure for other species in the community, such as mangrove forests, many coniferous forests, seagrass beds, and agroecosystems. Thus, particular concern for habitat-mediated stresses occurs for near-monoculture-dominated ecosystems. For example, mangrove trees provide a complex physical substrate for other plant and animal species, allowing the differentiation and development of a series of diverse ecological communities in particular niches within that ecosystem. Loss of the mangroves would consequently result in loss of habitat for many other species that might not be directly affected by the stress that destroys the mangroves, but that might have a particular importance for humans.

Another important mechanism for habitat-mediated indirect effects is the amelioration of physicochemical conditions by biota or other components of the ecosystem. For example, the presence of tree and shrub species in semi-arid environments can induce evapotranspiration, which in turn can increase precipitation regionally; hence, the biota act as an enhancing pump in the hydrological cycle upon which virtually all terrestrial ecosystems depend. Adverse impacts on this role can indirectly result in other ecological effects, as starkly demonstrated by the current positive feedback from desertification underway in sub-Saharan Africa; indeed, a major component of the drought and famines recently experienced is the human impacts on the environment through biomass harvesting for energy and food. Thus, the ecological effect of concern here (tree and shrub productivity) is of both *direct* importance (with respect to the economic value of this resource) and of *indirect*, habitat-mediated importance (with respect to inducing local- to synoptic-scale reductions in precipitation). Evaluation of this species-level effect becomes an evaluation of an ecological endpoint, determining the ability of that environment to support human life.

Species-level Endpoints: Ecological Role

Another species-level, indirect effect relates to the *ecological role* the affected species plays in the community, such as in maintaining the trophic structure of the community. Such critical species have been identified in several ecosystems; a well-known example is provided in Paine's work (1974, 1980) on keystone species, i.e., particular predatory invertebrates whose presence or absence is the determinant of the presence of other species in intertidal ecosystems. In other ecosystems, a few lower trophic-level species control the food availability for an entire suite of trophic levels directly or indirectly consuming that energy resource base.

Consequently, a species-level ecological endpoint might be the seagrass species *Thalassia*, which supports a diverse ecosystem through reliance on it as a detritus base and relatively stable substrate; another might be the wolves that control populations of herbivores through predation. Other examples include the alligator-controlled ecosystems in the Florida Everglades, known as "alligator holes," in which the alligator determines both the habitat structure and the trophic structure of the ecosystem.

Community-level Endpoints

Another type of ecological endpoint involves *community-level* issues. At this level of endpoint is the overall trophic structure per se, not as a mechanism for supporting a particular species of concern but as a characteristic of direct importance. Humans have come to value the diversity of ecosystems as having intrinsic worth, and consequently any change in species diversity constitutes an ecological change of concern. This value is reflected in some regulatory endpoints that specify maintenance of species diversity as an endpoint and a measure of overall ecological health. For instance, Section 403(c) of the Federal Water Pollution Control Act specifically calls for consideration of changes in species diversity (Harwell, 1984b). Similarly, Section 301(h) of the Federal Water Pollution Control Act indirectly calls for maintenance of species diversity through its "balanced indigenous population" endpoint as interpreted by regulations and litigation (Harwell, 1984a).

The issue of species diversity applies to ecosystem-scale biological units; but the idea also extends across landscape and regional units incorporating many different ecosystems and ecosystem types. Stresses that extend across this spatial scale can result in loss of species, or at least regional-scale loss of species. Consequently, a broader community-level concern arises, often termed a concern for *biotic diversity.* Thus, the elimination of large numbers of species in the tropics, primarily through human destruction of forests for biomass and for agricultural uses, is of immense importance because of the overall reduction in biotic diversity that is occurring essentially on a global scale.

Ecosystem-level Endpoints

The final level of ecological endpoints involves direct or indirect effects at the ecosystem level. Here the concern is for maintenance of processes that are of particular importance. Often, the roles that ecosystems perform in ameliorating environmental extremes are greatly underappreciated; but clearly, many instances exist of ecosystem processes providing tremendous economic or other societal benefit to humans. Examples include maintaining the biogeochemical cycles in wetlands to decrease the high amounts of nutrients in wastewater; maintaining a forest for water retention and floodcontrol; and maintaining dune ecosystems to protect coastal areas from storms.

Other important ecosystem-level endpoints relate to how changes in ecosystem processes cause other changes of concern to humans. In general, changes in biotic populations may not be important, as redundancy or other compensatory mechanisms may prevent adverse changes in ecosystem processes. But the converse is not true, and changes in ecosystem processes almost invariably result in changes in biological constituents.

Some ecological endpoints that need assessment are explicitly recognized in regulations, such as endpoints relating to water quality of surface waters and endpoints relating to habitat quality. There are no simple measures of such concepts as water or habitat quality, but often physical parameters (e.g., soil structure), chemical parameters (e.g., dissolved oxygen levels), or biological parameters (e.g., available habitat for waterfowl, or forage base for deer populations) can be used as surrogates for these ecosystem-level concerns.

DEFINING ECOSYSTEM RESPONSES TO STRESS

Frequency and Novelty of Stress

Qualitatively different responses by ecosystems to stress will depend on the frequency of occurrence and novelty of the disturbance (or closely analogous disturbance) in the evolutionary history of the ecosystem. Thus, the same disturbance can have dramatically different consequences on different ecosystems, e.g., fire affecting grassland ecosystems versus tropical rain forests. In the former case, fire is a natural part of the long-term biogeochemical cycles of the ecosystem, necessary to rejuvenate a biotic community that is adapted to survival or redevelopment after fire. In the case of a tropical rain forest, a fire would lead to extreme disruption of the physical habitat and nutrient reservoirs, and reestablishment of the biotic community would take a very long time, if it were possible at all. Similarly, a particular ecosystem will likely respond differently to different disturbances; for example, the grassland may do well in the presence of fire, but be devastated by overgrazing.

Time Scale of Stress

An ecosystem's response to stress must also be characterized in relation to the particular time scale of occurrence. Acute disturbances (i.e., those involving abrupt, large-magnitude changes in some characteristic of an ecosystem) can be exemplified by the removal of live biomass (Grime, 1979), or by the removal of total biotic material, including previously living material such as litter and detritus (Reiners, 1983). Large-magnitude changes typically involve alterations of species composition, as in the conversion of forests in Vietnam to grass and bamboo ecosystems following the spraying of defoliants (Tschirley, 1969). Elimination of sensitive species, reduction in pools of organic matter, and decreases in diversity have been observed by numerous researchers to occur simultaneously following acute disturbance (Weinstein and Bunce, 1981; Freedman and Hutchinson, 1980; Woodwell, 1970; and Gordon and Gorham, 1963).

Ecosystems often are adapted to cope with many types of natural disturbances, especially chronic stresses that are predictable (e.g., intertidal ecosystems adapted to diurnal and monthly cycles) or periodic (e.g., grassland ecosystems adapted to fire). When these stresses are small relative to the scale of the ecosystem, and when they have occurred commonly during historical development of the ecosystem, the disturbance often will be absorbed within the system structure, adding more heterogeneity but not changing the basic ecosystem functioning. For example, in northern temperate forests a response to wind-induced treefalls can be a formation of a shifting mosaic pattern in which patches of early-successional, mature, and late-successional vegetation composition can be found throughout the landscape (Bormann and Likens, 1979). Wind damage provides the mechanism for the replacement of established patches with early-successional patches that then go through the dynamical processes of maturing. However, chronic human disturbances that do not mimic the frequency of natural disturbances can alter the ability of a forest to absorb damage, and in turn alter basic properties of the ecosystem (Reiners, 1983).

Intensity of Stress

The intensity of the stress will affect the types of responses of the ecosystem. In general, sudden onset of an intense stress causes quite different effects on an ecosystem than a gradual, chronic stress of the same type. Sensitivity to intense stress may depend on the phenological state of the system and the degree of preadaptation. For example, sudden exposure to freezing temperatures in the middle of summer can severely affect a temperate forest ecosystem, whereas the same temperature decrease imposed gradually, over normal seasonal cycles, may cause no effect whatsoever (Harwell and Hutchinson, 1989). The ability of the ecosystem to compensate or acclimate to gradual or low-level stress can lead to a fundamentally different response. On the other hand, long-term exposure to chronic stress may be more consequential, such as prolonged flooding resulting in death to a riparian forest that might well accommodate a brief period of flooding. A stress that extends across a large spatial scale, such as a landscape, simultaneously affecting many different ecosystem units, is likely to be considerably more consequential and to cause considerably greater impediment to recovery than a disturbance of local extent. This effect is related to the availability of propagules to regenerate populations, the distance these propagules must travel, and the availability of refugia for organisms to escape the stress. Landscape-wide disturbances may affect the patterns of these factors.

Internal Stress Protection

Some ecosystems may be protected from stress-induced damage by internal processes. For example, microorganisms may be capable of biochemical degradation, converting toxics to non-toxic substances (Smith, 1980). In other systems, toxic materials are removed from sites of biological activity by the organisms themselves, which uptake compounds and sequester them in storage tissues (Amundson and Weinstein, 1980). Low doses of air pollutants can be absorbed by forests with no immediate changes, although the possibility exists that delayed damage could occur when accumulations reach toxic levels (Smith, 1980). However, ecosystems tend not to be adapted to accommodate rare, extreme events (e.g., freezing events affecting mangrove ecosystems) and may not be adapted to accommodate many anthropogenic stresses (e.g., inputs of toxic xenobiotic chemicals).

Summary

For novel stresses, such as many caused by human activities, stress/response relationships are quite unpredictable. The discussions here are concerned primarily with human-caused stresses on the environment, but the same issues apply to responses to non-human disturbances, such as storm events or invasions by exotic organisms. In summary, the effect of a stress on an ecosystem is a function of the frequency, duration, and intensity of the stress; its similarity to natural stresses to which the ecosystem has had a history of exposure; the spatial extent of the stress; and the natures of both the stress and the ecosystem.

MEASURING ECOSYSTEM RESPONSES TO STRESS

On what scale should ecosystem responses to stress be measured? Clearly, no single scale can be selected exclusively. For example, the population dynamics of forest soil bacteria are very rapid and may fluctuate considerably in time and space; by contrast, the population dynamics of the forest bears are long-term and cover a large spatial extent. Responses of these populations must be measured on very different time scales. If they are not, critical characteristics of population dynamics that occur out of synchrony with measurement intervals may be missed, and the response of the population thus misinterpreted.

Changes within a forest ecosystem can occur gradually over long time periods and may be affected by slowly changing external factors such as climate, which can operate over centuries to millenia. Whereas at one time/space scale the forest is perceived as essentially a steady-state ecosystem with characteristic structure and processes, at a larger scale the same system is perceived as highly variable, responding to climatic alterations over geological time scales. For example, as climate changes, a present temperate forest might become a tropical deciduous forest or perhaps the near monoculture of an arctic coniferous forest.

These characteristics indicate that an ecosystem can only be perceived and defined in an operational context (Levin et al., 1984; O'Neill et al., 1986), and the scales of examination are determined by the processes and biota of interest. Consequently, any effect on the ecosystem must also be examined operationally; there cannot be any single measure that alone adequately characterizes the status or health of an ecosystem.

Although the health of an ecosystem must be examined contextually, there are some general concepts that are germane. To characterize the response of an ecosystem to a stress, one must select specific *indicators* and establish specific norms for comparing those indicators to their non-stressed states (Table 2). An indicator is defined here to mean a measure that reflects some facet of the ecosystem, whether biotic or abiotic, structural or functional. Selecting appropriate indicators is not trivial and cannot be done generically. One measure of an ecosystem's health is the displacement of the selected indicators of the ecosystem compared to their baseline state (Figure 2). This change in indicators may be considered in an absolute or a relative sense, i.e., there might be some weighting of the relative importances of various indicators into a more inclusive, integrative measure. Also, there might be a focus on changes from a base state or a focus on the state of the system itself, e.g., it may be more relevant to know that primary productivity changed by 25% than to know what the actual productivity level is.

To appreciate what this means in a practical sense, consider the case of a forested ecosystem exposed to an anthropogenic stress, such as an input of a toxic chemical. By what measure does one decide if the ecosystem has been altered by this exposure? If the composition and productivity of the forest trees are unchanged, does that demonstrate no effect? How does one evaluate the importance of changes in deer populations; or increased incidences of tree disease and pest outbreaks; or changes in the relative abundances in the soil invertebrate species; or alterations in the rates of nitrogen release from decomposing leaf litter; or replacement of one bacteria species by another that performs the same functions; or mortality from the toxic chemical to a breeding pair of raptors or a breeding pair of protozoans? Which of these responses would constitute an adverse ecological effect from the toxic input? Obviously, there is a wealth of direct and indirect responses that can occur at various hierarchical levels of the ecosystem. It is also obvious that each are not of equal importance.

Resistance and Stability

How readily the ecosystem indicators change in response to a given disturbance is an inverse measure of the *resistance* of the ecosystem to that type of stress (following definitions in Webster et al., 1975; Harwell et al., 1978). Thus, a highly resistant ecosystem would change only slightly in response to the same stress that would cause major displacement in a low-resistance ecosystem. Note the reference to a specific type of stress here: all evaluations of an ecosystem's resistance (or other measures of response)

TABLE 2 Indicators of ecological effects.

- PURPOSES FOR INDICATORS
  - *intrinsic importance* - key: indicator is endpoint
    - e.g., economic species
  - *early warning indicator* - key: rapid indication of potential effect
    - use when endpoint is slow or delayed in response
    - minimal time lag in response to stress; rapid response rate
    - signal-to-noise low; discrimination low
    - screening tool; accept false positives
  - *sensitive indicator* - key: reliability in predicting actual response
    - use when endpoint is relatively insensitive
    - stress specificity
    - signal-to-noise high
    - minimize false positives
  - *process/functional indicator* - key: endpoint is process
    - monitoring other than biota; e.g., decomposition rates
    - complement structural indicators

- CRITERIA FOR SELECTING INDICATORS
  - *signal-to-noise ratio*
    - sensitivity to stress
    - intrinsic stochasticity
  - *rapid response*
    - early exposure; e.g., low trophic level
    - quick dynamics; e.g., short life span, short life cycle phase
  - *reliability/specificity of response*
  - *ease/economy of monitoring*
    - field sampling
    - lab identification
    - pre-existing data base; e.g., fisheries catch data
    - easy process test; e.g., decomposition, chlorophyll
  - *relevance to endpoint*
    - addresses "so what?" question
  - *monitoring feedback to regulation*
    - adaptive management

must be with respect to a particular stress, or combination of stresses, since the nature of the different ecosystems may evoke different responses from different stresses. In this light, one cannot accurately characterize a type of ecosystem as being intrinsically resistant; further, the level of resistance may be determined from measuring one indicator of the ecosystem, but differ in measuring another. How an ecosystem responds to perturbation and how readily it recovers, i.e., the *stability* of an ecosystem, like the ecosystem itself, can only be defined operationally.

Sensitivity

A similar concept to resistance is the idea of *sensitivity*. A sensitive ecosystem is one that responds readily to a particular stress; an insensitive ecosystem may be oblivious to the stress. Sensitivity is not identical to

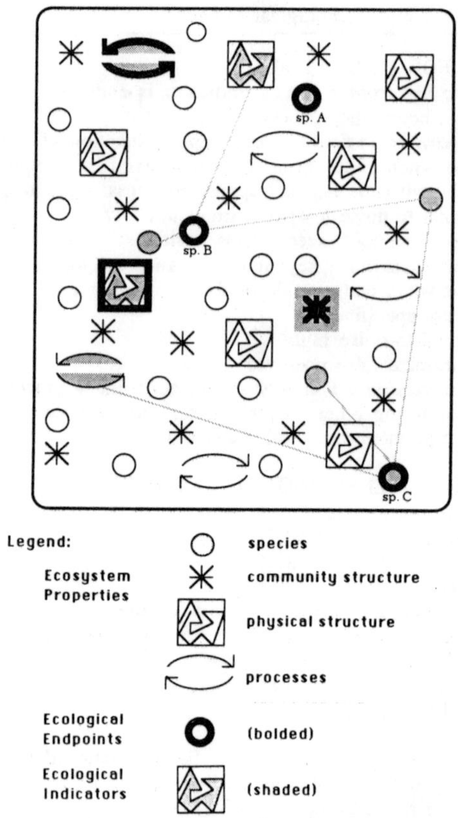

FIGURE 2 *Relationship among ecosystem properties, ecological endpoints, and ecological indicators*. Ecosystems are characterized by a variety of properties that exist across many space and time scales; included are species- and community-level properties, physical structure, and ecological processes. Stress on an ecosystem can change some or all of these properties. Ecological endpoints (shown in bolded figures) are those ecosystem properties (species, community, structural, or process) for which changes would have importance to humans and thus would represent changes in ecosystem health. Each ecological endpoint is measured or monitored by ecological indicator(s) (shown as shaded figures). Sometimes, the endpoint itself is its own indicator (as in the example here for the process, community, and structural endpoints, and for species A and C). Other endpoints are measured indirectly, as for species B and C, and their indicators are other species, community, structural, or process properties of the ecosystem.

resistance, although both measure how much an ecosystem is affected by a disturbance. But sensitivity also has a temporal component, and a system that responds more rapidly than another, or to lower levels of disturbance, is considered to be more sensitive.

Recovery and Resilience

Another concept of importance is *recovery*, i.e., how the ecosystem responds following removal of the stress. Again, there are two components, one related to how rapidly the ecosystem recovers, the other to how effectively the ecosystem recovers. The temporal aspect is characterized as the ecosystems' *resilience*, which is defined as the inverse of the length of time required for an ecosystem to return to near-normal. Note that one cannot define this as a complete return to a pre-perturbed state, because natural heterogeneity might preclude ever attaining that precise state; however, the time required for an ecosystem state to return to a point *within a specified range* of its pre-stress state could be a measure of resilience (Harwell et al., 1981).

One complicating factor sometimes considered is that the non-perturbed ecosystem may well not be at steady-state, even in the absence of human interference. Properties of the ecosystem may change over time. For example, diversity of a forest ecosystem will increase during the early stages of ecosystem development, decline in the middle stages of succession, and increase again during the later stages (Woodwell, 1970). In this case, the ecosystem is characterized by a moving set of values describing the trajectory of the undisturbed ecosystem. How resilient a non-steady-state ecosystem is to disturbance reflects the mechanism of *homeorhesis* of the ecosystem, i.e., feedbacks that tend to direct the ecosystem's state along a specific time sequence. The analog for the steady-state ecosystem responding to disturbance is *homeostasis*, a more commonly known term because of its applicability to physiological control of individuals, such as in maintaining a human's body temperature or blood pH. The human analog to homeorhesis is the developmental sequence and timing associated with embryology and with maturation of the individual into an adult.

Beyond the resilience issues are questions as to whether the ecosystem effectively ever will return to its pre-perturbed state or trajectory. It is possible that some complex ecosystems, when subjected to particular disturbances, will become irreversibly transformed into another system, having different components, steady-states, and dynamics; this is a well-known characteristic of many ecosystems. For example, deforestation in the coastal hills of Venezuela has changed soil structure, seed sources, and the local physical environment sufficiently that forests have not returned even after the areas were abandoned by humans. This phenomenon repeats the irreversible loss of the great forests in Britain during neolithic times, as humans cleared land for agricultural production and energy resources. Perhaps the above examples merely reflect an exceedingly long time period of recovery, and the ecosystem will *eventually* recover. Yet for practical purposes, these examples of ecosystem change are permanent.

Recovery of ecosystems is in part dependent upon the characteristics of the stress (i.e., the disturbance regime, including such factors as the nature of the stress, its frequency, duration, and intensity), and in part the history of the ecosystem (e.g., the level of preadaptation to disturbance, past history of disturbance, and susceptibility of organisms within the ecosystem). An ecosystem that has been subjected to repeated disturbances may tend to deteriorate over time because of loss of nutrient reserves or substrate. Examples of such deterioration can be found in the forests of the San Bernadino Mountains of California following periodic ozone exposures (Miller, 1973), and in salt marshes exposed to a series of oil spills (Baker, 1973).

Recovery from repeated stress may be rapid if most of the important species within the ecosystem complete their life cycles within the interim between disturbance events (Noble and Slatyer, 1980). Alternatively, for single disturbances of longer duration, recovery will be promoted if the important organisms within the ecosystem are capable of outlasting the toxicant by remaining in a latent or resting stage. For example, poor recovery has been noted in grassland systems exposed to oil because oil degradation proceeds slowly, and the actively growing portions of the grasses become directly exposed to the toxicant during their growth periods (Hutchinson and Freedman, 1978).

Characterizing recovery of ecosystems has the same problems as characterizing the ecosystem response to stress, specifically which indicator to examine. Is an ecosystem recovered when its pools of nutrients are back to the pre-stressed state; or when a specific species has reestablished its population at a particular density; or when the residues of a toxic chemical in sediments or in biological tissues have decreased to below some threshold? Just as an ecosystem functions and responds to stress at widely differing rates, hierarchical levels, and spatial extents, it also recovers differentially. There are substantial difficulties added in establishing an appropriate baseline for comparison with the stressed ecosystem, especially since when evaluating homeorhesis, one must not only have an adequate existing baseline but also a representation of what the ecosystem dynamics would have been had the ecosystem not been disturbed. Also, natural heterogeneity and fluctuations again raise the issue of detecting signals from among the noise of natural variations.

Summary

In summary, the health of an ecosystem is much too complex a concept to be quantified by a single measure. The multiscaled and multilayered nature of ecosystems establishes an almost infinite variety of ways of characterizing the ecosystem's state and relationship to some baseline condition.

Simple schemes to overcome this intrinsic complexity of ecosystems are by necessity simplistic and cannot be trusted. This is a true, but unfortunate reality to which society must accommodate; it means there can be no simple, generic answers to the complex problems of environmental protection and decision making. However, an operationally selected suite of indicators, chosen in the context of the ecosystem of interest and the regulatory and ecological endpoints of concern, can offer a reasonable and realistic approach to evaluating ecosystem response and recovery to stress (Figure 2).

SELECTION OF INDICATORS

Given that specific ecological endpoints need to be evaluated for a particular ecosystem and stress, the next step is to identify what indicators should be measured to detect potential changes in the ecosystem (Figure 1). Again there are innumerable components of the ecosystem that could be evaluated, but some careful thought can reduce these to a manageable set of indicators selected to optimize the detection of potential or actual changes in the selected ecological endpoint of concern (Table 2). The first approach to this process is to focus on the purposes of indicators.

Purposes of Indicators

Intrinsic Importance

Some indicators have *intrinsic importance*, such as when populations of direct human interest are measured directly. An example is the valuable striped bass population of the Hudson River and other estuaries of the east coast of the United States. Through the regulatory and, especially, the litigation process, measurements of this species have developed into the central concern for major human disturbances to the Hudson, such as involving thermal power plant siting, the management of the PCB-laden sediments in the river, and the recently resolved Westway Project in New York City (Limburg et al., 1986a, b). Striped bass populations became a primary indicator of ecological effects, especially through evaluations of population levels, age structures, recruitment rates, mortality rates, and migratory patterns. Many other examples of the endpoint itself being the indicator include: deer population levels, breeding success in bald eagles, productivity of Douglas fir stands, and harvested yields of shrimp. The common theme for this intrinsic importance criterion is some direct, usually economic, value of the species or processes.

Early Warning

But a big problem can develop if too much reliance is placed on just monitoring economically important species as the indicator of effects on ecological endpoints of concern. In many cases, by the time an effect shows up on the indicator, it is too late for effective management or mitigation. Thus, another category of indicators is warranted, i.e., *early warning indicators*.

The key characteristic of an early warning indicator is for it to respond rapidly to a stress. Often this criterion means the indicator needs to be exposed to the stress early in the introduction of the stress into the ecosystem. Further, the indicator needs to respond rapidly once exposed. Thus, there is both a time lag and a rate of response factor involved here.

Since the key issue is the *rapid* indication of a potential effect, the early warning indicator is a red flag hoisted to signal the need for closer examination of a potential problem. Consequently, the discrimination of the indicator can be rather low, i.e., it need not provide all the information needed to evaluate effects on the ecological endpoints of concern, and tight, causal relationships between the stress and the triggering of the early warning indicator are not required. Hence, this functions as a screening tool, where false positives are acceptable at a relatively high rate (i.e., having the flag go up even though further evaluation demonstrates no ecological effects of concern). Conversely, early warning indicators need to minimize false negatives; thus, they need to avoid missing a warning for a problem which is real. One way of enhancing this protective aspect is to incorporate more than one early warning indicator in an environmental protection scheme.

Reliability/Sensitivity

Another goal is for it to be a *reliable indicator*, with high capability in characterizing an adverse effect on an ecological endpoint of concern. Note that this category of indicators is focused on *actual* ecological effects rather than on *potential* ecological effects. Thus, the key issue is not the rapidity of response, but the reliability for characterizing changes in ecological endpoints. This type of indicator does require strong evidence of causal relationships with the stress, and the response should be relevant to the state of the ecosystem.

This type of indicator is used when the ecological endpoint itself is relatively insensitive to the stress, or when it is difficult to separate stress-induced changes from the normal variation that occurs over time and/or space. Stress specificity is essential here if the indicator can demonstrate causality needed to justify specific management or protection policies. Also, a criterion for this category of indicators is to minimize false positives, since incorrectly predicting unacceptable adverse impacts could lead to uneconomical overregulation.

Long-term indicators might be necessary to reflect alterations at large spatial/temporal scales, alterations that might not become evident by only examining short-term indicators. As an example, remote sensing of land-use patterns can illustrate loss of estuarine habitats in coastal regions of southwest Florida, alterations that might not be as clearly apparent by monitoring, say, the population levels of tarpon.

Process/Function

Indicators may be chosen to represent alterations in ecological functions and processes. Such *process indicators* may be the ecological endpoints themselves, but it is more likely that process indicators represent the potential for changes in other ecological endpoints of more immediate concern to humans. Note that process indicators are not excluded from also being early-warning indicators, or reliable indicators of change or of state.

Much has been made of the relative value of structural indicators (i.e., biotic indicators of population and community structures) as opposed to *functional indicators* (i.e., of ecosystem processes; see Kelly et al., 1987 and Kelly, 1989). Some authors have suggested that structural indicators, involving effects on biotic populations, are more sensitive and better early-warning indicators than functional indicators (cf., Schindler et al., 1985). Several reasons are offered for this generalization:

- ecological effects are first manifest as effects on individual organisms and subsequently on populations; thus, functional responses would imply prior associated changes in biotic populations performing those functions;
- there is often functional redundancy in ecosystems, so that effects on specific biota may not translate into functional effects; and
- recovery of biotic structure of an ecosystem often lags behind recovery of functional attributes.

However, there are instances where functional indicators respond at least as rapidly and sensitively to stress as structural indicators (Kelly et al., 1987). The point is not to prefer functional over structural indicators or vice versa, but rather, carefully to select functional indicators that can significantly enhance our ability to evaluate ecological responses to stress.

Criteria for Selection of Indicators

Sensitivity

Criteria can be listed for selecting a particular indicator to measure a specified ecological endpoint. One factor is the *sensitivity* of the indicator to stress, i.e., how large is the response of the indicator to a unit of stress. This measure of indicator resistance is important with respect to the normal *variation* that the indicator experiences over time and space in the absence of stress. These two factors, sensitivity and variation, combine to form the major determinants of the signal-to-noise ratio for the indicator. A high signal-to-noise ratio is required for sensitive, stress-specific indicators; a low signal-to-noise ratio is acceptable for screening indicators, especially involving inexpensive or easily measured variables.

As an example of this issue, consider the purported impacts on the striped bass population in the Hudson River ecosystem in comparison to the natural variability of that species. Having density-independent mechanisms as the primary control for these populations means a poor signal-to-noise relationship, and experts were able to argue effectively on both sides of the controversy concerning the presence or absence of demonstrated effects, compensatory mechanisms, and other issues. By contrast, consider the data on CO_2 concentrations in the atmosphere at the Mauna Loa observatory in Hawaii (Keeling et al., 1982). The annual cycle in CO_2 levels, related to seasonal turning on and off the primary production potential of the Northern Hemisphere, is clearly discernable; and it is superimposed over a rather constant, inexorable rise in the annually averaged CO_2 levels, reflecting effects from human inputs of CO_2 and human-caused destruction of primary production. Here the signal-to-noise ratios of both the long-term trend and the annual cycle are quite good, and the indicator is convincing.

Rapidity of Response

A second criterion relates to the *rapidity of response* of the indicator, especially with no time lag and a high rate of signal processing, as discussed previously. Early exposure is important; consequently, for some stresses, especially those transmitted through food webs, the rapidly responding indicator is likely to occur at lower trophic levels. Quick response also implies quick population dynamics, such as having a short life span or at least a short duration for one phase of the life cycle; for example, changes in phytoplankton are likely to occur much more rapidly than changes in whale populations.

Specificity

Another criterion is the *specificity* of the response indicator. High specificity may be critical to establishing causal relationships and, hence, appropriate management decisions. Conversely, broader response characteristics (i.e., low specificity) may be much more appropriate for screening indicators.

Ease/Economy of Monitoring

The criterion of the *ease and/or economy of monitoring* has historically been of special importance. In one sense, seeking the most economical indicator is of major benefit, in that a larger data base is likely to be developed. For example, historical records such as fisheries' catch data or board-feet of lumber harvested provide a major means for comparing current environmental conditions with those in existence prior to the development of stresses from industrial society. On the other hand, we seem to become too enamored with the ease or historical precedent for indicators; and we may find ourselves focusing efforts on large amounts of data with great precision, but with poor accuracy and little relationship to ecological endpoints of concern. This problem applies to laboratory testing (e.g., bioassays on easily maintained but ecologically insignificant species); field sampling (e.g., counting the 95th species in benthic samples even though looking at the top dozen or so will provide virtually all the relevant information); and pre-existing databases (e.g., fishery catches, where the endpoint is a poor indicator of anthropogenic stress on the environment because of poor signal-to-noise ratios).

Relevance

A final criterion is the degree of *relevance* of the indicator to the ecological endpoint of concern. Clearly, if the indicator itself is identical to the endpoint (e.g., the population levels of an endangered species), the relevancy is maximal. Otherwise, the more closely linked the indicator is to the ecological endpoint(s) of concern, the less difficult it is to answer the "so what?" question that often haunts demonstrations of environmental change. Process indicators tend automatically to be considered more relevant than, for example, sensitive species indicators, since loss of a species sensitive to stress, but otherwise not of particular note for humans or ecosystems, raises the "so what?" question (cf., Kelly et al., 1987).

SUMMARY

Ecosystems are complex and varied, multiscaled and multitiered, and subject to continuing change and adaptation. Consequently, a sophisticated approach is needed to characterize ecological effects from human activities, relying on a suite of ecological response/recovery indicators that reflect the status of the variety of facets about the ecosystem, or endpoints, of concern to humans. Focusing on these suites of indicators and endpoints can provide a systematic framework for incorporating scientific knowledge and understanding into a broader process of ecological risk assessment. It is unreasonable and futile to expect that a simple, generically applicable, single measure of ecosystem health can ever be realized. However, such a scheme is not required, and an ecological risk assessment methodology for enhancing environmental decision making is a reachable goal for ecological science.

Acknowledgment

This report is ERC-153a, Ecosystems Research Center, Cornell University. The ERC was established under a cooperative agreement between the U.S. Environmental Protection Agency (EPA) and Cornell. This chapter represents the views of the authors, and not necessarily the views of the EPA.

REFERENCES

Amundson, R.G., and L.H. Weinstein. 1980. Effects of airborne fluoride on forest ecosystems. Pp. 63-78 in the Proceedings of the Symposium on the Effects of Air Pollutants on Mediterranean and Temperate Forest Ecosystems. U.S.D.A. Forest Service, Pacific SW Forest Experiment Station, General Technical Report PSW-43, Berkeley, California.

Baker, J.M. 1973. Recovery of salt marsh vegetation from successive oil spillages. Environmental Pollution 4:223-230.

Bedford, B.L., ed. 1987. Modification of plant-pest interactions by air pollutants. ERC-117. Ecosystems Research Center, Cornell University, Ithaca, NY.

Bormann, F.H., and G.E. Likens. 1979. Pattern and process in a forested ecosystem. New York: Springer-Verlag.

Freedman, B., and T.C. Hutchinson. 1980. Long-term effects of smelter pollution at Sudbury, Ontario, on forest community composition. Canadian Journal of Botany 58:2123-2140.

Gordon, A.G., and E. Gorham. 1963. Ecological effects of air pollution from an iron-sintering plant at Wawa, Ontario. Canadian Journal of Botany 41:1063-1078.

Grime, J.P. 1979. Plant Strategies and Vegetation Processes. Chichester, U.K.: Wiley and Sons.

Harwell, C.C. 1984a. Regulatory Framework of the Federal Water Pollution Control Act, Section 301(h). ERC-29. Ecosystems Research Center, Cornell University, Ithaca, NY.

Harwell, C.C. 1984b. Analysis of Federal Water Pollution Control Act, Section 403: Ocean Discharge Criteria. ERC-26. Ecosystems Research Center, Cornell University, Ithaca, NY.

Harwell, C.C. 1989. Regulatory framework for ecotoxicology. Pp. 497-516 in Ecotoxicology: Problems and Approaches, S.A. Levin, M.A. Harwell, J.R. Kelly, and K. Kimball, eds. New York: Springer-Verlag.

Harwell, M.A., W.P. Cropper, and H.L. Ragsdale. 1978. Nutrient cycling and stability: A reevaluation. Ecology 58:660-666.

Harwell, M.A., W.P. Cropper, Jr., and H.L. Ragsdale. 1981. Analyses of transient characteristics of a nutrient cycling model. Ecological Modelling 12:105-131.

Harwell, M.A., T.W. Duke, and J.R. Kelly. 1985. Evaluation of proposed environmental scoring methodology. ERC-106. Ecosystems Research Center, Cornell University, Ithaca, NY.

Harwell, M.A., and T.C. Hutchinson, with W.P. Cropper, Jr., C.C. Harwell, and H.D. Grover. 1989. Environmental Consequences of Nuclear War. Volume II. Ecological and Agricultural Effects. Second Edition. Chichester, U.K.: John Wiley & Sons.

Harwell, M.A., and C.C. Harwell. 1989. Environmental decision making in the presence of uncertainty. Pp. 517-540 in Ecotoxicology: Problems and Approaches, S.A. Levin, M.A. Harwell, J.R. Kelly, and K. Kimball, eds. New York: Springer-Verlag.

Harwell, M.A., C.C. Harwell, and J.R. Kelly. 1986. Regulatory endpoints, ecological uncertainties, and environmental decision making. Oceans 86 Proceedings 3:993-998.

Hutchinson, T.C., and W. Freedman. 1978. Effects of experimental spills on subarctic boreal forest vegetation near Norman Wells, N.W.T., Canada. Canadian Journal of Botany 56:2424-2433.

Keeling, C.D., R.B. Bacastow, and T.P. Whorf. 1982. Measurements of the concentration of carbon dioxide at Mauna Loa, Hawaii. Carbon Dioxide Review 1982:377-382.

Kelly, J.R., T.W. Duke, M.A. Harwell, and C.C. Harwell. 1987. An ecosystem perspective on potential impacts of drilling fluid discharges on seagrasses. Environmental Management: 11(4):537-562.

Kelly, J.R. 1989. Ecotoxicology beyond sensitivity: A case study involving 'unreasonableness' of environmental change. Pp. 473-496 in Ecotoxicology: Problems and Approaches, S.A. Levin, M.A. Harwell, J.R. Kelly, and K. Kimball, eds. New York: Springer-Verlag.

Levin, S.A., K.D. Kimball, W.H. McDowell, and S.H. Kimball, eds. 1984. New perspectives in ecotoxicology. Environmental Management 8(5):375-442.

Limburg, K.E., S.A. Levin, and C.C. Harwell. 1986a. Ecology and estuarine impact assessment: Lessons learned from the Hudson River and other estuarine experiences. J. Environmental Management 22:255-280.

Limburg, K.E., M.A. Moran, and W.H. McDowell. 1986b. The Hudson River Ecosystem. New York: Springer-Verlag.

Miller, P.L. 1973. Oxidant-induced community change in a mixed conifer forest. Pp. 101-117 in Air Pollution Damage to Vegetation, J.A. Naegle, ed. Adv. Chem. Series No. 122, American Chemical Society, Washington D.C.

O'Neill, R.V., D.L. DeAngelis, J.B. Waide, and T.F.H. Allen. 1986. A hierarchical concept of ecosystems. Monographs in Population Biology 22. New Jersey: Princeton University Press.

Paine, R.T. 1974. Intertidal community structure: Experimental studies on the relationship between a dominant competitor and its principal predator. Oecologia 15:93-120.

Paine, R.T. 1980. Food webs: Linkage, interaction strength, and community infrastructure. J. Animal. Ecol. 49:667-685.

Rapport, D.J., H.A. Regier, and T.C. Hutchinson. 1985. Ecosystem behavior under stress. Amer. Nat. 125(5):617-640.

Reiners, W.A. 1983. Disturbance and basic properties of ecosystem energetics. Pp. 83-98 in Disturbance and Ecosystems, H.A. Mooney, and M. Godron, eds. New York: Springer-Verlag.

Schindler, D.W., K.H. Mills, D.F. Malley, D.L. Findlay, J.A. Shearer, I.J. Davies, M.A. Turner, G.A. Linsey, and D.R. Cruikshank. 1985. Long-term ecosystem stress: The effects of years of experimental acidification on a small lake. Science 228:1395-1401.

Smith, W.H. 1980. Air pollution: A 20th century allogenic influence on forest ecosystems. Pp. 79-87 in the Proceedings of the Symposium on the Effects of Air Pollutants on Mediterranean and Temperate Forest Ecosystems. U.S.D.A. Forest Service, Pacific SW Forest Experiment Station, General Technical Report PSW-43, Berkeley, California.

Tschirley, F.H. 1969. Defoliation in Vietnam. Science 167:779-786.

Webster, J.R., J.B. Waide, and B.C. Patten. 1975. Nutrient recycling and the stability of ecosystems. In Mineral Cycling in Southeastern Ecosystems, F.G. Howell, J.B. Gentry, and M.H. Smith, eds. ERDA CONF-740513, NTIS, Springfield, VA.

Weinstein, D.A., M.A. Harwell, and L. Buttel. (in prep.) A sensitivity analysis of a forest simulator: Methodology and application. ERC-147. Ecosystems Research Center, Cornell University, Ithaca, NY.

Weinstein, L.H., and H.W.F. Bunce. 1981. Impact of emissions from an alumina reduction smelter on the forests at Kitimat, B.C.: A synoptic review. Pp. 2-16 in the Proceedings of the 74th Annual Meeting of the Air Pollution Control Association, Philadelphia, Pennsylvania, June 21-26, 1981.

Woodwell, G.M. 1970. Effects of pollution on the structure and physiology of ecosystems. Science 168:429-433.

Measuring Function and Disfunction in Ecosystems

Alicja I. Breymeyer
Institute of Geography and Spatial Organization
Polish Academy of Sciences

Debates over the forms of environmental damage caused by increasing and hasty human activity often revolve around a central question: How should these deviations be measured? This is an especially difficult issue in moving from simple situations where injury can be easily observed to more complicated ecological systems, i.e., ecosystems composed of hundreds or even thousands of species which, within their trophic groups, are disposed to take over the niches and functions of eliminated competitors. These deviations should be recognized early, when adverse effects begin to occur in an ecosystem and when remediation is still possible; belated recognition of dying ecosystems signals both ecological disaster and the defeat of ecologists who have not foreseen them.

The behavior and condition of particular species do not necessarily reflect the condition of the whole ecosystem, unless the system is dominated by only one producer species. Except for agriculture, monocultures are rather rare today; however, in Europe, until quite recently, pine cultures had been maintained for several generations, i.e., for several hundred years. Over such a long period, a specific ecosystem based on only one producer species is formed. Such a system will be fundamentally altered if the main producer of organic matter (e.g., pine) is devastated, and only in such a case can the state of an ecosystem be evaluated through the condition of a single species. Usually, however, there are many producers in an ecosystem, and the state of the ecosystem cannot be recognized from the condition of only one species of plant or animal.

Some very specific types of ecosystems function with one or few consumers, such as the case of a tundra ecosystem described by Klekowski and Opalinski (1984). The tundra in the Fugleberget catchment on Spitsbergen, north of Norway, is supplied with allochthonous organic matter and nutrients from breeding colonies of marine birds feeding in the near-shore waters and nesting on land. The excrement of birds is distributed by surface waters and fertilizes a thin layer of soil. However, there are usually many consumers in an ecosystem, and their condition is not in itself a sufficient indicator of the state of the ecosystem. Similarly, particular plant species commonly used as bioindicators are not adequate to diagnose the condition of an ecosystem. They may indicate the contamination of air, soil, or water, and therefore, work as live instruments to measure concentrations of different chemical compounds (Chapter 14, this volume). However, they do not answer questions about the response of an ecosystem to these substances, nor do they characterize its condition.

ECOSYSTEMS: DEFINITION AND FUNCTION

In order to make further discussions more explicit, we should define what is meant by an *ecosystem* (Breymeyer, 1981b). An ecosystem is an ecological system in which organic matter produced within this system (or partially from outside) is transformed at successive trophic levels according to known and described patterns of matter cycling and energy flow. This definition is consistent with the hierarchical approach presented by O'Neill et al. (1986).

The ecosystem is treated as a "medium number system," consisting of a number of smaller functional components. Such an explicitly functional definition of an ecosystem makes its delimitation difficult; generally speaking, so-called "small cycles" of basic elements are included in this system as they are conditioned by biotic components of the ecosystem. "Large cycles" of energy, water, and elements are dependent on atmosphere, geology, and other physical and chemical factors. These are external and can be considered only as an environment in which the ecosystem functions.

The main inputs from this environment to the ecosystem are solar energy, O_2, CO_2, and water solutions of essential elements. Main outputs are gases, the energy dispersed in metabolic processes, and the remains of organic matter falling into the soil or sediments. Interpreted in this way, the boundaries of an ecosystem are designated in the diagram of carbon cycling (Figure 1). The ecological system operates within these limits; its environment is outside.

The creation and transformations of organic matter through a network of functioning trophic levels are the basic ecosystem processes. These are the production of organic matter, its consumption and incorporation into the bodies of successive consumers, and its breakdown by decomposers. This sequence of basic ecosystem processes included in the functional definition of an ecosystem is necessary for the self-restoration of an ecosystem and for its theoretically unlimited duration. Under conditions of increasing

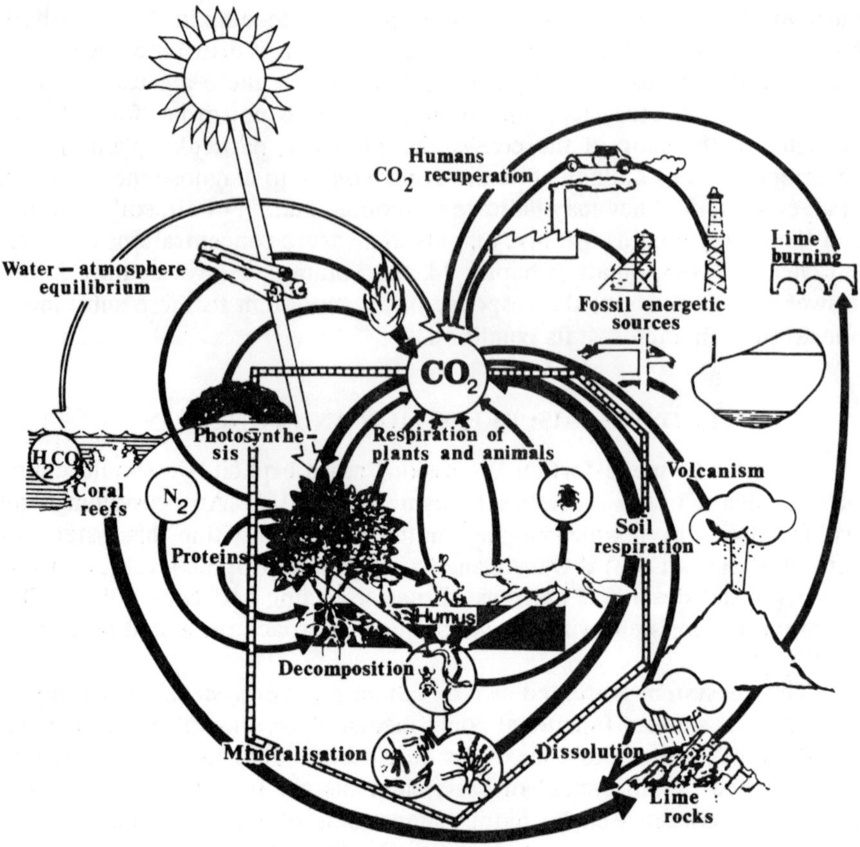

FIGURE 1 Carbon cycle and ecosystem boundaries. The "small" carbon cycle is framed by the ecosystem boundary (IIIIII); the "large" carbon cycle connects the ecosystem with the outside environment (Duvigneaud and Tanghe, 1962, adapted).

human impact, fewer ecosystems function solely on the basis of internal production of organic matter. Similarly, the process of decomposition, which releases nutrients necessary for biomass production and subsequent recycling, is often supplemented by intended or unintended fertilization from human activities. However, in spite of the fact that basic ecosystem processes are intensified or limited from outside, the fundamental ecosystem function is still the same: energy flow and cycling of matter.

The functional definition of an ecosystem provides a convenient basis to determine the limits of ecosystem capacity for pollutants and other stresses, i.e., the limits of ecosystem resilience can be estimated by the resilience of ecosystem processes. These processes can obviously be intensified or reduced; and the proportions among production, decomposition, and consumption can vary in each ecosystem. However, these processes must continue. If any of them disappear, the ecosystem by definition no longer exists. The disappearance of ecosystem processes is measurable, so the capacity (vulnerability) of an ecosystem can be measured as well.

How can we determine the level at which the rates and the character of ecosystem processes are altered from their "natural" state, resulting in disturbances in the functioning of the whole ecosystem? Comparative analyses provide a useful starting point. The natural range of rates of ecosystem processes should be determined in presently existing, ecologically stable ecosystems operating at different levels of production and decomposition. Comparative ecology of ecosystems provides the opportunity to utilize both older data from literature and recent information collected intensively at several ecological centers. Examples are the current studies of Berg et al. in Sweden, Meentemeyer and Box in the United States, Singh in India, and Parkinson et al. in Canada. In Poland, investigations of pine forests are carried out on a rather wide scale; data from these studies have yet to be published, but they are partly used in this chapter.

NATURAL INFLUENCES ON ECOSYSTEMS

Climate is the basic factor of the exterior environment that conditions the production and decomposition of organic matter in ecosystems. Materials from the International Biosphere Program (IBP) and the older studies compiled by O'Neill and DeAngelis (1981) show the dependence of forest productivity upon two basic components of climate: temperature and precipitation (Figure 2). These comparisons are based on a wide range of both elements of climate comprising the whole earth, so that the extreme values of forest productivity are represented. A similar, distinct pattern exists for the productivity of grassland ecosystems over a wide range of precipitation (Figure 3). The allocation of aboveground (green) and underground biomass in grassland ecosystems shows distinct dependence on the quantity of water available to the site (Figure 4): the fraction of underground biomass increases with decreasing precipitation.

Litter decomposition measured along the transect from northern Sweden to central Poland seems to react to lower average temperatures by decline in the decay rate (Figure 5). Stands examined along the transect were chosen with great care; pine forests of similar age growing on similar sandy soil were selected. The range of precipitation is not wide, although the two Spanish sites are situated in much drier conditions. The differentiation of temperature can be assumed to be the primary limiting factor, although the dependence is not simple, as can be seen in the variability of measurements.

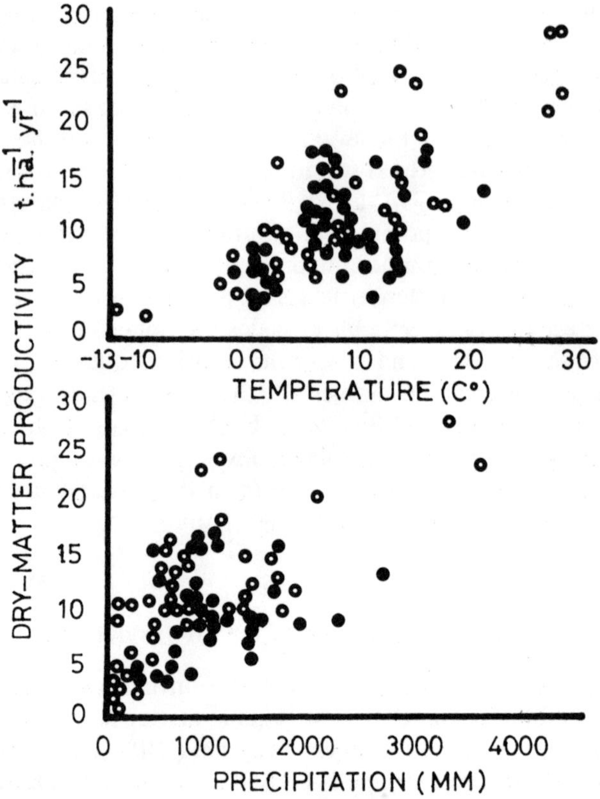

FIGURE 2 Forest productivity in relation to temperature and precipitation. Open and closed circles show two sources of information (IBP data after Reichle measurements, cited by Lieth, 1975).

Meentemeyer (1984) calculated the index of evapotranspiration and showed its correlation with decomposition. The question of the dependence of litter decomposition on its lignin content has provoked heated debate. Initial calculations made on measurements from the temperate zone suggest that the rate of litter decomposition could be precisely predicted by lignin content. Very high correlations of both these values were recorded in subsequent studies. This index was not applicable to a desert environment in which the rate of litter decay was found not to depend on lignin content (Schaefer et al., 1985).

Berg (1986) surveyed the research on the dependence of litter decomposition rate on its chemical composition, suggesting that different stages of decomposition are controlled by different factors and that there are "lignin-controlled" as well as "nitrogen- and phosphorus-controlled"

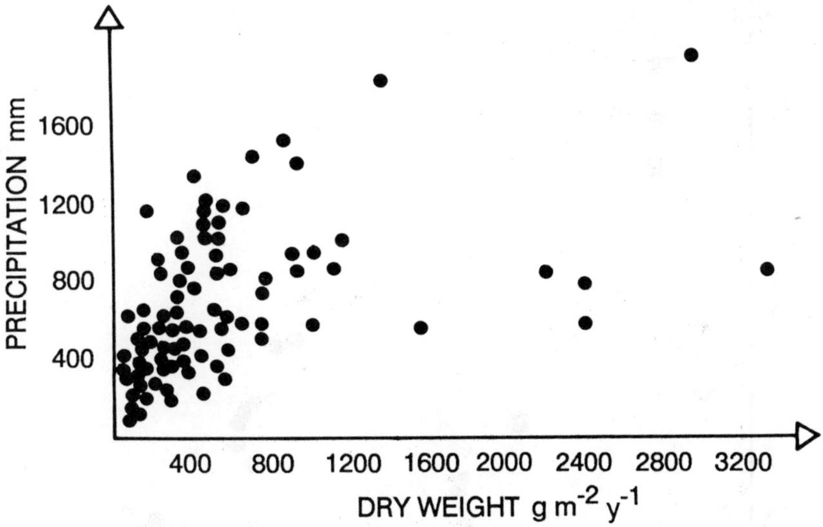

FIGURE 3 Annual precipitation and aboveground production of grasslands (Breymeyer, 1981).

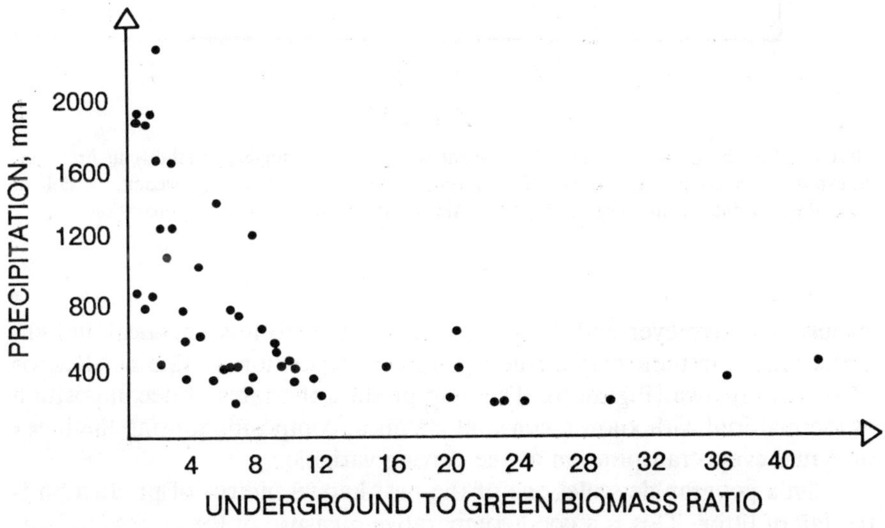

FIGURE 4 Ratio of underground/green biomass in relation to precipitation for various types of grasslands (Breymeyer, 1981).

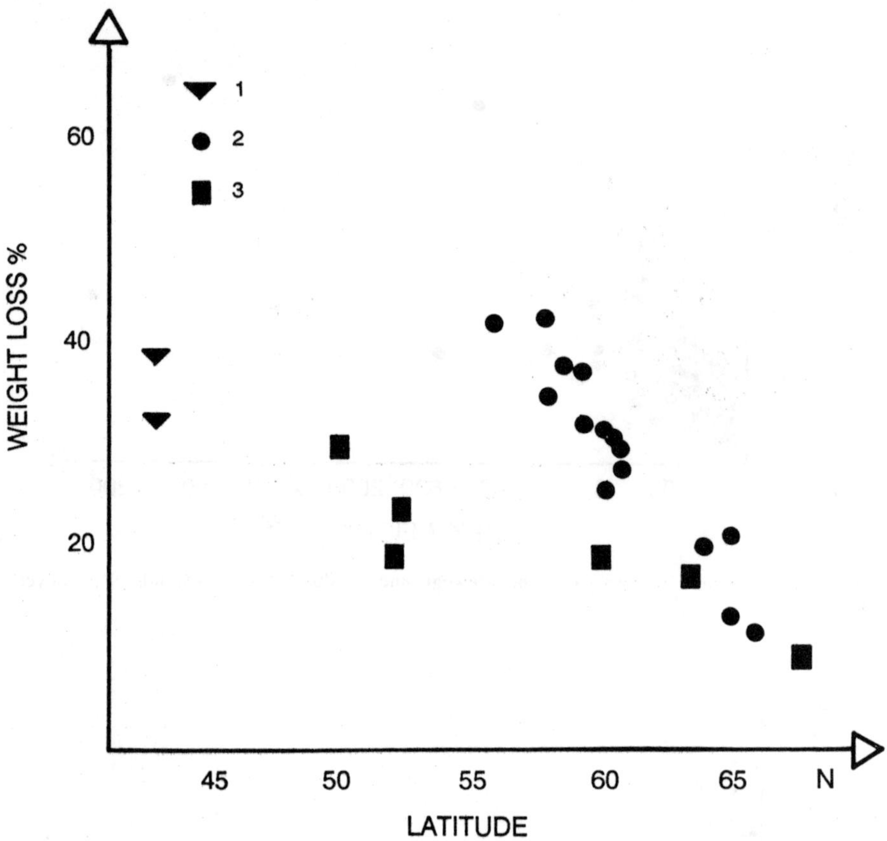

FIGURE 5 Litter decomposition as percent weight loss (litter-bag method) in first year of exposure in coniferous forests of different latitudes. 1=Spain, 2=Sweden, 3=Poland (according to data from Berg et al., 1983; Alvera, 1981; Breymeyer, in preparation).

stages. Meentemeyer and Berg (1986) attempted to link chemical and climatic factors in their map of the rate of decomposition of pine needles for all of Scandinavia (Figure 6). The map predicts the rates of decomposition of the material with known, constant chemical composition, using the index of actual evapotranspiration as the driving variable.

On a geographic scale, one of the best known indices of production is the fall of litter. This is a good comparative measure of forest productivity, though it is known that the correlation between litter fall and production becomes less direct in more productive forests (Figure 7). Nevertheless, measurement of litter fall is commonly used as a measure of forest production, particularly in monitoring programs (Breymeyer 1981, 1984) or

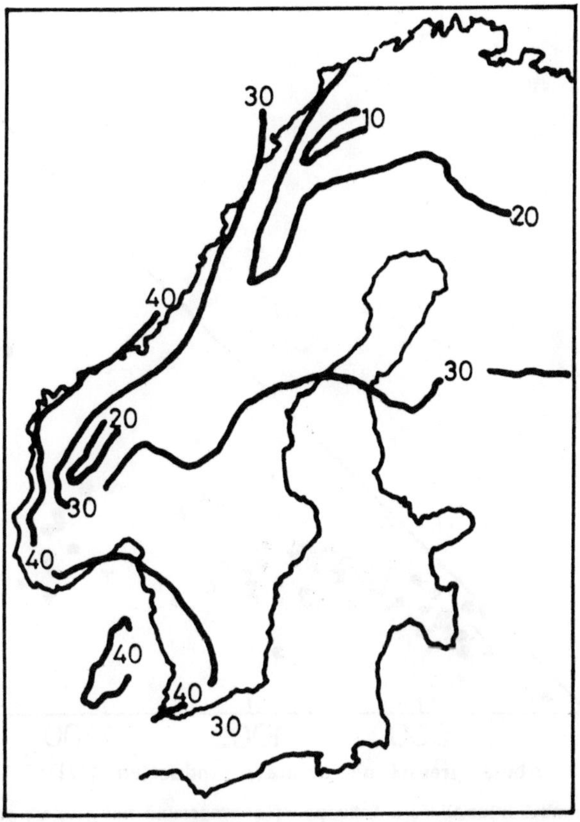

FIGURE 6 Isolines of decomposition rates for Scandinavia. The first-year needle-litter mass loss (%) was predicted on the basis of similar initial chemical composition of needles (N=0.41%, P=0.022%, lignin=25.7%) with changing actual evapotranspiration (AET) value (Meentmeyer and Berg, 1986, amended).

in modeling (Ajtay et al., 1979), since the complete estimation of forest productivity is quite laborious.

Theoretically, it can be assumed that all biomass produced in an ecosystem must at some time die and fall into the soil or sediments; error in this estimation results from ignoring the loss of biomass eaten by consumers, as some portion of organic matter produced is always consumed before it falls down. In forest ecosystems, this error can be neglected. According to the estimates of different authors, the consumption of crown herbivores does not exceed a few percent of a canopy biomass, except in conditions of pest outbreaks. Hence, litter fall—the measure of "annual green production" of a forest—can be assumed to be a good index of total aboveground production of a forest ecosystem. As an example, the production and production

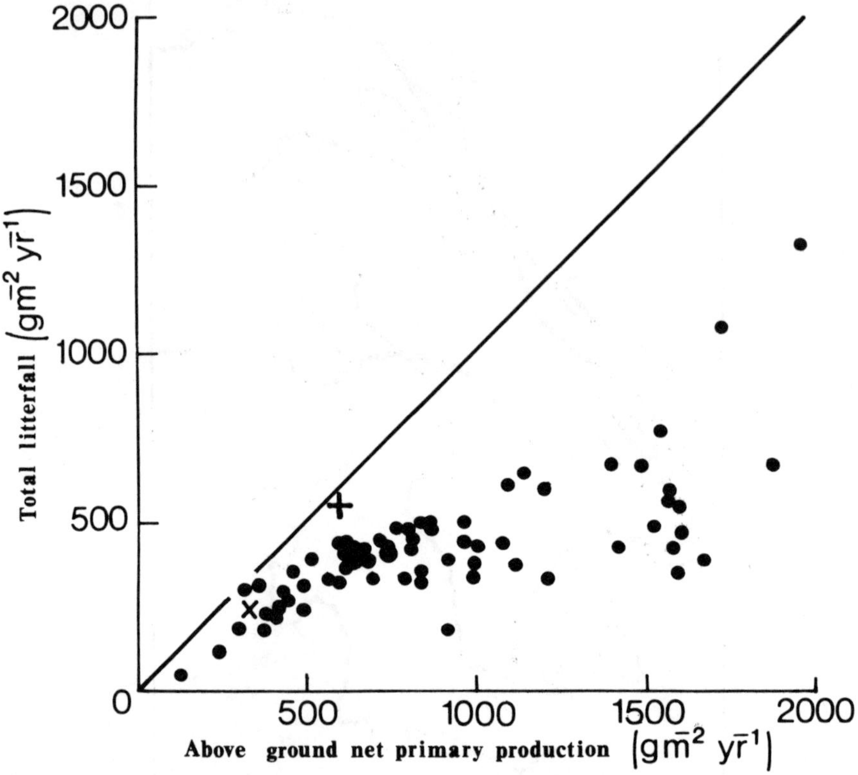

FIGURE 7 Annual litterfall as a function of aboveground net primary production. The solid line shows the situation when the 100% of production falls down (O'Neill and DeAngelis, 1981; Debazac, 1983; Grodzinski et al., 1984).

efficiency in four mixed-pine forests in Poland and the United States can be analyzed (Table 1). According to these measurements, it can be assumed that half of aboveground forest production falls to the soil annually; in the years when cones are produced, this percentage is considerably higher.

Let us analyze the dependence of litter fall on climatic conditions (Figure 8). (It is worth mentioning that only a small number of research centers have undertaken studies of complete forest productivity.) A large number of measurements show that the amount of organic matter delivered to the soil in forests decreases with nothward movement from the equator. This dependence is unquestionable within the gradient from 0° to 70°N latitude. However, at a smaller scale, such as the size of the territory of Poland (6° in latitude, 49° to 50°N), variance is relatively large and the data do not show any correlation with latitude.

TABLE 1 Production and production efficiency in four stands of mixed-pine forest.

MEASUREMENTS	FOREST STANDS 1	2	3	4
Tree production				
Tdw ha^{-1} y^{-1}	2.5	4.7	7.9	14.3
% of total:				
Leaves, fruits, twigs	40.4	53.7	24.7	57.4
Production efficiency (P/B)				
Trees	1 - - - 5%			
Leaves	50 - - - 53%			

SOURCE: Reiger et al., 1884; Whittaker and Marks, 1975.

On the basis of this information, one could conclude that the minimum north-south span for observing litter fall must be 20°. Does this also mean that the total production of these ecosystems reacts to changes in macroclimate only when the area comprises 20°? Some data suggest that such areas may be even more extensive. The measurements of wood production collected during the IBP are noted in Figure 8. Although the values maintain the general increasing trend from the north toward the equator, they show no such clear tendency between 30° and 60°N; in these latitudes, wood production may actually increase as one moves northwards. These measurements are not numerous, so it is not yet known whether this tendency is definitive or not. Nevertheless, it is an indication that climate regulates the wood production only over large areas.

Some idea of the variability of organic matter on a global scale is offered by the map by Olson et al. (1983) entitled "Major World Ecosystem Complexes Ranked by Carbon in Live Vegetation." The Eurasian continent is reproduced from this map in Figure 9. The biome of temperate needle-leaf forests is marked according to the UNESCO-Udvardy (1975) biome map. This biome covers four classes of the complexes defined by Olson et al. (1983): tundra; northern and maritime taiga; main and southern taiga; and cool conifer. According to the measurements collected by Olson et al., tundra is characterized by the smallest biomass (i.e., carbon accumulation) at 0-1 kg cm^{-2}. Northern and maritime taiga accumulates at 3-6 kg cm^{-2}; and the main and southern taiga and cool conifer complexes are both characterized by a wide range of carbon accumulation at 4-25 kg cm^{-2}. Zonal distribution of these ecosystem complexes is distinct, especially in the case of two taiga complexes. Further analysis of the map shows that this natural variability of needle-leaf forest takes place in the span of 20° North-South. Does this mean that living carbon in ecosystems is more

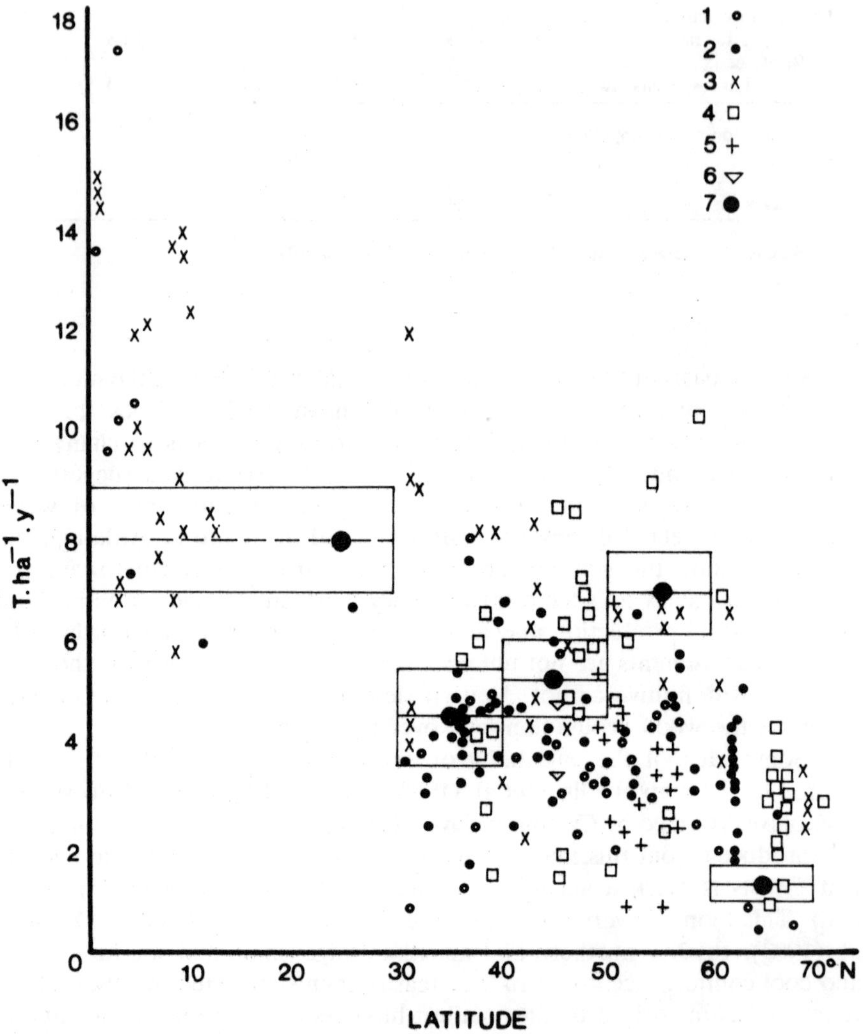

FIGURE 8 Annual litterfall in relation to latitude. 1=Bray and Gorham, 1984; 2=IBP Woodland Data Bank (Reichle, 1981); 3=Vogt et al., 1986 (broadleaved forest); 4=Vogt et al., 1986 (needle forest); 5=Breymeyer et al.(unpublished data); 6=Alvera, 1980; 7=means and standard errors of measurements of wood production according to Woodland Data Bank. Vogt et al. data are collected for northern and southern latitudes, the others for northern latitudes only.

sensitive to climate changes than is timber productivity described earlier (Figure 8), or are these differences only in methods of data interpretation?

ANTHROPOGENIC INFLUENCES ON ECOSYSTEMS

In addition to analyses of the natural variability of ecosystems, let us consider variability induced by humans by analyzing two examples of human-provoked disfunction in grassland ecosystems which were subjected to drastic changes in water conditions.

Irrigation

Bulla et al. (1981; in press) presented results of experimental irrigation of savanna in Venezuela. Water level was raised by several dozen cm over an extensive area by using dikes to retain the savanna rainwater that previously flowed to the Apure River. In this way, an experimental "modulo" was created that was covered by water over an entire year.

Grass production in the area increased quickly; the maximum standing crop exceeded 1,700 grams of dry weight per square meter (g dwm^{-2}), while previously it did not reach 900 g dwm^{-2}. However, satisfaction at the success of such a great and expensive experiment was disturbed soon: "as early as 2 or 3 years after the construction of the experiment, a thick layer of undecomposed material covered the ground" (Bulla et al., in press). When the layer of litter exceeded 1,000 g dwm^{-2}, it became evident that the pasture ecosystem had been transformed and was in danger of being turned into a kind of tropical peat bog. In the artificial conditions created in the modulo, the subsystem of decomposers was inefficient and unable to decompose the organic matter produced.

It has been calculated that even a small change in the decomposition rate results in a considerably high accumulation of litter. For example, the decrease in decomposition rate from 140 mg g^{-1} to 100 mg g^{-1} $month^{-1}$ doubles litter accumulated in the system. The authors built a predictive model, which indicated that the modulo ecosystem should become stabilized at a certain level of production/decomposition processes within six years.

Drainage

An opposite type of transformation in the functioning of an ecosystem can be observed in the large areas of peat bogs in Poland. In the postwar period, these bogs were drained in order to increase grass production. However, they were drained so carelessly that a rapid and irreversible increase in the rate of peat mass decomposition was provoked in some regions. The subsystem of decomposers stimulated by lowering the water

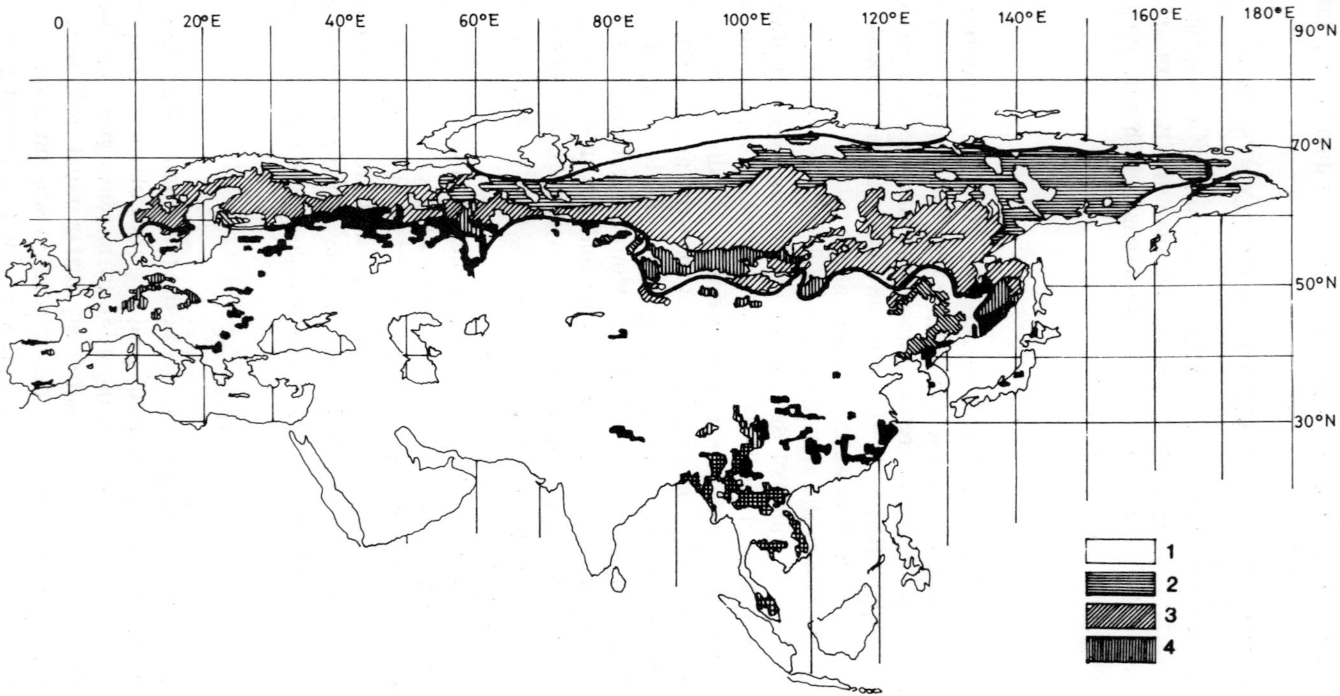

FIGURE 9 Carbon in live vegetation of coniferous temperate forest biome. 1=tundra, 0-1 kg c m^{-2}; 2=northern and maritime taiga, 3-6 kg m^{-2}; 3=main and southern taiga, 4-25 kg c m^{-2}; 4=cool conifer, 4-25 kg c^{-2} (Olson et al., 1983; amended by introducing the boundaries of biomes according to UNESCO-Udvardy map).

level and greater oxygen access worked so efficiently that dead organic matter disappeared from these systems.

Zurek (1987) compiled data from different authors which suggest that dried peat in Poland decreases at a rate up to 2 cm per year. Okruszko (1978) calculated that such mineralization causes the disappearance of 10 tons of peat mass per hectare per year on an average. This constitutes a great loss of organic matter, which is formed peculiarly and retains large amounts of water; it has been calculated that peat bogs, in their spongy mass of organic remains, retain more water than inland open water bodies (Okruszko, 1978).

Effect of Industrial Emissions on Grassland and Forest Ecosystems

Very large disturbances in the functioning of grassland ecosystems can be observed in Polish Silesia, which is heavily polluted by coal mining and other extractive industries. Local stimulation and growth of plant production have been recorded in the region; however, decomposition rates decreased at the same time. As a result, plant remains saturated with a range of chemical compounds from gas, dust, and water accumulate in the ecosystem and are consumed by detritivores; and biomass of this trophic group is thereby reduced.

It has been recorded that, under such conditions, a tendency toward diminution of body size occurs in different trophic groups of consumers, i.e., larger animals are replaced by smaller ones (Figure 10). This exchange is probably a favorable tactic for a species, as it results in faster metabolism and faster disposal of toxic substances from the population. However, for an ecosystem, this change is very expensive with respect to energy costs—the quantity of energy transformed in the ecosystems increases.

It has also been recorded that the percentage of predators in the group of all consumers increases. This change is also inefficient with regard to energy, as the controllers, which predators are in an ecosystem, need not occur in a great mass to act effectively. There is some evidence from other investigations that predators distinctly accumulate certain elements in their bodies (Table 2). Thus, the top invertebrate consumers accumulate chemicals in their tissue and, at the same time, they speed up their metabolic processes.

Let us return to the forest. In the vicinity of Pulawy, Poland, where the immense "Azoty" nitrogen fertilizer plant was built in 1966, distinct changes in the functioning of pine forests have been observed. The factory began to emit a whole complex of nitric and other compounds, which fell on the surrounding forests and caused their dramatic degradation. A very interesting sequence of symptoms was then observed. At the first stage,

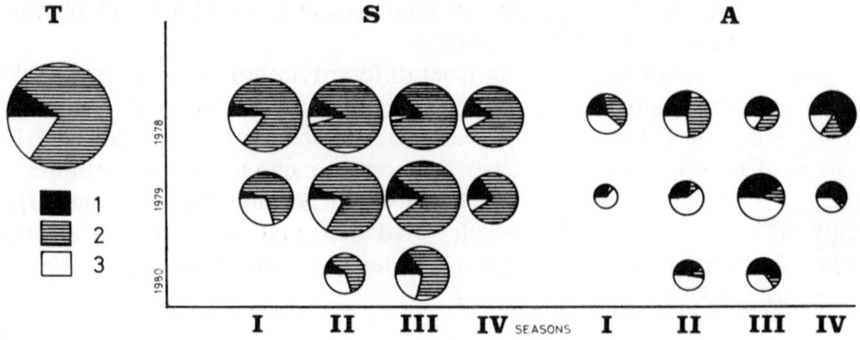

FIGURE 10 Biomass of various groups of soil fauna (g fw m^{-2}) in three grassland ecosystems: T=non-polluted meadow of medium fertility in Poland (data from literature), S=Szczyglowice meadow in Silesia (moderately polluted); A=industrial Aniolki meadow (heavily polluted). Circle size is proportional to the total biomass of soil fauna (for a non-polluted meadow the biomass 135 g fw m^{-2} was assumed). In each circle three groups of fauna are shown as percentages of total biomass: 1=microfauna, 2=oligochaeta, 3=other invertebrates. Samples were collected in four seasons and three years (spring, summer, fall, and winter 1978-1980) (Cianciara and Pilarska, 1984, amended).

TABLE 2 Increase in content of selected elements in the bodies of three main groups of consumers in a NPK fertilized meadow. Calculated on the basis of consumer biomass evaluated on fertilized and unfertilized plots (mg m^{-2}).

Invertebrate bodies from trophic groups	Elements analyzed	Fertilized Non-fertilized
Phytophagans	C	1.55
	N	1.70
	K+Na	1.72
Saprophagans	C	1.71
	N	1.69
	K+Na	2.00
Predators	C	2.73
	N	2.88
	K+Na	3.00

SOURCE: Mochnacka-Lawacz and Olechowicz, 1978.

there was a kind of nitric fertilization which caused substantial elongation of needles and sprouts (shoots), but also induced their flabbiness and early dropping from the trees (Jakubczak and Pieta, 1968). Several years later, the first portions of the forest began to die, and the border of living forest moved away from the factory. Investigations on the production and decomposition of organic matter in the area were carried out in the period 1980-1982. In successive zones moving away from the factory, a whole range of changes could be observed—a zone of bare soil; a zone of thick litter composed of very long needles and twigs; a zone of dwarf pines;

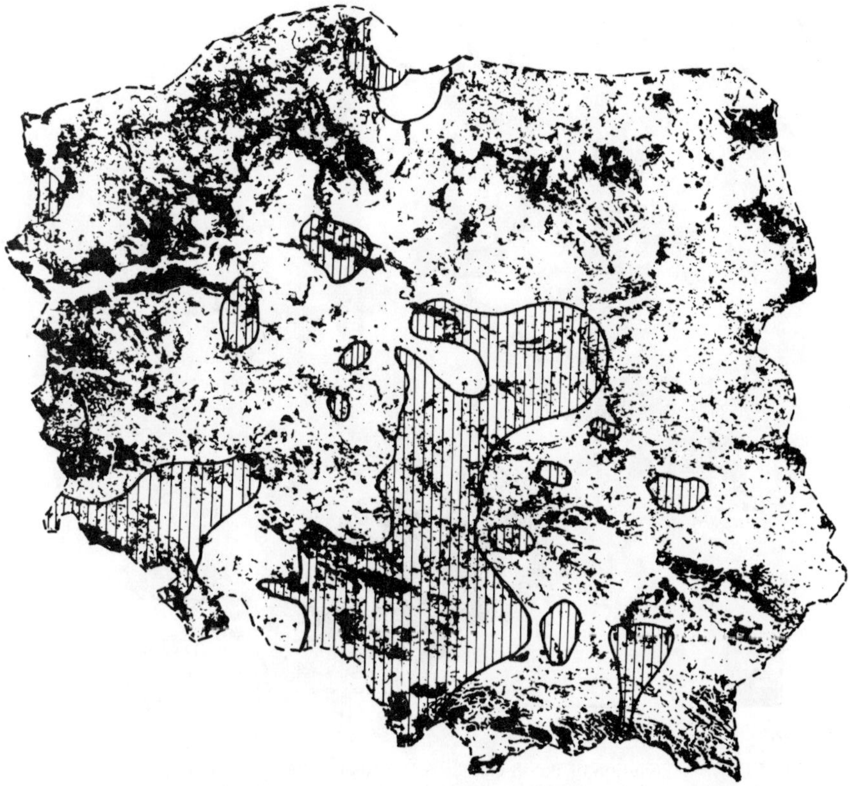

FIGURE 11 Distribution of coniferous and mixed forests and air pollution in Poland. The map was elaborated at the Institute of Geodesy and Cartography in Warsaw under the supervision of Andrzej Ciolkosz on the basis of *Landsat MSS* and *Salut-6* images acquired in 1978. The isolines delimit the area with SO_2 mean annual concentration equal/higher than 50 micrograms per m^3; dark areas show regions above 150 micrograms. Data on pollution from the map predicting air pollution in Poland in 1990 by J. Juda et al., 1982 (manuscript).

and the forest wall. Living but strongly damaged forest was characterized by already small organic fall from balding trees and a high rate of litter decomposition. Grabinska (dissertation, 1985) suggested that this was probably abiotic decomposition, as the litter was dissolved in the solution of nitric and sulfur compounds in rainwater.

Let us move on to the territory of Poland as a whole and the condition of coniferous forests as demonstrated by scientists and foresters. Satellite pictures of pine and mixed-pine forests in Poland are shown in Figure 11. The simulated isolines of contamination by sulfur oxides delimit the area of real danger for conifers as projected to 1990. Present evaluations by forestry services of the state of these forests are shown in Figures 12 and

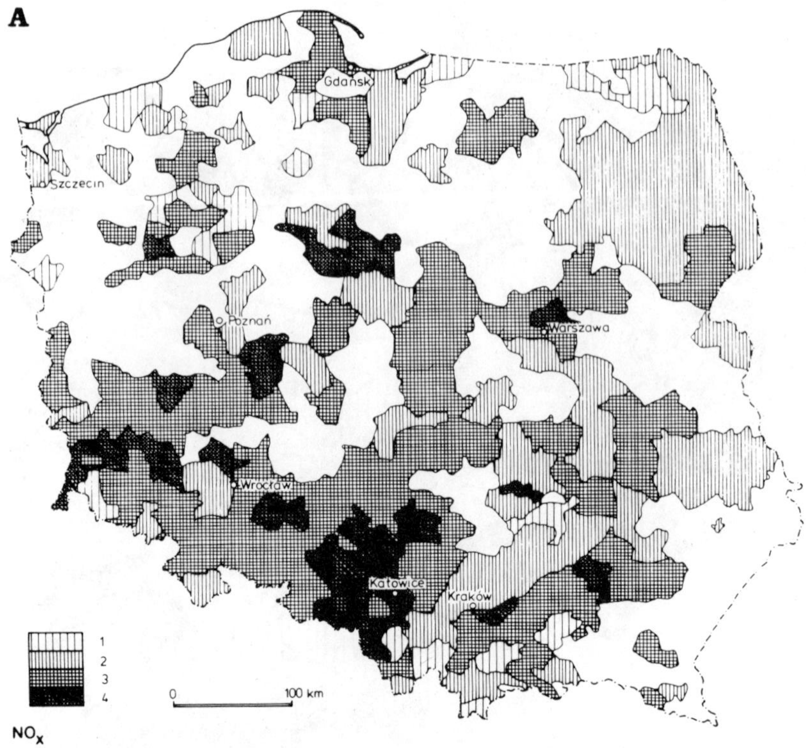

FIGURE 12a Pollution of Polish coniferous forests by NO_x in 1986 measured 1,800 points as dry sediments on specifically prepared filter paper exposed for 30 days in the canopy. Contamination expressed as mg m^{-2} d^{-1} in four classes: 1=0.000-0.200 mg; 2=0.201 - 0.5000 mg; 3=0.501-1.000 mg; 4=higher than 1.001 mg. A=winter pollution, B=summer pollution.

13. Comparison of winter (i.e., heating season) and summer contamination of forests provides a basis for some optimism: in many regions, forests are not as stressed during the warm season when heating is not used.

STATUS OF FOREST ECOSYSTEMS IN POLAND

According to maps of potential vegetation and the knowledge of geobotanists, Poland should be covered with a mixture of forest types. Due to various factors, however, Poland is covered primarily with coniferous forests, which constitute more than 75% of total aboveground biomass of forests in the country (pine alone covers 62%). Of the total production of

B

NO_x

FIGURE 12b

Polish forests, conifers make up 74%, and pine alone is 54.4% (Figure 14). Of the three species of conifers grown in Poland, pine shows the lowest productivity and yet provides the greatest wood production in the country, since pine forests occupy the largest area. Coniferous forests are considered to be the terrestrial ecosystems most imperiled by air pollution; they are affected particularly by sulfur and nitric oxides, whose concentrations in the air have recently increased considerably (Figure 12).

There is still no single answer as to which mechanisms are responsible for the dying of coniferous forests, but some are particularly worthy of analysis. Coniferous forests, and pine forests in particular, are in general ecosystems of cool climates and poor soils. Matter cycling is rather slow in these ecosystems, and the rate of organic matter decomposition is distinctly slower than in deciduous forests (Table 3). Interesting mechanisms for saving essential elements in these ecosystems are described in Figure 15.

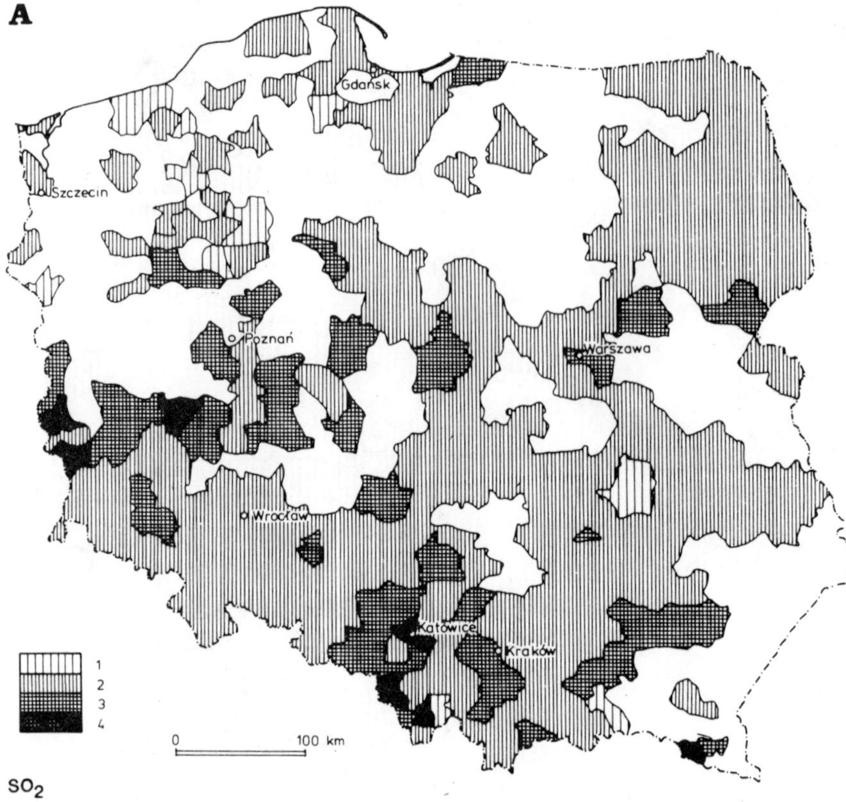

FIGURE 13a Pollution of Polish forests by SO_2. Description as for Figure 12. Classes of contamination: 1=0.000-10.000 mg m^{-2} d^{-1}; 2=10.001-30.000; 3=30.001-50.000; 4=above 50.000 mg. A = winter pollution, B = summer pollution.

Pine trees translocate some essential elements from the needles before leaf fall. In this way, the forest accumulates a larger portion of elements in plant tissue instead of losing them from litter and sandy soil. Moreover, conifers shed their needles two or three times slower than do other plants in similar climates. Consequently, pine forests can be considered to be the ecosystems that retain in their biomass a relatively large proportion of the elements taken from the abiotic environment.

This tendency to accumulate the elements which are difficult to acquire can be seen as an evolutionary strategy. However, this strategy may turn against coniferous ecosystems. When the chemical composition of the atmosphere and water changes rapidly in the surrounding environment,

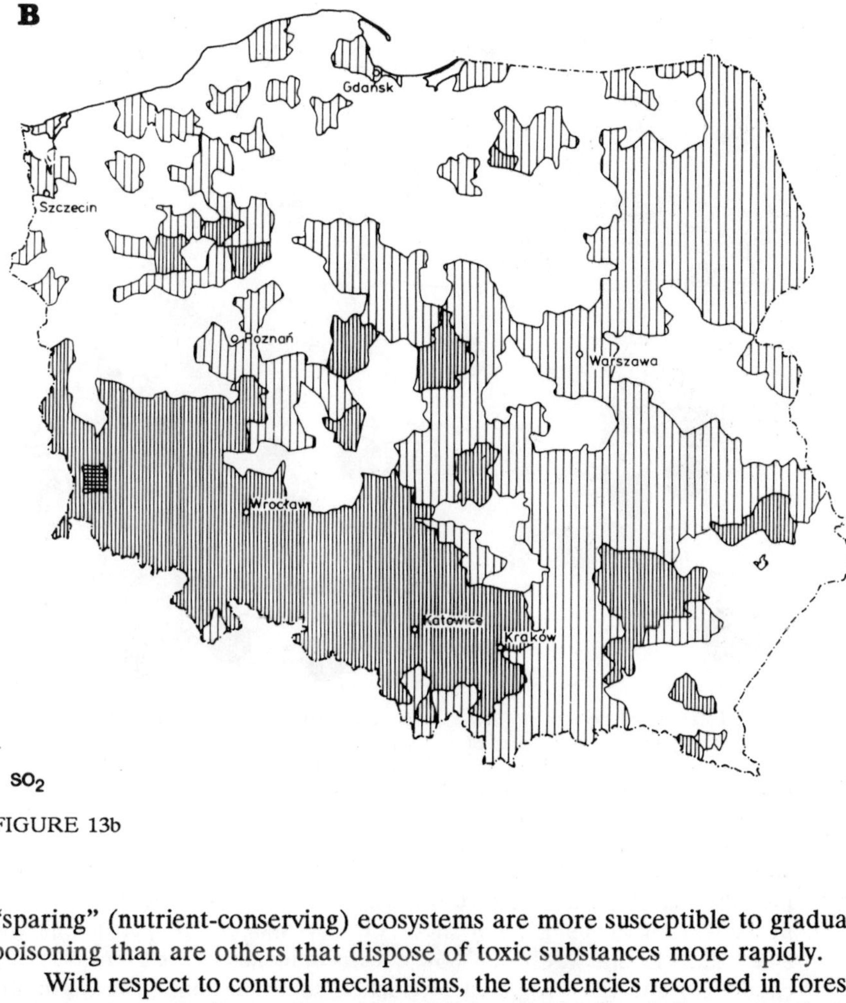

FIGURE 13b

"sparing" (nutrient-conserving) ecosystems are more susceptible to gradual poisoning than are others that dispose of toxic substances more rapidly.

With respect to control mechanisms, the tendencies recorded in forest ecosystems are similar to those observed in grasslands—in heavy stress situations, the number of species often decreases, the percentage of predators increases, and the dominance structure becomes more uneven (i.e., fewer species are dominant). These tendencies were recorded by Lesniak (in press) in his investigations of soil and epigeic (soil surface) invertebrates carried out in 52 forest districts distributed over the entire territory of Poland. In plots situated in pine forests of southwest Poland (which are heavily polluted), 71% of all predators were registered in epigeic fauna, while in northeast Poland (which is least polluted), the percentage of predators was only 28%

The regulatory mechanisms induced by an increasing share of predators

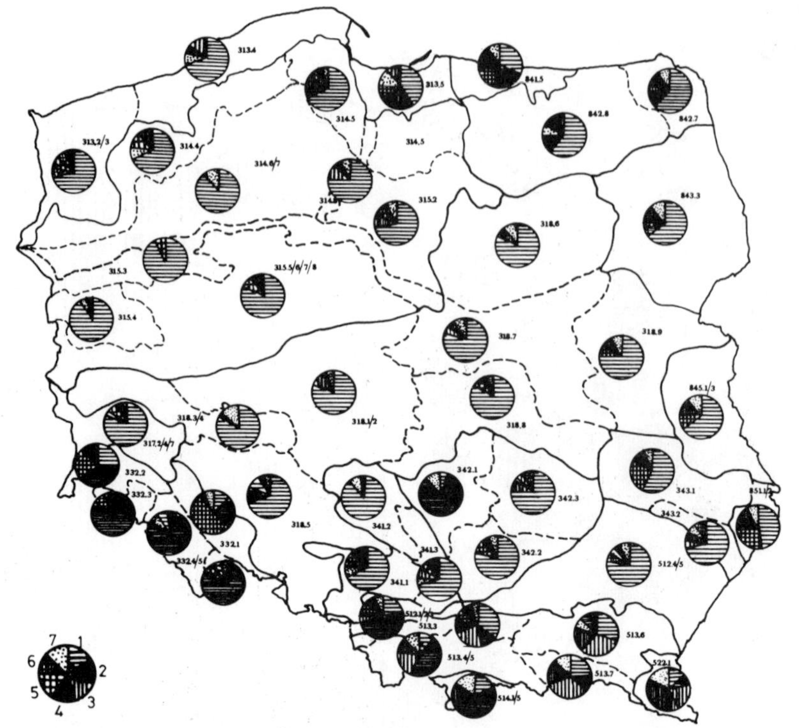

FIGURE 14 Composition of forests in geographical units in Poland. The portion of tree species expressed as percent of surface covered by this species. 1=pine; 2=spruce; 3=fir; 4=beach; 5=oak; 6=birch and hornbeam; 7=alder and aspen (Trampler et al., 1982, prepared from Forestry Service Data).

and their increased metabolism rate are strengthened in stress conditions; perhaps this is the first distinct, repeatable reaction of an ecosystem noted in the case of changing conditions. Is the ecosystem trying to defend and adapt itself in this way? And how does it influence the pattern of matter circulation and the budget of production/decompostion? These are still open questions, but the answers must come soon as the demands of society and the interest of scientists grow simultaneously.

EVOLUTION OF ECOSYSTEMS

Considering the functioning and the character of ecosystems as defined above, we often pause on the question of their evolution: How do ecosystems make evolutionary progress since they do not inherit their

TABLE 3 Weight loss of litter in central Poland. Litter bags exposed on 28 stands (10-30 bags per stand) during warm season. Decomposition rate expressed as percent of initial weight.

Site	Forest type	Number of stands	Daily decomposition rate (%)
Bolimow	Pine and mixed	6	0.08-0.30
Malogoszcz	Pine	7	0.10-0.24
Pulawy	Pine	5	0.20-0.24
Bialoleka Dw.	Pine and mixed	2	0.22
	Coniferous	20	0.08-0.30
Bialoleka Dw.	Alder	2	0.34
Bialoleka Dw.	Birch	1	0.44
Bialoleka Dw.	Oak hornbeam	5	0.51
	Broadleaved	8	0.34-0.51

SOURCE: Unpublished data, Breymeyer and Grabinska.

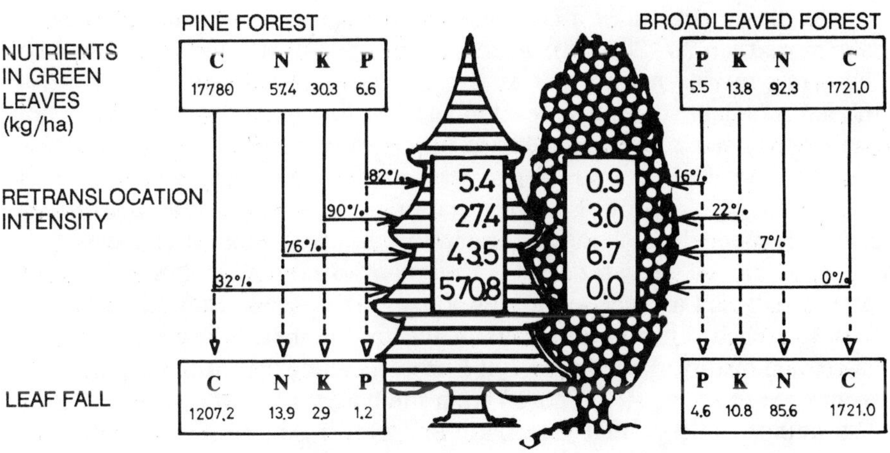

FIGURE 15 Pattern of nutrient transfer to soil in pine and alder forests (Stachurski and Zimka, 1981; modified).

characteristics as do species? They have no mechanisms of heredity, since they are not organisms provided with sets of features coded in their genes. However, it appears that ecosystems adjust so as to permit the recurrence of the composition of species (or the communities of ecologically redundant species) worked out during long-term evolution. And perhaps ecosystems evolve by even simpler means than the do cogeneric populations.

If, for example, a sand dune is colonized, the first group of colonizers is selected; there are only a small number of pioneer species that are fit for living on dune sand. They prepare the environment for succeeding species, which in turn do the same for the next generation. A layer of soil appears and it grows thicker, pine seedlings attach in it, and the succession proceeds in the same way, only to reach the climax stage as one of the forest associations common to the climatic zone. Just as in the case of populations where conditions select the fittest of overabundant progeny, it can be assumed in the case of an ecosystem that only certain groups of the overabundant seeds, spores, or juvenile forms of plants and animals which expand in different ways and are ready to colonize a given area will be accepted there. These sets of species or communities:

- accept the physical conditions of an area;
- can produce organic matter or feed on organic matter produced there; and
- can defy competition or other ecological dependences which grow in number and intensify as the colonization of niches increases.

In the conditions of a cool temperate climate, forest succession must take more than 200-300 years to reach a climax forest stage (assuming that the forest attains maturity at an age of about 100-150 years, and all of the intermediary stages can develop and last for shorter periods). It is not proven how much time is needed to develop the complex trophic structure registered in contemporary forest stands. Probably many generations of trees produce the pool of organic matter in the forest soil as detritivores gradually enter the soil and put in motion the circulation of elements.

Can this be recognized as a beginning of ecosystem function? And how many years are needed to develop sets of species ready to replace or complete each other in all points of the trophic net? Kornas (1972) discussed the historical background of European and North American conifers (genus *Pinus*, *Picea*, *Abies*, *Larix*) from the boreal zone of coniferous taiga. The author believes that Eurasiatic and North American lowlands were colonized by taiga-type forest not earlier than in the Holocene. The taiga trees have had their long, early history in primary centers in the mountains of Eastern Asia and Pacific North America, and they dispersed in lowlands after Pleistocene glaciation (before the Tertiary Holarctic realm divided). Moreover, there is some proof that pairs of close, vicariant taxa of American and European flora still show distinct ecological correspondence:

> Nearly all series of closely related taxa in the forests of temperate Eurasia and North America consist of ecologically corresponding components which have similar ranges of tolerance, occupy similar habitats, and grow in strictly analogous plant communities. The ecological constitution of such groups must be very ancient and very rigid. Even those nemoral taxa which are separated

into morphologically distinct species have undergone no noticeable ecological changes since their isolation some 10-15 million years ago (Kornas, 1972).

Thus, contemporary boreal coniferous forest ecosystems began their evolution in the Holocene 10,000 years ago, but they consist of some very conservative taxa which did not change ecological habits for millions of years.

REFERENCES

Ajtay, G.K., P. Ketner, and P. Duvigneaud. 1979. Terrestrial primary production and phytomass. Pp. 129-181 in The Global Carbon Cycle, B. Bolin, R.T. Degnas, S. Kempe, P. Ketner, eds. SCOPE 13 (Wiley & Sons).

Alvera, B. 1981. Decomposition de hojas en un pinar altoaragones. Ann. Edafol. Agrobiol. Madrid 40:37-46.

Berg, B., P.E. Jansson, and V. Meentemeyer. 1984. Litter decomposition and climate-regional and local models. Pp. 389-404 in State and Change in Forest Ecosystems, G.I. Agren, ed. Report 13 Swed. Univ. Agric. Sci.

Bray, J.R., and E. Gorham. 1964. Litter production in forests of the world. Ad. ecol. Res. 2:101-157.

Breymeyer, A. 1981a. Comparative studies on trophic structure of grassland ecosystems. Wiad. Ekol. 27:116-147 (in Polish with summary in English).

_____. 1981b. Monitoring of the functioning of ecosystems. Environmental Monitoring and Assessment 1:173-183.

_____. 1984. Ecological monitoring as a method of land evaluation. Geographia Polonica 50:371-384.

_____. 1986. Pine forests in Poland, their productivity, distribution, and degradation. Geographia Polonica 52:101-109.

Bulla, L., J. Pacheco, and R. Miranda. 1981. Ciclo estacional de la biomasa verde, muerta y raices en una sabana inundata de estero en Mantecal (Venezuela). Acta Cient. Venezolana 31:339-344.

Bulla, L., J. Pacheco, and G. Morales. In press. Seasonally flooded meotropical savanna closed by dikes. In Damaged Grasslands, A. Breymeyer, ed. Elsevier.

Cianciara, S., and J. Pilarska. 1983. Numbers and structures of soil fauna communities in tow meadows in Upper Silesia. Ekol. Pol. 31,4:947-997.

Debazac, E.F. 1983. Temperate broad-leaved evergreen forest of the Mediterranean region and Middle East. Pp. 107-122 in Temperate Broad-Leaved Evergreen Forests, J.D. Ovington, ed. Elsevier.

Duvigneaud, P., and M. Tanghe. 1962. Ecosystems et biosphere. Travaux du Centre d'Ecologe Generale. Bruxelles.

Grabinska, B. 1985. The rate of organic matter decomposition and stands characteristics in coniferous forests of Central Poland. Dissertation in Polish. Inst. of Geography, Polish Academy of Sciences. Warsaw.

Grodzinski, W., J. Weiner, and P.F. Maycock, eds. 1984. Forest ecosystems in industrial regions. Springer Verlag.

Jakubczak, A., and J. Pieta. 1968. Some morphological variation of the needles and shoots of Scotch pine (*Pinus sylvestris L.*) in the polluted surroundings of nitrogen plant at Pulawy. P. 68 in the Proceedings of the VI International Conference, "Influence of air pollution on forests". Research Laboratory GOP, Polish Academy of Sciences.

Juda, J., S. Chrosciel, J. Jedrzejowski, M. Nowicki, and W. Jaworski. 1982. The influence of fossil fuel combustion on environment in the perspective of the year 2000 in Poland. Typescript in Polish.

Klekowski, Z.R., and K.W. Opalinski. 1984. Matter and energy flow in Spitzbergen tundra. Wiad. Ekol. 30:143-166 (in Polish with summary in English).

Kornas, J. 1972. Corresponding taxa and their ecological background in the forest of temperate Eurasia and North America. Pp. 37-59 in Taxonomy, Phytogeography, and Evolution, D.H. Valentine, ed. Academic Press.

Lesniak, A. 1988. Monitoring of soil fauna on 52 stands of coniferous forest in Poland. Institute of Zoology, Polish Academy of Sciences. Warsaw (manuscript in Polish).

Lieth, H. 1975. Modelling the primary production of the world. Pp. 237-242 in Primary Productivity of the Biosphere, H. Lieth and R.H. Whittaker, eds. Springer Verlag,.

Meentemeyer, V. 1984. The geography of organic decomposition rates. Ann. Assoc. American Geographers 74(4):551-560.

Meentemeyer, V., and B. Berg. 1986. Regional variation in rate of mass loss of *Pinus silvestris* needle litter in Swedish pine forests as influenced by climate and litter quality. Scand. J. For. res. 1:167-180.

Mochnacka-Lawacz, H., and E. Olechowicz. 1978. The role of selected heterotrophic groups in the nutrient economy of a meadow ecosystem. Pol. Ecol. Stud. 4(1):243-257.

Molski, B.A., and W. Dmuchowski. 1986. Effects of acidification on forest and natural vegetation, wild animals, and insects. Pp. 29-51 in Acidification and its Policy Implications, T.S. Schneider, ed. Proceedings of the International Conference, Elsevier Studies in Environmental Science 30.

Okruszko, H. 1978. Meliorations and changes in the environment. Pp. 38-65 in The Role of Meliorations in Management of Natural Environment. Proceedings of the Conference, IMUZ-Falenty (in Polish).

Olson, J.S., J.A. Watts, and L.J. Allison. 1983. Carbon in live vegetation of major world ecosystems. Oak Ridge National Laboratory.

O'Neill, R.V., and D.L. DeAngelis. 1981. Comparative productivity and biomass relations of forest ecosystems. Pp. 411-449 in Dynamic properties of forest ecosystems, D.E. Reichle, ed. Boston: Cambridge University Press.

O'Neill, R.V., D.L. DeAngelis, J.B. Waide, and T.F.H. Allen. 1986. A hierarchical concept of ecosystems. Monographs in Population Ecology 23, Princeton University Press.

Reichle, D., ed. 1981. Dynamic properties of forest ecosystems. Boston: Cambridge University Press.

Rieger, R., S. Grabczynski, S. Orzel, and J. Ramier. 1984. Primary production in the Niepolomice forest ecosystems. Pp. 70-78 in Growing Stock and Increasing of Tree Stands in Forest Ecosystems in Industrial Regions, W. Grodzinski, J. Weiner, and P.F. Maycock, eds. Springer Verlag.

Schaefer, D., Y. Steinberger, and W.G. Whitford, 1985. The failure of nitrogen and lignin control of decompositoin in a North American desert. Oecologia 65:382-386.

Stachurski, A., and J. Zimka. 1981. The patterns of nutrient cycling in forest ecosystems. Bull. Acad. Pol. Sci., CI. II, 29, 3-4:141-147.

Trampler, T., A. Girzda, and E. Dmyterko. 1986. Differentiation of biomass and production of coniferous forests in Poland. Institute of Geography and Spatial Organization. Warsaw (typescript in Polish).

Trampler, T., E. Dmyterko, and B. Lonkiewicz. 1987. State and prognosis of forest damages in Poland. Bull. of Forestry Institute 3,87:10-18 (in Polish).

Udvardy, M.D.F. 1975. A classification of the biogeographical provinces of the world. IUCN occasional paper No. 18. Morges, Switzerland.

Vogt, K.A., C.C. Grier, and D. J. Vogt. 1986. Production, turnover, and nutrient dynamics of above- and belowground detritus of world forests. Advances in Ecologial Res. 15:303-377.

Whittaker, R.H., and P.L. Marks. 1975. Methods of assessing terrestrial productivity. Pp. 55-118 in Primary productivity of biosphere, H. Lieth and R.H. Whittaker, eds. Springer Verlag.

Zurek, S. 1987. Peat deposits in Poland on the background of European peat zones. Dok. Geogr. 3:85 (in Polish with summary in English).

Air Pollution Impacts on Forests in North America

ANN M. BARTUSKA
U.S. Forest Service
U.S. Department of Agriculture

No other impact of air pollution has so mobilized public concern in North America as has the effect on forests. During the 1980s, funding for air-pollution research on forests increased tenfold. "Acid rain" was the primary issue early in the 1980s when effects on lakes and streams were the focus; now, as concern increases about effects on forests, studies are focusing on acidity (sulfur and nitrogen) in association with photochemical oxidants (especially ozone) and other gases. This chapter presents our current understanding of the impact of air pollution on the productivity and health of forests in various parts of North America, and discusses the major uncertainties still remaining in this area of research. This chapter will not review the substantial literature dealing with effects near major point sources of sulfur dioxide or fluoride (e.g., smelters), but instead will concentrate on regional air pollution.

EFFECTS OF AIR POLLUTION ON FORESTS: CASE STUDIES

Forest decline is a term used to refer to a general decrease in health and vigor leading to tree mortality over a large geographic region. Declines are not new phenomena, and have been widely reported in the literature. Some declines are caused by a complex interaction of multiple abiotic and biotic factors (Manion, 1981). Others are caused by only one or two casual factors (Benoit et al., 1982). A classic example was the dieback of birch which occurred between 1930 and 1950 in southeastern Canada and the northeastern United States. In this example, higher than average soil temperatures in summer apparently resulted in extensive root mortality. The weakened trees were subsequently attacked by foliar and bark-boring

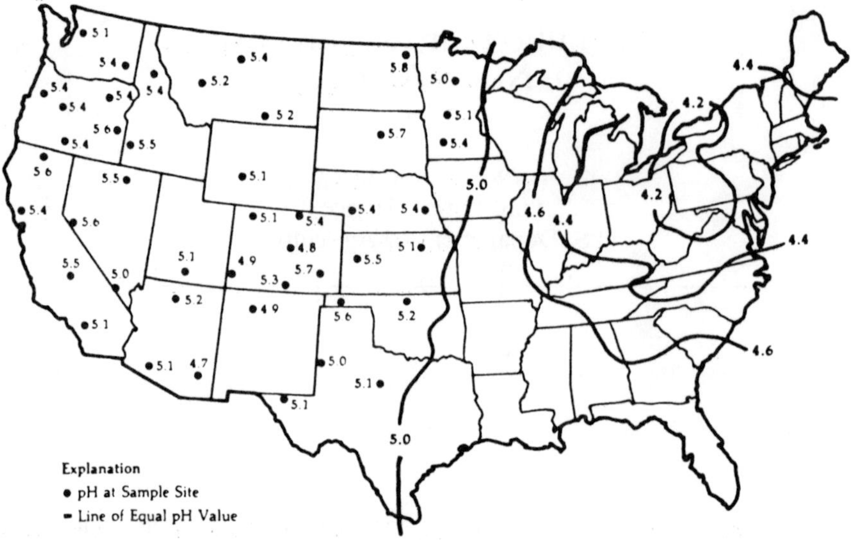

FIGURE 1 The 1986 distribution of precipitation-weighted pH (NAPAP, 1986).

insects—*Armillarea mellea* and various viruses—leading to widespread mortality. Clearly, there was a multiplicity of causes. Whether atmospheric deposition plays a role in forest decline can only be assessed within the context of the numerous other stress factors that affect forests.

While there have been numerous studies on the effects of a specific air pollutant on a specific plant, the complex nature of declines can only be assessed through an intensive, multilevel investigation of the community at risk. Due to this complexity, it is not surprising that we have few clear-cut examples of regional declines where air pollution has been a contributory factor (Figures 1 and 2). In only two cases has air pollution been implicated as a cause of damage: the mixed conifer forests of the San Bernardino mountains of southern California and the Eastern white pine (*Pinus strobus* L.) in eastern North America. A third and more recent decline—red spruce (*Picea rubens* Sarg.) at high elevations in the eastern United States—has been suggested as being due to atmospheric deposition in association with natural causes. Finally, recent growth reductions in southern pines has led to speculation that regional air pollution is a primary cause. These four cases will be discussed in detail below (Figure 3).

San Bernardino Mountains

Foliar damage to a variety of western conifers was observed in the San Bernardino Mountains of southern California in the early 1970s. These

FIGURE 2 Average of maximum 7-hour daily mean concentration (ppb) of O_3 during 1984 growing season at selected rural sites in the United States (NAPAP, 1987).

FIGURE 3 Areas where air pollution is suspected to impact forest trees.

effects provoked an extensive interdisciplinary study of this mixed conifer forest ecosystem (Miller et al., 1977; Miller, 1983).

Eighteen vegetation study sites were established along a geographical gradient of visible injury in the San Bernardino Mountains. The frequency and intensity of visible injury in the forest types occupying the transect was monitored from 1973 to 1976. The five forest types identified along this gradient include: ponderosa pine (*Pinus ponderosa* Dougl. ex Laws.); ponderosa pine/white fir (*Abies concolor* [Gord. & Glend.] Lindl. ex Hildebr.); ponderosa pine/Jeffrey pine (*P. jeffreyi* Grev. & Balf.); Jeffrey pine/white fir; and Jeffrey pine.

The symptoms visible on these species are very characteristic of what is termed "typical symptoms of ozone exposure," and include chlorotic mottling, necrotic tipburn progressing from the tip to the base, and premature needle abscission (Miller et al., 1982). The trees showing the most visible injury were ponderosa pine and Jeffrey pine. The symptoms on these trees appeared to worsen in degree and magnitude over time, i.e., some of the trees showing chlorotic symptoms in 1973 showed symptoms of necrosis by 1976, and a few of the trees showing necrosis and needle drop in 1973 were dead by 1976. Overall, the accumulated mortality of ponderosa pine and Jeffrey pine varied with forest type, but ranged from 0 to 6% for the 18 plots from 1973 to 1978. Average accumulated mortality rates by type were 4% for ponderosa pine, 4% for ponderosa pine/white fir, 6% for ponderosa pine/Jeffrey pine, 1% for Jeffrey pine/white fir, and 0% for Jeffrey pine (Taylor, 1980).

The visible symptoms and mortality were associated with decreased photosynthetic capacity, suppressed radial growth, and reduced nutrient retention and availability. There was also evidence of decreased cone and seed production, increased litter accumulation, and increased infestation by pests and pathogens such as the pine beetle (*Dendroctonus brevicomus*) and root rot caused by *Fomes annosus*. The long-term effects expected for this system result in shifts from a pine overstory to a self-perpetuating ozone-tolerant community of oak and shrub species.

The evidence is strong for ozone as the agent causing foliar injury and other types of injury. The San Bernardino National Forest (SBNF), which is located at the east end of the South Coast Air Basin (including Los Angeles), has been exposed to increasing annual dosages of air pollutants from heavy urban and industrial development over the last three to four decades. Photochemical oxidants have been carried by marine air currents to the mixed conifer forests since at least the early 1950s, when signs of injury were detected in ponderosa pine. As human population growth in the Basin has continued, both pollutant concentrations and the extent of the affected geographic area has increased (Miller, 1977; Taylor, 1980).

Thirteen air monitoring sites accompanied the vegetation study sites along the geographical gradient of visible foliar injury (Miller, 1983). The natural background ozone (O_3) concentration in the study areas was 30 to 40 parts per billion (ppb); this concentration is comparable to that in other mountainous regions in the United States. The first signs of injury of ponderosa pine needles were associated with a 24-hour average

O_3 concentration exceeding 50 to 60 ppb during the months of May through September. When the average ranged from 100 to 120 ppb, injury to sensitive species such as ponderosa pine was severe and ecosystem functions were affected (Miller, 1983). More recently, Peterson et al. (1987) demonstrated a significant decrease in Jeffrey pine growth north of Los Angeles in the Sierra Nevadas of California. This growth decrease has been linked to ozone exposure and the appearance of visual damage symptoms, and is alarming due to the high aesthetic value of the area.

Eastern White Pine

Needle-blight injury on Eastern white pine has been reported for decades; however, it was not determined until the 1960s that the injury was caused by ozone (Berry and Hepting, 1964). The conditions of Eastern white pine in areas of the southeastern United States mimic those of western conifers in the San Bernardino Mountains. In addition to the symptoms being similar, the short-term effects on the biological, chemical, and physical properties of the trees appear to be analogous.

Foliar injury was visible on Eastern white pines in the Blue Ridge Mountains of southern Virginia (Skelly et al., 1972; Stone and Skelly, 1974; Phillips et al., 1977a,b; Benoit et al., 1982). The pines showed typical ozone symptoms which varied in intensity within the species. Trees were classified as sensitive, intermediate, or tolerant based on needle length, needle retention, and presence of foliar injury. Eleven, 67, and 22% of the white pines surveyed were rated sensitive, intermediate, and tolerant, respectively (Skelly et al., 1979).

Nine Eastern white pine trees were sampled in Tennessee, three in each sensitivity class, and cored for radial growth analysis. Similar growth trends were found in the pines of tolerant and intermediate class, while a steady decline in growth rate (approximately 70% over 15 years) was observed in pines of the sensitive class (McLaughlin et al., 1981). This study also detected altered allocation of carbon, resulting in a reduction of photosynthate to the boles and roots. Radial growth decline was also evident in the trees sampled in the Blue Ridge Mountains and was correlated to changing local air quality (Benoit et al., 1982). Skelly and Stone (1972) also noted a 66% decrease in height growth. The rate of decline of these white pine will be furthered by bark beetle infestation and root disease such as *Verticicladiella procera* (Skelly et al., 1983).

McLaughlin et al. (1982) suggest the following sequence of events and conditions afflicting Eastern white pine in the southeastern United States:

1. premature senescence and loss of older needles at the end of the growing season;

2. reduced storage capacity in the fall and resupply capacity in the spring to support new needle growth;
3. increased reliance of new needles on self-support during growth;
4. shorter new needles resulting in lower gross photosynthetic productivity;
5. higher retention of current photosynthate by foliage resulting in reduced availability of photosynthate for external usage (including repair of chronically stressed tissues of older needles);
6. premature casting of older needles.

The net result of this condition is a reduction in the total amount of photosynthesizing tissue and carbohydrates available for growth and maintenance. This sequence of events was derived from continuous observation of Eastern white pine (*Pinus strobus*) in the Cumberland Plateau region of eastern Tennessee. Perhaps more importantly, these stages also are appearing in loblolly pine following ozone exposure, and may reflect a common pattern of tree response.

Concentrations of ozone in the study areas of the Blue Ridge Mountains repeatedly reached 40-60 ppb, with peak episodes ranging from 100 to 200 ppb (Skelly et al., 1983). Foliar injury occurred most frequently on Eastern white pine and milkweed and was absent on characteristically insensitive species. The observations of foliar injury corresponded with periods when ozone concentrations were the highest. The Eastern white pines that did not show visible injury did appear to exhibit growth effects (Skelly et al., 1983).

High-Elevation Red Spruce

Although evidence of injury has been recorded in high-elevation spruce-fir forests throughout the eastern United States, the patterns of symptom expression vary a great deal, and the linkage to air pollution is uncertain. In the northern Appalachian forests, Siccama et al. (1982) noted a reduction in density of all size classes of red spruce in higher elevational zones (>760 m) in the Green Mountains of Vermont. Similar observations on the decline of spruce in the Adirondacks were reported by Raynal et al. (1980), as well as in the White Mountains of New Hampshire, where a decline in number and basal area of this species was noted (Siccama et al., 1982).

The major visible foliar symptoms are loss of needles from the tips of the branches and from the apex of the crown, without the pronounced chlorosis indicating nutrient deficiency and with only slight macroscopic structural damage, i.e., necrotic spotting on the older foliage (Friedland et al., 1984a). Field and microscopic observations indicate that cold and/or winter moisture stress are responsible for the loss of the young foliage

(Friedland et al., 1983b). Mature red spruce showed foliar loss from the tips of the crowns down and from the tips of the branches inward, similar to foliar loss exhibited on younger trees (Friedland et al., 1984a). Balsam fir (*Abies balsamea* [L.] Mill.) appears vigorous in that the symptoms of stress and necrosis exhibited by the spruce have not yet been noted. However, measurement of average annual increment on balsam fir shows a decrease similar to that of spruce (Siccama et al., 1982). The symptoms cited above are typical of declining red spruce in the northern Appalachians.

The spruce-fir forests of Camels Hump in the Green Mountains of Vermont show the decline symptoms typical of the northern region. Camels Hump is comprised of a complex of schist overlain by glacial till. On the lower slopes (< 750 m), northern hardwoods with occasional conifers are found. The upper slopes (750-1,210 m) are primarily red spruce-balsam fir, with occasional white birch (*Betula papyrifera* var. *cordifolia*). The upper slopes are characterized by shallow acid till and very acidic Spodosols or Histosols (pH in H_2O 3.0-4.5). Average annual rainfall is 1,200-2,000 mm, increasing with elevation, and there is prolonged contact with cloud moisture (70-120 days per year). Visible characteristics of decline include the loss of younger foliage, severe frost damage (i.e., browning of foliage), and damage to needles at the cellular level, including to the chloroplasts and tonoplasts, which is typical of the damage to high-elevation spruce-fir in the north.

The decline of red spruce, which is a long-lived, shade-tolerant species, is not the anticipated pattern based on its known ecological strategies and former abundance. Significant in the assessment of this decline is the fact that the pattern is recorded for both larger trees (30 m in height) as well as saplings (2 m in height in the substrate). Equally important is the fact that these symptoms occur throughout all the elevational gradients and across all age and size classes (Siccama et al., 1982).

In the southern Appalachians (i.e., Virginia, North Carolina, and Tennessee), some of the spruce-fir forests are in a severe state of decline (Bruck, 1984). Here, symptoms include the loss of older foliage, proceeding from the inner to outer portions of the crowns. Chlorosis on the needles is apparent. Also, the appearance of epicormic branches has been noted on both the red spruce and the Fraser fir (*Abies fraseri* [Pursh] Poir.) on the branches and stems.

Mount Mitchell is located in western North Carolina and the first symptoms of decline in the south were noted here (Bruck, 1984). At 2,038 m, it is the highest peak in the eastern United States and in the Appalachians. The bedrock is Precambrian metamorphic, gneiss and schist, with fine granitoid layers. Slopes of 20-60% are common, but some flat areas occur as draws and on crests. Where soils have coarse textured, parent material and a stable surface, Spodosols may be likely; where soils are of medium textured parent material with unstable substrates, Umbrepts are likely. The climate is characterized by high precipitation: 1,800 mm throughout the year with extensive periods of cloud cover and relatively high input of total precipitation from cloud moisture.

The survey of Mount Mitchell in 1984 showed that trees aged 45-85 years, primarily red spruce, averaged 75-90% defoliation characterized by loss of the oldest needles, often leaving a chlorotic tuft of needles at the branch ends. Most trees at or above 1,900 m exhibited some form of growth reduction beginning in the early 1960s. The spruce stands below 1,800 m also have begun to show signs of defoliation. Foliar analyses of needles collected during the survey showed no unusual concentration of macro- or micronutrients or trace elements. However, anticipated decreases of nitrogen (N) content with flush age for red spruce were not observed. Instead, N content of the 1983 needles was actually greater than the 1984 or 1982 needles (Bruck, 1984).

More recent surveys completed in 1985 and 1986 have shown continued deterioration of the spruce-fir stands in the Mount Mitchell area. A concomitant study of cloud and rain chemistry in the area is showing signficantly greater deposition of pollutants than at lower elevations; however, cause and effect cannot be attributed to any specific factor(s) at this time.

Another important factor is the extensive Fraser fir mortality throughout the southern Appalachians due to the balsam woolly adelgid (*Adelges piceae* Ritz). The loss of fir has significantly altered the microclimate of these sites as the crown has broken up, increasing light penetration to the forest floor. Whether this structural change has affected the red spruce is still conjectural; however, it is an important natural stress factor which must be considered (Bartuska and Medlarz, 1986). Recent hypotheses suggest that air pollution in the region is acting as a predisposing stress on Fraser fir, increasing the fir's susceptibility to adelgid attack (Hain and Arthur, 1985). However, the link is circumstantial at present, and underscores the point that high elevation ecosystems are subject to many co-occurring stresses. Therefore, it may not be reasonable to expect that one factor can be isolated as the ultimate cause of mortality.

Southern Pines

Of the various case studies discussed in this chapter, the role of air pollution in the southern commercial forest region (specifically, Piedmont and Coastal Plain) of the United States is the most uncertain, but also the most important with regard to potential economic impact. Data from ten-year remeasurements of Forest Inventory and Assessment (FIA) plots in Alabama, Georgia, South Carolina, and North Carolina indicate that there has been a decrease in the rate of diameter growth in some portions of the southern pine region (Sheffield et al., 1985). These U.S. Department of Agriculture Forest Survey data suggest that commercially important loblolly (*Pinus taeda* L.), shortleaf (*Pinus echinata* Mill.), and slash pine (*Pinus elliottii* Engelm.) forests in the Piedmont and low elevation mountain regions of these states grew less rapidly—by about 16-20% in diameter during the past 5 to 10 years—than during the previous 10 years. Shortleaf pine also showed a decrease in growth in the Coastal Plain, although loblolly and slash pine showed a volume increase of 17% and 49%, respectively.

The cause of the growth reduction has been attributed to increased hardwood competition, loss of the "old field condition," an increase in pine mortality due to pine-beetle outbreaks, and the natural aging of stands. However, in evaluating any of these factors, air pollution can not be ignored. The average growing season rainfall pH is < 4.6, with events as low as pH 3.6 (NADP, 1988). Perhaps of greater importance with respect to southern forests, the Piedmont and Mountain regions of the southeastern United States showed higher hourly O_3 levels than most other regions (Lefohn and Pinkerton, 1988). The influence of these air pollution factors on tree or forest response is still unknown and is being studied extensively.

RECENT FINDINGS AND UNCERTAINTIES

From 1985 to the present, significant research efforts have been underway to study the effects of atmospheric deposition and ozone exposure on forest species, with support from federal and state agencies, forest industry, and other institutions. While much of the data from this research is still being evaluated, some significant advances in understanding have been made. The purpose of this section is to provide some more recent research results which perhaps bring us closer to evaluating the link between air pollution and forest health.

Western Conifers

While the San Bernardino forests of southern California have long been known to be impacted by air pollution as described earlier in this chapter, recent research also is focusing on other areas which are, or have the potential to be, at risk. These areas are: Puget Sound in Washington State; the Sierra Nevada in California; and the Front Range of the Colorado Rockies.

- Sites in the southern Sierra Nevada have shown recent reduced growth rates from increment core analysis and visible ozone damage. Reduced growth was an isolated occurrence among sites; the region as a whole showed no growth decline for ponderosa pine.

• Comparative analysis of cloud water and rainfall indicates that clouds can have 3-10 times more hydrogen (H^+), nitrate, and sulfate than precipitation in many parts of the western United States. These results are consistent with similar observations on cloud water chemistry in the eastern part of the country.

Spruce-Fir

• Measurement and analysis of increment cores collected from dominant and codominant red spruce and balsam fir trees at permanent plot locations throughout New York and New England have documented that there has been a significant decline across the region in the rate of individual tree growth since about 1960 for red spruce, and since about 1970 for balsam fir; however, all other major forest tree species are currently growing at rates that equal or exceed the rates prior to 1960 (Hornbeck et al., 1987).

While these analyses initially focused on the general question of growth reduction without attempting to answer questions of cause, recent analyses of both red spruce and balsam fir data have indicated that the declines should have been expected due to a natural stand growth phenomena. Stands of red spruce and balsam fir growing at elevations below 700 m are experiencing reduced growth as would be expected because of natural factors of stand development. Although atmospheric deposition impacts on growth rate cannot be completely ruled out by these analyses, the involvement of such pollutants must be considered minimal. This position is now widely accepted by other scientists, many of whom continue to corroborate these findings with other data.

• A field study at Whitetop Mountain, Virginia, confirmed that wet acidic deposition leaches basic cations from red spruce foliage. A comparison of cloud water and throughfall chemistry showed that there was an exchange of incoming H^+ ions with foliar cations, mainly Ca^{2+} and Mg^{2+}. This leaching increased as cloud pH decreased. In addition, a depletion of NH_{4+} in throughfall appeared to indicate the occurrence of direct foliar nitrogen uptake.

The significance of these results is unclear. Foliar leaching is a normal process that usually occurs to some extent. It has not been determined whether the leaching losses observed here are excessive with respect to internal nutrient balances. Additional research is needed to better quantify nutrient leaching and to determine its significance as related to growth and productivity.

• Controlled laboratory and field studies have not demonstrated a significant ozone effect on red spruce seedlings, although ozone stress may alter winter hardiness. In contrast, exposure to sulfate acidic mist produced foliar symptoms rapidly on red spruce seedlings. Results indicate that spruce is particularly vulnerable to injury from acidic mist when sulfate concentrations are high, and there are repeated opportunities for drying of liquid on foliage.

• Recent hypotheses have suggested that a primary mechanism of pollutant impact on red spruce is through an alteration in the cold-hardening process, leading to increased winter injury. Several studies are ongoing, so that these data are only preliminary; however, results to date indicate that most hardiness is strongly negatively influenced by acid mist treatments. Further research is needed to better define the linkage between acidic deposition and winter damage.

Southern Pines

• A dendrochronological analysis of over 2,000 increment cores from dominant and codominant naturally regenerated loblolly pine trees on typical Piedmont sites in North Carolina, South Carolina, and Georgia has led to the conclusion that these trees are growing today at less than two-thirds the annual rate that equivalent trees in stands on the same sites were growing 35 years ago. Analysis of the data—using a modeling technique that allowed for the quantitative analysis of the impact of various stress factors on growth, including site, quality, stand-density changes, and climatic impacts—revealed that an additional unexplained reduction in radial growth remained.

• A number of coordinated projects have demonstrated that families of loblolly pine (*Pinus taeda*) differ significantly in response both to simulated acid rain and to ozone, although the differences were most striking for ozone. An examination was made of 100 half-sib families of loblolly pine, selected for superior growth under ambient conditions, from throughout the southern commercial pine region. There were significant differences in response to acidic rain and to ozone among these families. Response to acidic rain treatments (pH 3.5, 4.3, and 5.2) was variable; some experiments showed negative effects, others showed positive, while still others showed no effect at all. In a two-year field study, the acid treatment affect did not appear until 18 months with rain exposure, at which point growth stimulation due to the pH 3.5 treatment was measured. These results emphasize the need for long-term research under controlled conditions to better quantify responses to acid deposition.

• In contrast, loblolly pine response to ozone was significant after only 6 months. Higher concentrations of ozone resulted in decreased height and diameter growth. These same saplings showed marked reduction in photosynthetic capacity and quantum yield. Similar studies with slash (*Pinus elliotii*) and shortleaf (*Pinus echinata*) pines are showing responses after eight months of treatment, consistant with the loblolly pine response. Further evaluation of the mechanism of impact is ongoing and must be expanded to include trees at all stages of development, from seedling to mature tree. There appears to be a strong link between ozone exposure and changes in carbon fixation and allocation. The net effect appears as a reduction in productivity without visible symptoms—a response only detectable with long-term repeated measurements.

CONCLUSION

Our knowledge concerning the impact of regionally dispersed (as opposed to locally dispersed) air pollutants on forest ecosystems clearly is incomplete. Improving this knowledge base is complicated by an absence of visible symptoms in many cases. The ability to perform research which links physiological or nutritional changes with growth appears to be a more useful approach to evaluating air pollution impacts on trees in the short-term. Ultimately, the response which society should consider is forest growth and productivity changes, which can only be detected with long-term measurements.

A major challenge will be in providing sufficient information in the short term to assist policy makers in the decision process, coupled with the understanding that only through long-term studies will uncertainty be reduced. Because multiple causal factors are probably involved in the response, the general public, policymakers, and scientists must recognize that if a large number of interactive factors contribute to a widespread change in forest conditions, even a well-funded research program may not produce a definite cause-and-effect relationship in forest ecosystems. We may have to be satisfied with a substantial body of circumstantial evidence with which to make decisions.

Acknowledgement

Sections of this document have been adapted from the Forest Response Program (FRP) Plan, the 1987 FRP Annual Report, the briefing document on the FRP prepared for the EPA Science Advisory Board, and the 1989 FRP Accomplishments Report. The contributions of the FRP management are appreciated.

REFERENCES

Bartuska, A.M., and S.A. Medlarz. 1986. Spruce fir decline: Air pollution related? Pp. 55-73 in the Proceedings of the Appalachian Society of American Foresters Annual Meeting. Raleigh, North Carolina, January 30-31, 1986.

Benoit, L.F., J.M. Skelly, L.D. Moore, and L.S. Dochinger. 1982. Radial growth reductions of *Pinus strobus* L. correlated with foliar ozone sensitivity as an indicator of ozone-induced losses in eastern forests. Can. J. For. Res. 12:673-678.

Berry, C.R., and G.H. Hepting. 1964. Injury to Eastern white pine by unidentified atmospheric constituents. Forest Science 10:2-13

Bruck, R.I. 1984. Decline of montane boreal ecosystems in central Europe and the southern Appalachian Mountains, TAPPI Proceedings, pp. 159-163.

Eagar, C. 1984. Review of the biology and ecology of the balsam woolly aphid in southern Appalachian spruce-fir forests. Pp. 36-50 in The Southern Appalachian Spruce-Fir Ecosystem: Its Biology and Threats, P.S. White, ed. USDI-National Park Service, Research/Resource Management Report SER-71.

Friedland, A.J., R.A. Gregory, L. Karenlampi, and A.H. Johnson. 1984a. Winter damage to foliage as a factor in red spruce decline. Can. J. For. Res. 14:963-965.

Friedland, A.J., A.H. Johnson, and T.G. Siccama. 1984b. Trace metal content of the forest floor in the Green Mountains of Vermont: Spatial and temporal patterns. Water Air Soil Pollution 21:161-170.

Hain, F.P., and F.H. Arthur. 1985. The role of atmospheric deposition in the latitudinal variation of Fraser fir mortality caused by the Balsam woolly adelgid *Adelges piceae*: A hypothesis. Zeitschrift für angewand te Entomologie 99:145-152.

Hornbeck, J.W., R.B. Smith, and C.A. Federer. 1987. Growth decline in red spruce and balsam fir relative to natural processes. Pp. 425-430 in Acidic Precipitation: Part II, H.C. Martin, ed. Boston: D. Reidel Publishing Company.

Lefohn, A.S., and J.E. Pinkerton. 1988. High resolution characterization of ozone data for sites located in forested areas of the United States. J. Air Pollut. Control Assoc. 38:1504-1511.

Manion, P.D., 1981. Tree Disease Concepts. Englewood Cliffs, New Jersey: Prentice-Hall, Inc.

McLaughlin, S.B., R.K. McConathy, D. Duvick, and L.K. Mann. 1982. Effects of chronic air pollution stress on photosynthesis, carbon allocation, and growth of white pine trees. Forest Science 28(1):60-70.

Miller, P.R. 1983. Ozone effects in the San Bernardino National Forest. P. 161 in Air Pollution and the Productivity of the Forest, D.D. Davis, A.A. Millen, and L. Dochinger, eds. Izaak Walton League and Pennsylvania State University.

Miller, P.R., R.N. Kickert, O.C. Taylor, R.J. Arkley, F.W. Cobb, Jr., D.L. Dahlsten, P.J. Gersper, R.F. Luck, J.R. McBride, J.R. Parmeter, Jr., J.R. Wenz, M. White, and W.W. Wilcox, Jr. 1977. Photochemical oxidant air pollutant effects on a mixed conifer forest ecosystem. Annual Progress Report, 1975-1976. EPA-600/3-77-104. U.S. Environmental Protection Agency.

National Acid Precipitation Assessment Program. 1986. Annual Report to Congress. pp. 163.

National Acid Precipitation Assessment Program. 1987. Interim assessment: The causes and effects of acidic deposition, Volume III. Atmospheric processes, chapter 5. Acidic deposition and its gaseous precursors. pp. 116.

Peterson, D.L., M.J. Arbaugh, V.A. Wakefield, and P.R. Miller. 1987. Evidence of growth reduction in ozone-injured Jeffrey pine in Sequoia and Kings Canyon National Parks. J. Air Pollut. Control. Assoc. 37:906-912.

Phillips, S.O., J.M. Skelly, and H.E. Burkhart. 1977a. Growth fluctuation of loblolly pine due to periodic air pollution levels: Interaction of rainfall and age. Phytopathology 67:716-720.

Phillips, S.O., J.M. Skelly, and H.E. Burkhart. 1977b. Eastern white pine exhibits growth retardation by fluctuating air pollutant levels: Interaction of rainfall, age, and symptom expression. Phytopathology 67:721-725.

Raynal, D.J., and D.C. LeBlanc. 1985. Acidic deposition and effects on forests in the United States. In the Proceedings of the 12th Energy Technology Conference, Washington, D.C., March 25-27, 1985.

Sheffield, R.M., N.D. Cost, W.A. Bechtold, and J.P. McClure. 1985. Pine growth reductions in the southeast. USDA Forest Service: Southeastern Forest Experiment Station, Resource Bulletin SE-83:131.

Siccama, T.G., M. Bliss, and H.W. Vogelmann. 1982. Decline of red spruce in the Green Mountains of Vermont. Bull. Torrey Bot. Club 109:163-168.

Skelly, J.M., L.D. Moore, and L.L. Stone. 1972. Symptom expression of Eastern white pine located near a source of oxides of nitrogen and sulfur dioxide. Plant Disease Reptr. 56:3.

Skelly, J.M., Y.S. Yang, B.I. Chevone, S.J. Long, J.E. Nellessen, and W.E. Winner. 1983. Ozone concentrations and their influence on forest species in the Blue Ridge Mountains of Virginia. Pp. 143-159 in Air Pollution and the Productivity of the Forest, D.D. Davis, A.A. Millen, and L. Dochinger, eds. Izaak Walton League and Pennsylvania State University.

Stone, L.L., and J.M. Skelly. 1974. The growth of two forest tree species adjacent to a periodic source of air pollution. Phytopathology 64:773-778.

Air Pollution and Forest Health in Central Europe: Poland, Czechoslovakia, and the German Democratic Republic

STEFAN GODZIK
Institute of Environmental Engineering
Polish Academy of Sciences

JADWIGA SIENKIEWICZ
Forestry Research Institute
Warsaw

Forest injury close to air pollution sources has been known in Central Europe for more than 100 years (Schroeder and Reuss, 1887; Reuss, 1893). Air pollutants—gaseous as well as particulate—are thought to be responsible for damage to a large part of the forested area in Poland, with both forest managers and forest scientists recognizing the situation as very serious (Grzywacz and Prusinkiewicz, 1986). Inventories carried out in the late 1960s for the Upper Silesian industrial region of Poland have shown major changes in wooded area, tree species composition, and their growth parameters (Godzik and Harabin, 1968). Since that time, significant changes in the extent of forest injuries have occurred not only in Upper Silesia, but in almost all European countries. Regional-scale air pollution has been implicated as a contributing factor in forest deterioration as damage has become evident outside industrial regions, especially in the mountainous part of Central Europe and in countries where emission of air pollutants from industrial sources has been relatively low (e.g., Switzerland, Scandinavia).

Beginning in the early 1980s, symptoms of declining tree health have been observed in Central European forests, increasing rapidly both in impacted area and in intensity. Considerable forest damage has been reported in Austria, the Federal Republic of Germany (FRG), Switzerland, and to some extent in Czechoslovakia, the German Democratic Republic (GDR), and Poland (Liefgreen, 1985; Figure 1). Recent forest damage in Central Europe is perceived as a real threat not only to the forest ecosystem, but to the environment in general.

The aim of this chapter is to present the most recent data available concerning forest injury and decline in Czechoslovakia, the GDR, and Poland. These three countries have been chosen because extensive damage

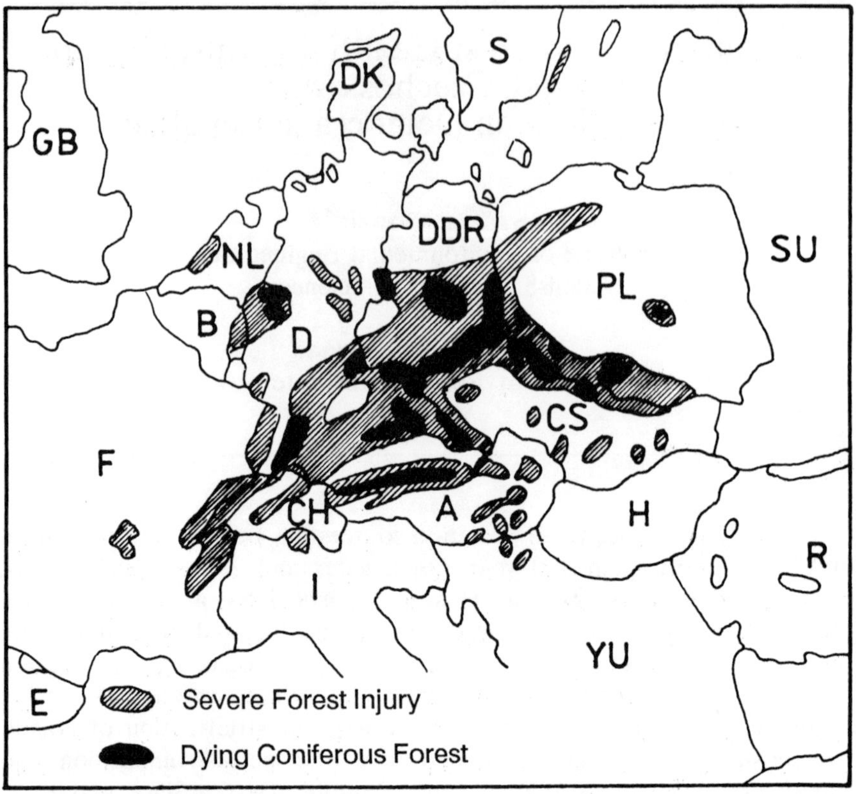

FIGURE 1 Distribution of forest injuries in some European countries (Wentzel, 1987). A=Austria; B=Belgium; CH=Switzerland; CS=Czechoslovakia; D=FRG; DDR=GDR; DK=Denmark; E=Spain; F=France; GB=Great Britain; H=Hungary; I=Italy; NL=Netherlands; PL=Poland; R=Romania; S=Sweden; SU=USSR; YU=Yugoslavia.

of forests has been associated with regional industrial pollutants of similar origin. These areas are characterized by a preponderance of coniferous forest, dominated by *Picea abies* (Norway spruce), *Pinus silvestris* (Scots pine), with *Fagus silvatica* (beech) and *Quercus robur* (pedunculate oak) as important broadleaved species.

Furthermore, fossil fuels such as bituminous and lignite coals are major energy sources for power generation and heating in the region, making transboundary transport of air pollutants more than a theoretical problem. Sulfur dioxide is the pollutant of greatest concern, although the effects of ozone and oxidants are poorly documented in contrast to the situation in the United States. Because of the greater amount of data available to the authors, the situation in Poland will be discussed in more detail and will serve as a basis for comparison with the other countries.

TABLE 1 Emission of sulfur dioxide in Czechoslovakia, GDR, and Poland.

Country	Area (x 1,000 km^2)	EMISSION OF SULFUR DIOXIDE Total TG per a	Metric tons per km^2
Czechoslovakia	127.9	4	36.9
GDR	108.3	3.2	25.0
Poland	312.7	4.3	13.7

SOURCE: National Program, 1988.

AIR POLLUTION

The total amount of sulfur dioxide emitted in Poland and Czechoslovakia calculated by unit area is shown in Table 1. The distribution of emissions sources within Europe has been taken from the 1979 OECD report. Figure 2 presents the distribution (isolines) of sulfur dioxide concentration in Poland, based on modeling. The mean values of sulfur dioxide and fluorine deposition in Polish forests, measured by using a surface active monitoring technique (with potassium carbonate as active substance), are shown in Figures 3-6 (Dunikowski et al., 1988, 1989).

Conclusions concerning the severity of the situation differ to some extent, depending on the criteria for evaluation and on the total amount of sulfur dioxide emitted or calculated by an unit area (Table 1). Poland is the largest "producer" of sulfur dioxide if the total amount is taken into account; however, if calculated by unit area, it is the smallest. It would appear that the latter conclusion is closer to reality.

The uneven distribution of fossil fuel deposits and their use for power generation is largely responsible for the very diversified situation within each of these countries (Figures 1 and 2). Due to emission source characteristics and meteorological conditions, the ground level concentration and, subsequently, its impact on the forest may differ both in severity and location. The role of sulfur dioxide is emphasized here, but it is by no means the only pollutant to cause adverse effects on forests. Also contributing are nitrogen oxides from coal-fired power stations and hydrogen fluoride (HF) from those using lignite as fuel (Dominok, 1984). Poor control of emission of particulate is another contributing factor if deposited not only on the plant surface but on the soil as well.

On a regional scale, the major pollutant is sulfur dioxide. On a local scale, even where other pollutants may be responsible for very serious effects, sulfur dioxide can not be ignored as contributing element. In addition to several nitrogen compounds, including fertilizers, sulfur dioxide is emitted and reacts with ammonia coming from the same source.

In the GDR, several types of pollutants have been recognized as

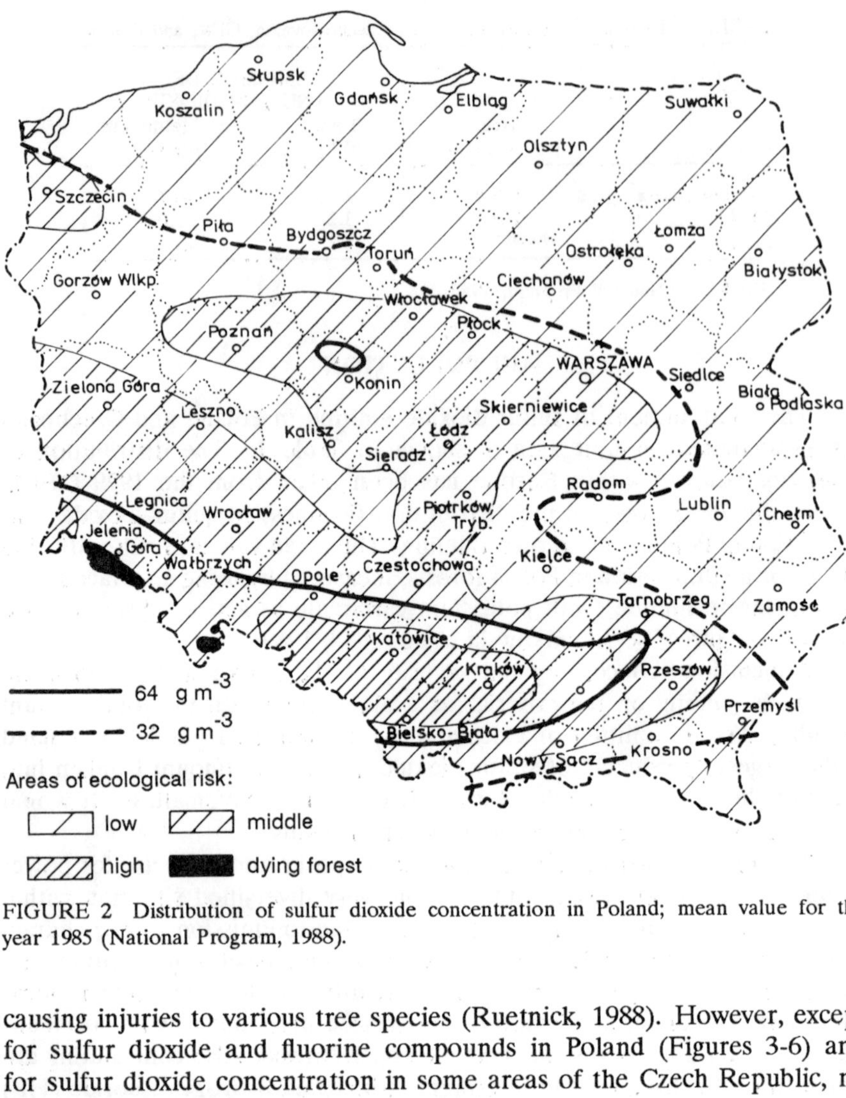

FIGURE 2 Distribution of sulfur dioxide concentration in Poland; mean value for the year 1985 (National Program, 1988).

causing injuries to various tree species (Ruetnick, 1988). However, except for sulfur dioxide and fluorine compounds in Poland (Figures 3-6) and for sulfur dioxide concentration in some areas of the Czech Republic, no similar data are available for the GDR.

For the Czech Republic the following mean concentrations of sulfur dioxide are available (given in grams of pollutant per cubic meter of air):

- Ore Mountains: 100 μg m^3
- Izerskie Mountains: 40 μg m^3
- Beskidy Mountains: 20 - 30 μg m^3.

These figures represent daily values but from long-term measurements (Materna, 1986). According to Kanok (1987), the emission of sulfur dioxide

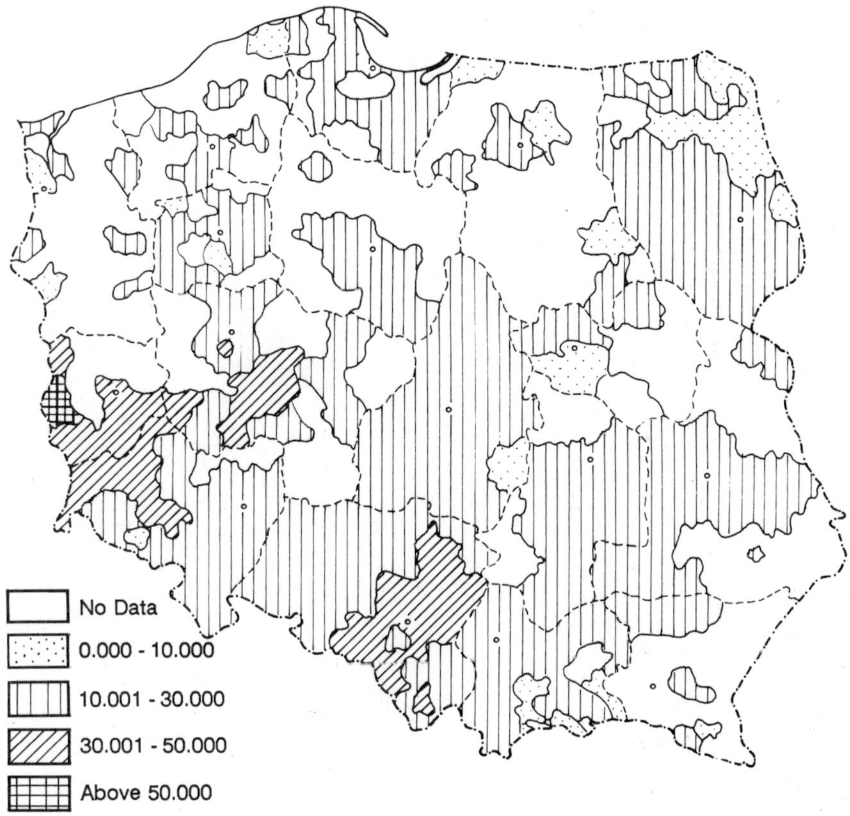

FIGURE 3 Sulfur dioxide deposition rate in Polish forests in winter. (Dunikowski et al., 1988). Values in mg per m^2 per day in increasing order of line density.

for the North Moravian area—which includes the Beskidy region—was estimated to be 221,000 metric tons per year. The amount of the same pollutant coming from industrial sources across the border in Poland (i.e., Upper Silesia) has been estimated at 780,000 metric tons per year (GUS, 1984). No similar data are available to the authors for other regions of Czechoslovakia or the GDR.

The distribution pattern of sulfur dioxide concentrations shows that the national standard (64 $\mu g\ m^{-3}$) for this pollutant is exceeded in the southern part of Poland, from the border with the GDR (west from Jelenia Gora) along the border with Czechoslovakia to Bielsko-Biaca (Figure 2). Data from measurements of the deposition rate of SO_2 and F^- carried out in about 2,000 forest locations (Figures 3-6) show a similar pattern to the one presented in Figure 2 for SO_2 concentration.

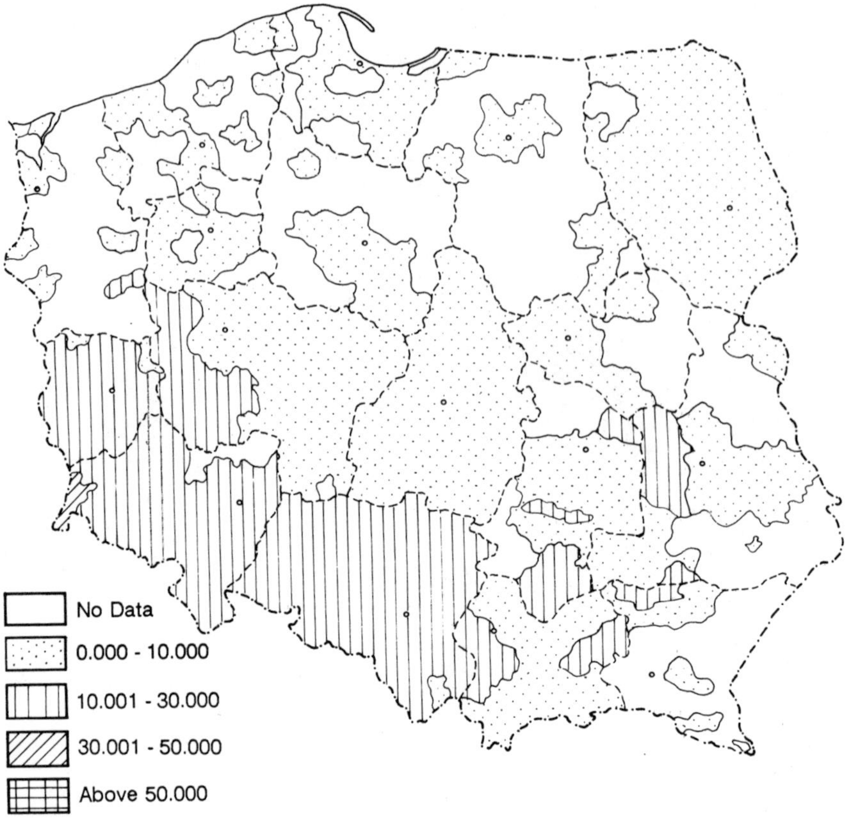

FIGURE 4 Sulfur dioxide deposition rate in Polish forests in summer (Dunikowski et al., 1989). Explanation as in Figure 3.

During the winter months, the deposition of SO_2 and F^- is generally higher than in summer (compare Figures 3 & 5 and 4 & 6). The highest dry deposition of these pollutants has been found in areas close to major air pollution sources as shown in the left corner of Figures 3-6, west of Jelenia Gora, where pollutants from all three countries are present (Praglowski, 1986). The amount of sulfur dioxide emitted from the Polish power station at Turoszow alone has been estimated at 160,000 metric tons per year (GUS, 1984). No official data for power stations located in Czechoslovakia and the GDR are available.

The second most polluted area is the region where the Ostrava (North Moravian) region and the Upper Silesian industrial region share a border (see Figure 2, south of Katowice). Similarities in distribution patterns of

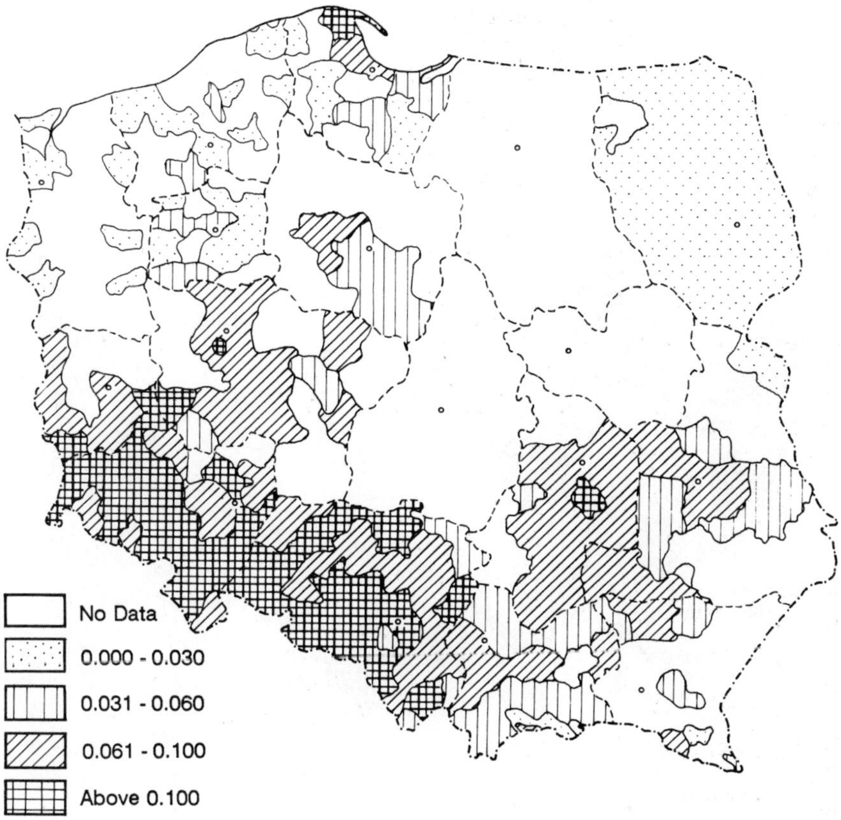

FIGURE 5 Fluoride deposition rate in Polish forests in winter (Dunikowski et al., 1988). Values in mg per m^2 per day in increasing order of line density.

sulfur dioxide and fluorine in both seasons may suggest that they are coming from similar sources.

FORESTS AND FOREST INJURY IN CENTRAL EUROPE

Information about forest composition in selected Central European countries and the most recent data from inventories of forest damage are shown in Table 2. Overall, forests cover about one-third of the area of Central Europe, and the quantity of wood harvested in the region constitutes roughly one-fifth of the total European wood harvest per annum (Liefgreen, 1985). As already mentioned, Scots pine and Norway spruce dominate over broadleaved species in the region, with pine growing mostly in the lowlands (i.e., in the GDR and Poland) and spruce in the mountain

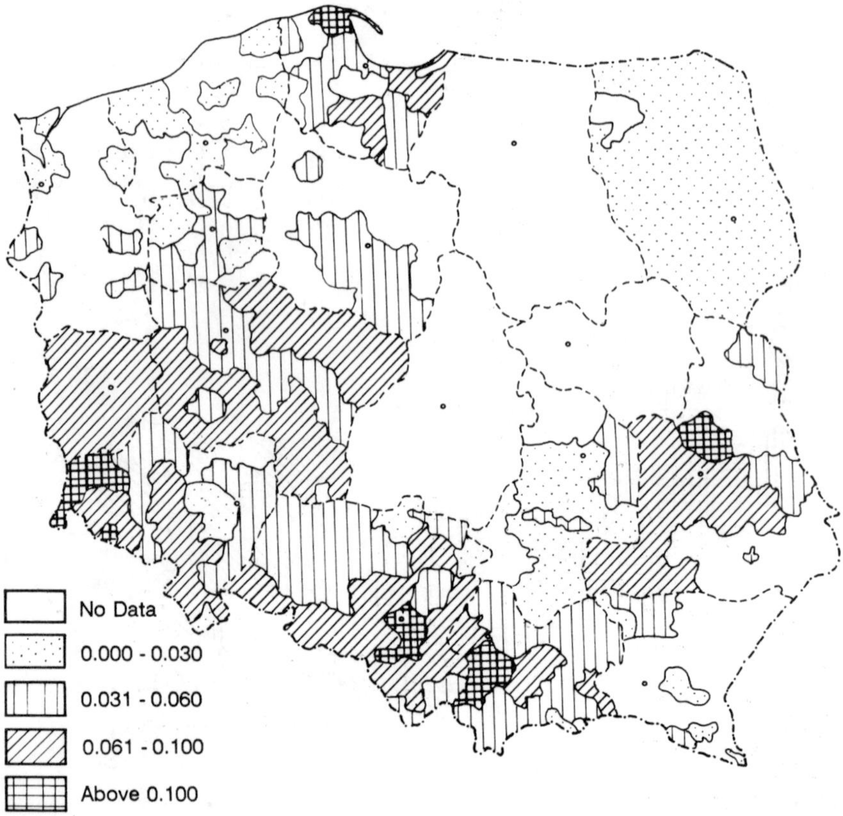

FIGURE 6 Fluoride deposition rate in Polish forests in summer (Dunikowski et al., 1989). Explanation as in Figure 5.

regions. In Czechoslovakia, spruce constitutes the main forest species in the Czech Republic, whereas beech dominates in the Slovak Republic (Czechoslovak Forestry, 1972).

Forests in Central Europe, for the most part, have been planted and are intensively managed. Coniferous species have been planted for commercial purposes during the last 200 years, replacing the original mixed-hardwood forests. This mass introduction of conifers rendered these forests more vulnerable to stresses including pathogens, insect epidemics, extreme climatic events, and nutritional problems. Air pollution can be seen as an additional stress factor contributing to the already weakened condition of the forests, especially conifers.

Data concerning forest injury (Table 2) generally are compiled only for whole countries; therefore, a more detailed description of their distribution

TABLE 2 Results of forest damage survey in some European countries; defoliation symptoms in all species and ages.

Country	Forest x 1,000 hectares Total	Conif.	Broadl.	Percent of total injured area
Austria	3,754	3,040	714	33.5
Czechoslovakia	4,578	2,942	1,636	52.3 *
GDR	2,955	2,275	680	37.0
FRG	7,360	5,078	2,282	52.3
Poland	8,654	6,895	1,759	57.6 **
Switzerland	1,186	777	409	56.0

*Conifers only
**Courtesy of Forest Inventory and Geodesy Bureau 1989.

SOURCE: UN-ECE, 1988.

by province or district within each country is often not available. However, more detailed information for Czechoslovakia and Poland can be found in several papers.

For the past several years, the most severe occurences of forest damage have been observed in Czechoslovakia in the Ore Mountains, and in Poland in the Izerskie Mountains (Figures 1 and 2). In each of those areas, about 30,000 hectares of spruce forests have been cut down since the damage started (Materna, 1986; Rykowski, 1987). The area of forest decline on the Polish side is expanding eastward toward the Beskidy Mountains (Figure 2). Similar phenomena have been observed in Czechoslovakia (Materna, 1986).

Based on data for the year 1986, 37.2% of the forests in the North Moravian area has been injured (Kanok, 1987), which is more than twice the area reported damaged in 1981 (Pokorny, 1984). A similar figure (36%) has been obtained for the forest on the Polish side of the border based on the inventory from 1983 (Godzik, 1987). In both cases, assessments of the impact have been carried out using methods which originated in each country. Official figures for forest areas injured in these three countries ranged from 7% for Poland and 10% for Czechoslovakia to 13% for the GDR. However, it is unlikely that figures obtained for this area during the most recent inventory carried out in all European countries using the same method will be significantly higher (Table 2). The criteria and methods applied in the inventory accepted by the European countries (Manual, 1986) are not designed to determine the specific cause of injury.

In the region delineated by these three countries, air pollution has been implicated in forest damage measurements. Areas of most severe

TABLE 3 Concentration of some heavy metals in needles and leaves from area surrounding a zinc smelter

	Concentration, ppm					
	Plot I *			Plot II **		
Plant species	Zn	Pb	Cd	Zn	Pb	Cd
Pinus silvestris	495.9	436.4	8.9	261.0	276.0	4.2
P. nigra	781.8	635.9	14.1	179.7	92.4	2.6
P. strobus	944.2	592.1	15.8	434.4	202.7	6.8
Betula verrucosa	2,780.1	535.9	15.7	1,709.3	293.0	8.8

Distance from the source:
* = 0.8 km
** = 2.0 km

SOURCE: Hawrys et al., 1986.

and widespread injury are found in the vicinity of heavy industrial regions (Materna, 1986; Ruethnick, 1988; Program, 1985). Data from the most recent assessment are based on the survey of 23,500 permanent plots in Poland, 2,500 plots in the GDR, and 164 plots in Czechoslovakia (Table 1) (National Program, 1988). According to data obtained from measurements of air pollutants in forests, the most likely compound responsible for most of these injuries is sulfur dioxide, although nitrogen oxides and fluorine may also be contributing factors. On a local scale, however, pollutants other than sulfur dioxide may cause very severe damage to the forest, e.g., in the vicinity of the nitrogen fertilizer plant in Pulawy, or the zinc smelter at Miasteczko Slaskie.

Injuries to forests around the Pulawy nitrogen fertilizer plant have increased at a very uneven rate (Kowalkowski, 1985), but are now visible at a distance of more than 60 kilometers downwind from the source. Because of the injury, all trees have been cut down within a zone of 1.5 km from the plant. Soil analysis has shown a very high input of nitrogen; however, it is not clear which compound (i.e., urea, ammonia, nitrogen oxides, ammonium nitrate, sulfur dioxide) was the most harmful. Ammonium sulfate produced as the result of interaction between sulfur dioxide and ammonia has not been taken into consideration so far. However, this compound is known to cause injuries to pine trees in the Netherlands (Boxman and Roelofs, 1986; Roelofs and Boxman, 1986).

A different combination of pollutants is emitted from the zinc smelter: sulfur oxides (di- and trioxides), sulfuric acid, and heavy metals (e.g., zinc

(Zn), lead (Pb), and cadmium (Cd)). The occurrence of forest injury from these elements is increasing at an estimated 250 meters per year in the downwind direction (Rostanski et al., 1985). Very high concentrations of all these metals have been found in both plants and soils around the smelter (Hawrys et al., 1986; Rostanski et al., 1985). Depending on distance and direction from the source of pollution, the concentrations of Zn, Pb, and Cd in the 1-7 cm soil layer were 1,200, 600, and 23.5, respectively, at a distance of 0.8 km, and 124, 160, and 4.5 ppm at a distance of 2.0 km (Hawrys et al., 1986). Concentrations of some heavy metals in needles or leaves in the locations mentioned above are shown in Table 3. A significant amount (i.e., up to 90%) of these metals can be removed using such unconventional methods as organic solvents and ultrasounds, but their concentrations in the tissue still remain higher than in samples from outside this region (Godzik et al., 1979).

Long-term exposure of pine populations to air pollutants and other environmental stress factors has led to differences in their genetic characteristics when compared to populations from a lightly polluted location (Prus-Glowacki and Nowak-Bzowy, 1987). Three generations of *Pinus silvestris* populations from two locations were compared, i.e., from near Poznan as a reference and from the Upper Silesian industrial region as a polluted site. In some cases, some alleles and genotypes showed reciprocal tendencies in these two populations, e.g., malic acid dehydrogenase (MHD) 3 allele and MHD 13 genotype. In the control population, the number of trees carrying the alleles increased while decreasing in the population from the polluted location.

Frequency of MHD alleles may constitute a good index of selective pressure in *Pinus silvestris* populations under sulfur dioxide stress as shown by Mejnartowicz (1983). An increase in average heterozygosity level and genetic polymorphism from the embryo group to the group of maternal trees has been found. In the population from the reference site, an increase was observed by 5% of heterozygosity level, and by 10% of genotypic polymorphism upon transition from the age group 15-30 years to the group of maternal trees (i.e., above 30 years old). In the population from the polluted site, heterozygosity level and genotypic polymorphism increased most rapidly—by 10% and 20%, respectively—upon transition from embryos to the youngest indidviduals. This probably reflects the differences in selective pressures, which are stronger for the population at the polluted site. An inbreeding index shows that the population from the reference location remained in Hardy-Weinberg equilibrium, while the population from the polluted site demonstrated an excess of homozygotes in all age groups (Prus-Glowacki and Nowak-Bzowy, 1987).

A reduction of heterozygosity level by 30% and of genetic polymorphism by 20% has been observed in seedling populations of a known genetic

TABLE 4 Concentration of some heavy metals (ppm) and sulfur (%) in pine [*Pinus silvetris*] needles and oak leaves [*Querus robur*] from Niepolomice Forest.

	CONCENTRATION			
PLANT SPECIES	Cd	Pb	Zn	S
Pinus silvestris	1.76	23.11	57.36	0.177
Quercus robur	1.19	14.35	45.88	0.203

SOURCE: Grodzinska, 1984, after Grodzinski et al., 1984

composition in experiments which are currently in progress (Prus-Glowacki, 1988). Similar differences among populations of *Arabidopsis thaliana* from Upper Silesia and other parts of Poland have been found by Kilian (1987).

As far as the impact of air pollution on forest ecosystems is concerned, the most extensive studies thus far were carried out in the Niepolomice Forest, the results of which were presented by Grodzinski et al. (1984). This lowland forest of pine (*Pinus silvestris*, comprising 71% of the total) and oaks (*Quercus robur* and *Q.petrea*, comprising 17% of the total) covers an area of 110 km and is located in southern Poland, east of Krakow.

The mean annual sulfur dioxide concentration for the entire study period was 20 $\mu g\ m^{-3}$, ranging between 10.25 and 11.98 $\mu g\ m^{-3}$ during the warm season, and between 36.35 and 37.85 $\mu g\ m^{-3}$ during the cold season of the year. The total deposition of sulfur (S) was 60.3 kg ha^{-1} per year, with 37.8 kg ha^{-1} by dry deposition. The injuries to pine trees have been estimated as medium to severe, depending on the location within the forest. The accumulation of Cd, Pb, and Zn has been estimated to be 0.3, 13.9, and 16.6 kg ha^{-1}, respectively, while the yearly input of these metals has been estimated at 0.015, 0.315, and 1.231 kg ha^{-1}, respectively.

Concentrations of these elements in needles of *Pinus silvestris* and leaves of *Quercus robur* are presented in Table 4. Under these conditions, the photosynthesis of pine needles was reduced by 13 and 18% for first- and second-year needles, respectively. In the Niepolomice Forest, the average net biomass production of deciduous stands was low, and that of pine stands was definitely lower when compared to European temperate lowland forests. The reduction of pine diameter growth has been estimated to be between 20 and 30%. According to the authors, both long-term effects of air pollution and lowering of the water table are responsible for these effects (Grodzinski et al., 1984).

Forest decline in Upper Silesia has been investigated since the early 1970s (Wolak, 1971). Extensive studies carried out in this region have resulted in the description of a typical pattern of forest vegetation transformations induced by pollutants. In heavily contaminated areas, the secondary succession of vegetation—which sets in as soon as the original conifer forests begin to die—has lead to the formation of new, less complex, and more stable plant communities. These new communities, developed and living under conditions of pollution stress, represent a state of *sui generis* ecological equilibrium with the most obvious regularity, i.e., the higher the degree of air pollution, the lower the vegetation. Thus, in the industrial region of Upper Silesia, in locations once overgrown with conifer or mixed deciduous forest, a stable zonation of vegetation has been established in relation to the degree of pollution from sulfur dioxide, nitrogen oxide, heavy metals, etc. (Wolak et al., 1977; Wolak, 1979). Under conditions of severe air pollution, an industrial desert occurs, followed by a grass zone with communities dominated by gramineous species, such as *Calamagrostis epigeioso* or *C. villosa.* Along with the decreasing gradient of air pollution, a zone of shrubby trees is formed with a zone of stunted forest closing the series. Thus, the original forest communities are being replaced by communities of grassland or sand dunes, which constitutes the reversal of natural succession of vegetation (Rostanski et al., 1985; Sienkiewicz, 1986).

SUMMARY

Data presented in the previous section illustrate the seriousness of the state of forest health in Central Europe. The importance of air pollutants, particularly dry and wet deposition, as a cause of forest injury and forest decline in Czechoslovakia, the GDR, and Poland is apparent. In all locations where injuries are moderate to severe, not only are the ambient air concentrations of phytotoxic agents high, but the concentration of heavy metals in soils is elevated simultaneously with a marked decrease in pH (Pelisek, 1986; Pokorny, 1984, 1987; Vesely, 1987; Hawrys et al., 1986). Populations of soil microorganisms also undergo changes; however, decline of trees occurs earlier than the collapse of soil microflora (Lettl, 1986, 1987). No correlation has been found between the pH of the soil and the severity of tree injury in mountain areas of Czechoslovakia (Pokorny, 1984; Kratochvilova and Marek, 1986). However, the lowest soil pH was found in the Ore Mountains, where injuries are most severe (Pokorny, 1984).

A correlation has been found between the elevation above sea level and the degree of tree injury (Ibid., Kwapis, 1988). Although the air pollution differs both quantitatively and qualitatively, the symptoms of forest injury are very similar throughout Europe, and do not differ markedly from those described as "classical" (e.g., Kandler, 1985; Wentzel, 1987). In addition, the severity of the deterioration in forest health does not correlate with the deposition of certain pollutants. For example, the concentration of some elements in spruce needles from some locations in Poland and the FRG does not reflect both the severity and extent of the damage to forests in these countries (Table 5). Similar deposition of sulfur in the Izerskie

TABLE 5 Concentration of sulfur and F- in spruce needles (*Picea abies*) from some injured forests in Poland and Federal Republic of Germany

	LOCATION			
ELEMENT	Beskidy Mts (PL)	Izerskie Mts (PL)	Sauerlach (FRG)	Donaustauf (FRG)
S (%)	0.168	0.228	0.097	0.097
F- (μg/g)	8.26	11.69	3.0	5.0

SOURCE: Godzik and Krause, 1987, from Godzik, 1987.

Mountains and the Beskidy Mountains of Czechoslovakia in the range of 60 kg ha per annum as wet deposition (Forest...,1988; Godzik and Szdzuj, 1988) has led to ecological disaster in the Izerskie Mountains, but not in the Beskidy Mountains to date. This seems to support the thesis that no single cause/effect explanation applies in the case of observed forest damage (Cowling et al., 1986; Godzik, 1984, 1987; Manion, 1981; Schutt, 1988). Consequently, the authors are also of the opinion that multiple factors, including various types of air pollutants, are responsible for the forest injury and decline.

REFERENCES

Boxmann, A.W., J.G.M. Roelofs. 1986. Some physiological effects of NH_{4+} and Al_{3+} on pine forest ecosystem. Pp. 407-414 in Wissenschaftliches Symposium zum Thema Waldschaeden Neue Ursachen Hypothesen. UBA Texte 19/86. UBA Berlin.

Cowling, E., B. Krahl-Urban, C. Schimansky. 1986. Wissenschaftliche Hypothesen zur Erklaerung der Ursachen. Pp. 120-1125 in Waldschaeden Kernforschungsanlage Juelich, E.H. Papke, B. Krahl-Urban, K. Peters, C. Schimansky, eds.

Czechoslovak Forestry. 1972. The Ministry of Forests and Waters CSR, and the Ministry of Forests and Waters SSR. State Agricultural Publishing House, Prague.

Dominok, B. 1984. Fluoride and environment: Situation in the district Cottbus. Fluoride 17:193-196.

Dunikowski, S. 1988. The state and perspectives of monitoring of the forest environment. Pp. 1-7 in Conference on Forest Decline in Poland. Committee for Forest Sciences of the Polish Academy of Sciences and Forestry Research Institute, Warsaw. February 1988 (in Polish).

Dunikowski, S., J. Wawrzoniak, A. Liwinska, J. Malachowska, D. Kowalska. 1988. Measurment of air pollution in the forest: Technical monitoring. Report for the winter season 1987-1988. Forestry Research Institute. Warsaw (in Polish).

Dunikowski, S., J. Wawrzoniak, A. Liwinska, J. Malachowska, D. Kowalska. 1989. Measurment of air pollution in the forest: Technical monitoring. Report for the summer season 1988. Forestry Research Institute, Warsaw (in Polish).

Forest Monitoring and Management Program in Sudety Mountains. 1988. Forestry Research Institute, Warsaw (in Polish).

Godzik, S. 1984. Air pollution problems in some central European countries: Czechoslovakia, the German Democratic Republic, and Poland. Pp. 25-34 in Gaseous Air Pollutants and Plant Metabolism, M.J. Koziol and F.R. Whatley, eds. Butterworth, London.

Godzik, S. 1987. An attempt to evaluate the effects of air pollutants on forests and soils. Ochrona Powietrza 6:156-160 (in Polish).

Godzik, S., Z. Harabin. 1968. Waldbestaende im Raume Katowice und Myslowice zu Ende des XIX Jahrhunderts und gegenwaertig. Pp. 355-369 in Referaten der VI Interantionalen Arbeitstagung Forstlicher Rauchschadensachverstaendiger. Katowice

Godzik, S., T. Florkowski, S. Piorek, and M.M.A. Sassen. 1979. An attempt to determine the tissue contamination of *Quercus robur* L. and *Pinus silvestris* L. foliage by particulates from zinc and lead smelter. Environm. Pollut. 18:97-106.

Grodzinski, W., J. Weiner, and P.F. Maycock. 1984. Forest ecosystems in industrial regions. Berlin: Springer Verlag.

Grzywacz, A., and Z. Prusinkiewicz, eds. 1986. Trends in development of forest sciences till the year 2000 and in subsequent years. Postepy Nauk Rolniczych 2-3:86, 229-250.

Hawrys, R., J. Chlodny, J.Zwolinski, I. Matuszczyk. 1986. Determination of tree species composition for afforestation of areas under air pollution stress. Documents. Forestry Research Institute, Warsaw (in Polish).

Kandler, O. 1985. Immissionen—versus Epidemie—Hypothese. Pp. 19-59 in Theorie und Praxis auf der Suche nach Antworten, G. von Kortzfleisch, ed. Waldschaeden. Muenchen-Wien: R. Oldenbourg Verlag.

Kanok, F. 1987. Economic consequences of air pollution in north Moravian state forests. Lesnictvi 33:607-620 (in Czech).

Kilian, A. 1987. Genetic effects of environmental pollution on the model plants. Ph. D Thesis. Silesian University, Katowice (in Polish).

Kowalkowski, A. 1985. Fundamentals of forest management in areas infuenced by air pollutants from the Pulawy nitrogen fertilizer plant. Forestry Research Institute, Warsaw (in Polish).

Kratochvilova, I., and M. Marek. 1986. The influence of air pollution on the soil reaction and some characteristic of spruce stands in Jizerske hory. Lesnictvi 32:1069-1080 (in Czech).

Kwapis, Z. 1988. Air, soil, and foliage pollution of spruce stands at Ustron forest management area. American-Polish Seminar on Long-Term Monitoring of Forest Ecosystems. Jaszowiec, 1988 (in Polish).

Lettl, A. 1986. Development of microbial populations in forest soils exposed to SO_2 emissions. Lesnictvi 32:593-610 (in Czech).

Lettl, A. 1987. Internal relations in the microbial populations of the soil of spruce, European mountain ash, and birch stands in the area with air pollution stress. Lesnictvi 33:769-786 (in Czech).

Liefgreen, D. 1985. Long range air pollution: A threat to European forests. Unasylva 37, No. 149:14-25.

Manion, P.D. 1981. Tree Disease Concepts. Englewood Cliffs, New Jersey: Prentice Hall Inc.

Manual on methodologies and criteria for harmonized sampling, assessment, monitoring, and analysis of the effects of air pollution on forests. 1986. Hamburg/Geneva: Programme Coordinating Centres/UN-ECE.

Materna, J. 1986. An outlook of the influence of air pollution on forests in the Czechoslovak Socialst Republic. Lesnictvi 32:319-328 (in Czech).

Moseholm, L., B. Anderssen, and I. Johnsen. 1985. Acid deposition and forest decline in central Europe. The Nordic Council of Ministers.

National Program of Natural Environment Protection Till the Year 2010. 1988. Draft. Ministry of Environmental Protection and Natural Resources, Warsaw (in Polish).

Organization for Economic Cooperation and Development. 1979. Programme on Long Range Transport of Air Pollutants. OECD, Paris.

Pelisek, J. 1986. The harmful effects of emissions on forest soils and stands. Lesnictvi 32:87-88 (in Czech).

Pokorny, P. 1984. To the problem of forest soil acidification in mountain regions of the Czech Socialist Republic (CSR). Pp. 81-91 in the Proceedings of the Symposium on Air Pollution and Stability of Coniferous Forest Ecosystems. Ostravice, Czechoslovakia, October 1-5, 1984.

Pokorny, P. 1987. Contamination of forest soils by air polluting substances. Lesnictvi 33:267-277 (in Czech).

Praglowski, R., et al. 1986. Integrated air pollution measurments at the Rozdroze Izerskie. Jelenia Gora cit. acc. to Forest Monitoring Program in Sudety (in Polish).

Program 1985. Hazards to the forest environment in Poland: 1985. Ministry of Forestry and Wood Industry, Main State Forests Management, Forestry Research Institute. Warsaw (in Polish).

Prus-Glowacki, W. and Nowak-Bzowy. 1987. Demographic processes in *Pinus silvestris* populations from regions under strong and weak anthro-opression (in press).

Prus-Glowacki, W. 1988. Direction and dynamic of adaptative processes in populations of forest trees (*Pinus silvestris*) in air polluted areas. Report. Institute of Environmental Engineering, Polish Academy of Sciences, Zabrze (in Polish).

Reuss, C. 1983. Rauchbeschaedigung in dem von Thiele-Winkler'schen Forstreviere Myslowitz-Kattowitz. Goslar.

Roelofs, J.G.M., A.W. Boxmann. 1986. The effect of airborne ammonium sulfate deposition on pine forests. Pp. 415-422 in Wissenschaftliches Symposium zum Thema Waldschaeden "Neue Ursachen-Hypothesen." UBA Texte 19/86. UBA Berlin.

Rostanski, K., ed. 1985. Plant communities of forests under air pollution stress. Report. Katowice.

Ruethnick, R. 1988. Die Aufgaben der Forstwirtschaft bei weitren Verwirklichung der Beschluesse des XI Parteitages der SED. Sozialistische Forstwirtschaft 9/88, pp. 257-265.

Rykowski, K. 1987. Conservation of forests under conditions of growing threats to the environment. Pp. 375-386 in the Proceedings of the Workshop on Forest Decline and Reproduction: Regional and Global Consequences, L. Kairiukstis, S. Nilsson, and A. Straszak, eds. IIASA Laxenburg.

Schroeder J. von, and C. Reuss. 1983. Die Beschaedigung der Vegetation durch Rauch und Oberharzer Huettenrauchschaeden. Zentralantiquariat der Deutschen Demokratischen Republic, Leipzig 1986.

Schutt, P. 1988. Der Wald Stirbt an Stress. Ullstein. Frankfurt M./Berlin.

Sienkiewicz, J. 1986. Effect of heavy metals industry on plant communities. Sci.Total Environm. 55:339-349.

Trampler, T. 1988. Evaluation of forest injury using biomonitoring methods. Pp. 8-18 in Conference on forest decline in Poland. Committee of Forest Sciences, Polish Academy of Sciences. Forestry Research Institute, Warsaw (in Polish).

Trampler, T. 1989. Forest injury in Poland. Report. Forestry Research Institute. Warsaw (in Polish).

UN-ECE. 1988. Forest damage and air pollution. Report for the 1987 Forest Damage Survey in Europe. GEMS, Geneva.

Vesely, J. 1987. The infuence of emissions on the chemical composition of forest soils. Lesnictvi 33:385-398.

Wentzel, K-F. 1987. Waldschaeden—was ist wirklich neu? Mensch and Umwelt, September 1987, pp. 19-28.

Wolak, J. 1971. Studies on the industrioclimax in Poland. Proc. XV IUFRO Congr. Wien.

Wolak, J., ed. 1977. Relationship between increase in air pollution toxicity and elevation above ground. Report. Forestry Research Institute, Warsaw.

Wolak, J. 1979. Reaction des Ecosystems a la Pollution Subnecrotique. ECE symposium on the effects of airborne pollution on forests. Warsaw, Poland, August 20-24.

Impacts of Air Pollution on Agriculture in North America

WALTER W. HECK
Agricultural Research Service
U.S. Department of Agriculture

This chapter highlights our current understanding of the effects of anthropogenic air pollutants on agriculture in the United States and assesses their impact on crop productivity. Information from a cross section of crop-oriented review articles has been used to help define the problem (Altshuller and Linthurst, 1984; Dochinger and Seliga, 1976; Evans, 1984; Guderian, 1985; Heck, 1984; Heck and Brandt, 1977; Heck et al., 1977, 1982a, 1982b, 1983a, 1984a, 1984b, 1984c, 1986; Irving, 1987; Jager and Klein, 1980; National Academy Sciences, 1977; Roberts, 1984; Shupe et al., 1983; Shriner et al., 1980; Treshow, 1984; Unsworth and Ormrod, 1982; U.S. EPA, 1978, 1982a, 1982b, 1986; and Winner et al., 1986).

Air pollution effects from point sources were first recognized in the early 1900s. Research was directed at recognizing injury symptoms and assessing losses in productivity in accordance with the severity of symptom development. Since that time, many chemicals have been identified as atmospheric contaminants that can injure or damage crops at some range of exposure concentration and time (Table 1).

Sulfur dioxide (SO_2) and fluoride (generally as hydrogen fluoride or HF) gases from point sources caused injury to vegetation around the turn of the century. Ducktown (Copper Hill), Tennessee in the United States, and Sudbury, Ontario in Canada still show marked effects from SO_2 releases from early smelter operations. Today, the installation of emission control systems and the use of high stacks has generally eliminated extensive injury to vegetation near sources of SO_2. However, high-capacity power stations with tall stacks have increased the distribution of lower SO_2 concentrations over larger areas. Thus, SO_2 is now considered a regional problem, although direct effects on agricultural production are poorly defined.

TABLE 1 Phytotoxic air pollutants, in order of importance to crop systems.[a]

Pollutant	Primary or Secondary	Form	Major Source(s)
O	Secondary	Gas	Atmospheric transformation (associated with automotive emissions, NO_2, hydrocarbons)
SO_2	Primary	Gas	Power generation, smelter operations
NO_2	Primary and secondary	Gas	From direct release and atmospheric transformation (high-temperature combustion, from NO); fertilizer production
HF	Primary	Gas-particulate	Superphosphate, aluminium smelters
C=C[b]	Primary	Gas	Combustion, natural
PAN-Oxid.[c]	Secondary	Gas	Atmospheric transformation (automotive emissions, NO_2, hydrocarbons)
NO	Primary	Gas	Combustion, natural
Cl_2	Primary	Gas	Spills, manufacture
HCl	Primary	Gas	Burning of plastics
Toxic elements	Primary	Particulate	Smelters, combustion processes
NH_{32-}	Primary	Gas	Feedlots, natural
SO_4	Secondary	Aerosol	Atmospheric transformation (SO_2)
NO_3-	Secondary	Aerosol	Atmospheric transformation (NO_2)
H_2S	Primary	Gas	Paper production, natural, geothermal
CO_2	Primary	Gas	Combustion, natural
UV-B!	Primary	Radiation	Natural, stratospheric O_3 depletion

[a] This list is not meant to be complete but represents the most important air pollutants with respect to terrestrial plant systems. Several of those low on the list have been poorly studied and may be more important than currently thought.

[b] Ethylene

[c] Peroxyacetyl nitrate-oxidant

SOURCE: Heck, 1982.

Fluoride (as HF) injury to vegetation was described by the turn of the century, but did not become a major problem until aluminum smelting and super-phosphate production increased in the 1940s. Fluoride symptoms are well characterized and a spectrum of sensitivities both between and within species is known. Foliar analysis is an acceptable diagnostic tool since fluoride accumulates in plant tissues. However, natural distribution of fluoride complicates the diagnosis. Shupe et al. (1983) present a comprehensive treatment of fluoride research.

Photochemical air pollution injury was first described in the Los Angeles area in 1944 and was evident over large segments of California by 1950. Components of photochemical oxidants (primarily ozone or O_3) injure and damage crops in many areas of North America. Ozone, peroxyacetyl nitrate (PAN), and nitrogen dioxide (NO_2) are phytotoxic components of the photochemical complex.

Ethylene is a major petrochemical and a by-product of combustion and of plant metabolism. It is a phytotoxic hydrocarbon gas and contributes to the formation of photochemical oxidants. Ethylene from anthropogenic sources probably contributes to crop losses.

EFFECTS OF AIR POLLUTANTS OTHER THAN OZONE ON AGRICULTURAL PRODUCTIVITY

Many phytotoxic air pollutants have caused serious damage around point sources. Others are ubiquitous across the continent of North America and are perceived as major problems. Fluoride is one of the most phytotoxic of the first group, while hydrogen chloride and hydrogen sulfide are less toxic. PAN is more phytotoxic than O_3 and is widespread (in low concentrations) around metropolitan areas. Nitrogen dioxide has been studied principally in combination with O_3 and SO_2 because alone it is not particularly phytotoxic at current ambient concentrations. However, it is a primary actor in ozone formation.

The air pollutants currently of greatest national concern are O_3, SO_2, NO_2, and the transformed products of SO_2 and NO_2 (SO_4^{2-} and NO_3^-) that are largely responsible for acidic precipitation. In North America, O_3 has the greatest effect on agricultural productivity because it is found in damaging concentrations in most sections of the continent. Sulfur dioxide is released from point sources and may, over weeks or months, injurc sensitive vegetation. Its primary impact may occur when it is associated with O_3 or NO_2. The importance of NO_2 in terrestrial ecosystems is as an ingredient in photochemical reactions (forming O_3) and when present in association with SO_2 or O_3. These three pollutants (O_3, SO_2, and NO_2) are critical components in the formation of acidic precipitation.

Space does not permit a discussion of the many air pollutants that have damaged plants due to accidental releases or on a fairly local basis. However, a brief discussion of NO_2, acidic deposition, and SO_2 are included in this section. The importance of ozone is recognized and treated in a separate section along with assessment methods.

Nitrogen Dioxide (NO_2)

Nitrogen dioxide can be used as a nitrogen source by plants. The first published study using ambient levels of $^{15}NO_2$ (0.097, 0.152, or 0.325 ppm NO_2 for three hours) reported a linear relation between exposure concentration and uptake for snapbean; essentially all the nitrogen was metabolized (Rogers et al., 1979). Although several studies have reported greater NO_2 uptake at night, Yoneyama et al. (1979) found that night absorption was only about 14% of that absorbed during the day.

Both nitrogen oxide (NO) and NO_2, which are associated with the production of high levels of greenhouse CO_2, substantially decreased growth and productivity of greenhouse crops (Mansfield et al., 1982). This raises serious concern for greenhouse operators who add CO_2 in the greenhouse.

Acidic Deposition

The effects on crops of acidic precipitation (often incorrectly called "acid rain," but encompassing all forms of precipitation) was highlighted in the United States in an international symposium (Dochinger and Seliga, 1976). Acidic deposition includes both wet and dry deposition of acidifying substances such as SO_2 and NO_2. Acidic precipitation is used to separate the transformed products (SO_4^{2-} and NO_3^-) from gaseous SO_2 and NO_2; the term "acid rain" denotes only that specific form of acidic precipitation. Vegetation growing where acidic fogs or clouds are common may also be affected.

The pollution control strategy of using tall stacks to reduce ground-level concentrations of sulfur and nitrogen oxides near fossil fuel combustion sources is a contributing factor to the long-range transport of acidic substances. The transformed products may be deposited hundreds of kilometers downwind in precipitation.

Acidic precipitation may affect the growth, reproduction, quality, and/or yield of agricultural crops (Heck et al., 1984a). Direct effects include changes in leaf surface morphology, foliar nutrient leaching, uptake of additional sulfur or nitrogen, or changes in metabolic function or reproductive processes; with perennial plants, the effect may be cumulative across growing seasons. Indirect effects include altered physicochemical characteristics of soils (e.g., water-holding capacity), nutrient availability, availability of toxic elements, and susceptibility of plants to biotic and other stresses.

Research efforts in the early 1980s developed around field studies utilizing rain exclusion systems to permit growth of plants under field conditions. Evans et al. (1986) reported significant acidic rain effects on 'Amsoy 71' over five consecutive growing periods; but no effects on other soybean cultivars were reported. DuBay and Heagle (1987) reported no effects on growth or yield from rain acidities as low as pH 2.7 (using similar field protocol) for the cultivar 'Forrest.' Banwart et al. (1987) found no effect on two field corn cultivars at present acidity levels, although a significant reduction in yield was found in one cultivar at a pH of 3.0. Pell et al. (1987) detected no effects on two potato cultivars at pH treatments as low as 2.8.

Acidic precipitation may affect plant systems, but the evidence is limited. It seems safe to suggest that we have not succeeded in developing an experimental approach that will permit the identification of small impacts.

Likewise, the presence of increasing nitrogen and sulfur with decreasing pH (increasing H^+) probably confounds experimental results.

Sulfur Dioxide (SO_2)

Sulfur dioxide was extensively studied in the first half of the century (Heck and Brandt, 1977). Interest was renewed by the increase in power generation and the subsequent increase in SO_2 emissions. In addition, results of long-term assessments around point sources convinced scientists that earlier perceptions of SO_2 thresholds needed to be reexamined.

The effects of SO_2 may be visible or subtle. Acute injury results in cell plasmolysis with foliar lesions, often bleached white; chlorosis may occur. Chronic injury may initially disrupt cellular activity, followed by chlorosis or other pigment changes that may lead to cell death. Chronic injury patterns are generally not characteristic of SO_2 and may be confused with symptoms caused by diseases, insects, other stresses, as well as normal leaf senescence.

The mechanism of crop response to SO_2 has been studied in several ways. Sensitivity differences in four cultivars of cucumber were related to SO_2 uptake (stomatal activity), but leaves of different sensitivity on the same plant involved a biochemical or developmental resistance mechanism related to the formation and loss of hydrogen sulfide (H_2S) (Bressan et al., 1978; Sekiya et al., 1982). Evidence for the photodetoxification of SO_2 was associated with increased injury to plants on which stomata remained open during dark exposures to SO_2; a strong case was presented for sulfite as the primary phytotoxicant for acute plant injury (Olszyk and Tingey, 1984). Plant physiological and biochemical processes are probably more important controllers of plant resistance to SO_2 (tolerance) than is control of gas entry via the stomata (avoidance).

Many biological and physical factors affect the response of plants to SO_2, including genetic, biological, environmental, and chemical. Research of a genetic nature has concentrated on the determination of relative sensitivities of species and their genotypes, and on the effects on pollen and pollen germination (Heck et al., 1986). Mexican bean beetle larva fed preferentially on soybean foliage exposed to chronic concentrations of SO_2 (Hughes et al., 1983). Environmental factors include light, temperature, humidity, CO_2, freezing, soil moisture, and soil nutrition. Several of these have been studied in combination. Plants are generally more sensitive to SO_2 as light intensity, wind speed, temperature, and humidity increase; elevated CO_2 levels protect plants; and freezing may increase plant sensitivity, while low soil moisture tends to make plants more resistant.

The effect on plants of mixtures of pollutants (i.e., SO_2 and O_3) is due primarily to the O_2 component and is discussed below. Research has shown that mixtures of SO_2 and NO_2 can cause interactive effects

TABLE 2 Growth and yield of selected crop species in response to sulfur dioxide exposures.

Test Plant	Sulfur Dioxide Exposure Characteristics	Results
Six species	Constant and stochastic concentration, 0-0.20 ppm (Greenhouse)	Constant SO_2 concentration underestimated effects compared to the time series treatments; excellent discussion of time series concept; interpretation of results difficult.
Barley	0.010, 0.023, 0.038 and 0.058 ppm mean over growing season (Field, open release system)	Controls not defined but three lowest SO_2 concentrations gave about 5, 20, and 12% yield increase respectively; the high SO_2 reduced yield about 18% (McLeod et al., 1986).
Barley	0.04 to 0.20 ppm mean concentration across growing season; a two year field study; open release of SO_2	Yield reduction found in both years; when average SO_2 during fumigation was used in regression analysis found a 2.2% loss per 0.01 ppm of SO_2 (Baker et al., 1986).
Rice (3cvs)	0.05 to 0.20 ppm, 24 hr per day, 5 da per week, 15 weeks; in pots in open-top chambers	Yield at highest SO_2 compared to lowest was reduced from 11 to 29% in the three cultivars (Kats et al., 1985).
Lolium perenne	0.012 to 0.029 ppm winter mean at 4 selected sites with differing SO_2 concentrations (field-correlational)	Two sites with lowest mean SO_2 showed same total dry wt yields; two highest sites showed a -19 and -54% reduction respectively. The site with highest SO_2 also suffered winter injury (Ashenden, 1987).

SOURCE: Heck et al., 1986. Individual references, not included here, are found in the original table.

on sensitive plant types. An interesting biochemical explanation for the synergistic action of SO_2 and NO_2 on several grass species involves the inability of the plant to detoxify nitrite in the presence of SO_2 (Wellburn et al., 1981).

Examples of the effects of SO_2 on carbon translocation and partitioning and on plant growth and yield are shown in Table 2. Generally, assimilates move to developing leaves rather than to roots under low SO_2 stress. Root growth is generally reduced more than shoot growth and occurs at relatively low SO_2 concentrations. The results support the contention that plants are sensitive to low SO_2 concentrations ($<$ 0.10 ppm), when exposed continuously.

Dose/response studies using an open-air SO_2 release system have simulated exposures of soybean to SO_2 near point sources. Soybean yield was decreased by periodic SO_2 exposures after flowering to doses of approximately 10 to 15 ppm-hours (Sprugel et al., 1980). These dose statistics

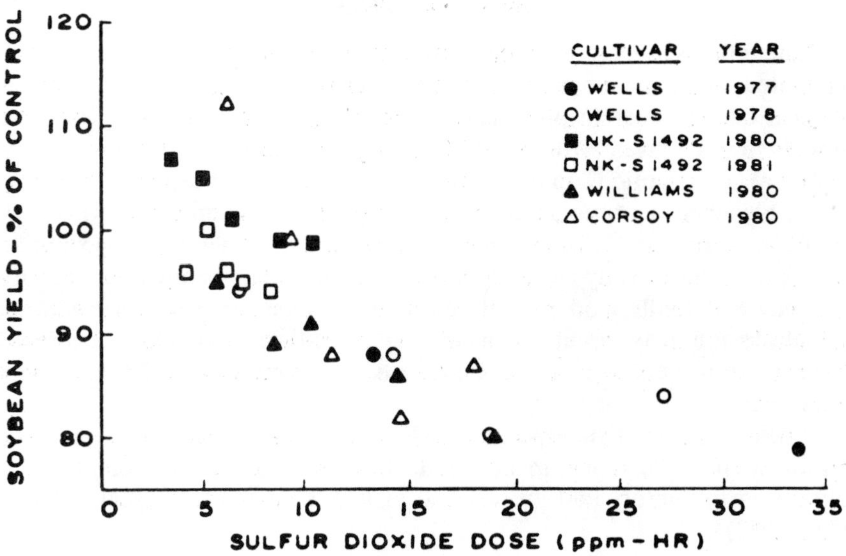

FIGURE 1 Effects of periodic SO_2 exposures in open air on the yield of soybeans (Heck et al., 1986).

were products of mean exposure durations of 2.5 to 4.2 hours, mean concentrations of 0.12 to 0.31 ppm, and 19 to 25 exposures. Doses in the 5 ppm-hour range were either stimulatory or inhibitory. Maximum peak-to-mean SO_2-concentration ratios were about 2.5. Figure 1 shows results of these studies.

Daily four- or seven-hour exposures of cotton and tomato (Heck et al., 1983b) or soybeans (Heagle et al., 1983) in open-top field chambers demonstrated that SO_2 concentrations, which are likely to occur regionally in the United States, probably do not cause decreased yield. A similar conclusion was reached in a review by Roberts (1984) on the effects of SO_2 on plant productivity, which included results from open-air studies. However, emissions of SO_2 near point sources can cause decreased yield in sensitive crop species.

EFFECTS OF OZONE ON AGRICULTURAL PRODUCTIVITY

This section presents summary information on the effects of O_3 on crop growth and productivity. It also provides background information on plant response, including some review of field research that supports efforts to assess O_3 dose, and crop yield responses for assessment purposes.

Symptomatology

The effects of O_3 on crops are either visible (i.e., morphological, pigmented, chlorotic, or necrotic foliar patterns resulting from major physiological disturbances in plant cells) or subtle (i.e., measurable growth or physiological changes without visible injury that may affect yield, or reproductive or genetic crop systems). Ozone injury often appears as flecks (small, bleached necrotic areas) or stipple (small pigmented areas) on upper foliar surfaces. Chlorosis may be associated with acute exposure to O_3. Chronic injury may cause chlorosis or other color or pigment changes that may eventually lead to cell death; early senescence with or without leaf abscission may occur. Chronic injury patterns are easily confused with symptoms caused by diseases, insects, other stresses, and normal leaf senescence.

From a practical standpoint, visible foliar injury is the only conclusive way to identify O_3 injury in the field. Research has shown relationships between visible injury and growth and yield for many plant species (Heck et al., 1977).

Physiological and Biochemical Effects

Ozone enters leaves through stomata. Any stress causing stomatal closure reduces O_3 uptake and thus protects the plant by avoidance. The fate of O_3 after entry into plant leaves is not known. Some fraction of O_3 may pass through the cell membrane (Mudd et al., 1984); O_3 may react with protein or lipid membrane components (Heath, 1980); or free, radical products of O_3 activity within the substomatal cavity may react with membrane components (Grimes et al., 1983). Regardless of the exact mechanism of action, the cell membrane is probably the site of initial O_3 reaction.

Factors Affecting Plant Response

Many biological and physical factors are known to affect the response of plants to O_3, including *genetic* (e.g., cultivar and species differences, effects on reproductive structures, inheritance of sensitivity); *biological* (e.g., plant diseases, insects); *environmental* (e.g., climatic [temperature, light, humidity] and edaphic [nutrition, soil moisture]); and *chemical* (e.g., herbicides, insecticides, special additives), as well as other factors.

Genetic

Research of a genetic nature has concentrated on the determination of relative sensitivities of species and their genotypes, and on the effects on pollen and pollen germination. Studies on the heritability of O_3 resistance suggest that O_3 resistance is heritable. Results of cultivar screening suggest that, when dealing with extremes of sensitivity, cultivars maintain the same relative separations when longer-term exposures are performed under both field and controlled conditions (Heck et al., 1988c).

Biological

Interrelationships between biological stresses (e.g., insects and diseases) and plant response to O_3 must be understood as part of a crop-loss assessment effort. Available information regarding effects of O_3 on plant parasites suggest that obligate fungal parasitism is generally inhibited and that some facultative parasites may benefit; these effects are probably indirect through the host. Foliar parasites may provide localized protection of the host from O_3 injury. Review articles by Heagle (1982) and Lawrence (1981) cover much of the research on plant/parasite interactions.

Environmental

Relationships between climatic or edaphic factors and plant response to pollutant exposure are discussed in several O_3 reviews (Heck et al., 1977; U.S. EPA, 1986). Generalizations on the modifications of plant response to O_3 by climatic factors are difficult because of the known exceptions. Evidence suggests that lower light intensity during growth, higher light intensity during exposure, higher growth temperatures, and higher growth and exposure humidities increase the sensitivity of many species to O_3. Information suggests that increased O_3 sensitivity, although related to stomatal conductance, reflects changed physiological conditions within the plant. Environmental conditions in the field can affect stomatal control of plant response the same day, and physiological control of plant response will start the following day. A two- to four-day period is usually necessary before physiological conditions that affect plant response to O_3 will completely reflect the changed environmental conditions.

Tingey et al. (1982), using a uniform water stress, reported that O_3 sensitivity in bean decreased with increasing plant water stress; protection occurred in one day at the highest stress, and recovery occurred within six days. Beans treated with a chemical to induce stomatal opening in water-stressed plants were as sensitive to O_3 as non-water-stressed plants (Tingey and Hogsett, 1985). Results suggest that O_3 protection in water-stressed plants is caused by stomatal control and not biochemical control. Heggestad et al. (1985) found that soybean growing in open-top chambers were more sensitive to ambient concentrations of O_3 under a small soil-moisture stress than when adequate moisture was available. Because of the

TABLE 3 Plant response to O_3 and SO_2 mixtures.

Test Plant	Exposure Information	Plant Response to the Mixture
Bean, snap	0.20 ppm each gas; 7 hours/ day, 4 days; varied salinity	Stomatal conductance: synergistic (variable); foliar injury: antagonistic; growth: additive; effect changed with salinity.
Bean, field	0.05-0.30 ppm O_3; 0.04 ppm SO_2; 4 hours	Net photosynthesis: additive or antagonistic, depending on O_3 concentration.
Tomato	0.005 to 0.468 ppm SO_2; 0.015 ppm or 0.056 ppm O_3; 5 hr/ day, 5 da/wk, 57 days; field study with open top chambers	Ripe fruit decreased 16% by O_3 (low SO_2 treatment), 18% by SO_2 (low O_3 treatment); 32% in high O_3 - high SO_2 treatment; additive response (Heggestad et al., 1986).
Lettuce, radish	0.4 ppm O_3, 0.8 ppm SO_2; 6 hours	Use of covariates increased precision for lettuce and radish; lettuce growth and injury effects antagonistic; radish was additive.
Potato	Four O_3 concentrations filtering of ambient O_3; 0.1 ppm SO_2; for 6 hours/day, 255 hours	Reductions in various growth and yield parameters were additive.
Soybean	0.06 or 0.08 ppm O_3, 0.06 or 0.11 ppm SO_2; 5 hours/day, days; in open field facility	Both O_3 and SO_2 caused decreases in a number of yield measures; mixture responses were additive.
Soybean	0.04 to 0.08 mean ppm O_3, 0.00 to 0.11 mean ppm SO_2; 5 hr/day, 16 days, during pod fill; linear gradient system in field.	Ozone caused 26% seed yield reduction, SO_2 a 6% reduction with no significant interactions (Reich and Amundson, 1984).

SOURCE: Heck et al., 1986. Individual references, not included here, are found in the original table.

importance of soil water stress in agricultural areas, it is imperative that we try to understand this interaction and include it in any assessment effort.

Pollutant Interactions

Plants are often more severely affected by mixtures of O_3 with other pollutants than by O_3 alone (Lefohn and Ormrod, 1984; Reinert, 1984). In general, mixtures tend to give a greater-than-additive (synergistic) response when the concentrations are below those causing visible effects from pollutants singly; concentrations around the injury threshold tend to produce an additive response; and concentrations above threshold tend to cause a less-than-additive (antagonistic) response. The term synergistic, although statistically appropriate, is not necessarily biologically appropriate since response functions may not be linearly related to pollutant concentration. Table 3 is a summary of results from selected studies using O_3 with SO_2.

TABLE 4 Growth and yield of selected crop species in response to ozone exposure.

Test Plant	Ozone Exposure Characteristics	Results
Bean, lima (8 genotypes)	Comparison of charcoal filtered (< 0.02 ppm mean daily max) and non-filtered (< 0.06 ppm mean daily max) greenhouses.	Yield reductions of 3.4 to 68.5% were found across the 8 genotypes (Meredith et al., 1986).
Bean, snap	0.30, 0.60 ppm; 1.5 hour, 2 times; 6 growth stages; harvest 7 days after exposure or at fresh harvest (controlled)	Reduced relative and absolute growth rates, pod production, nodulation, and fixed nitrogen; magnitude varied with O_3 concentration and growth stage.
Tomato, *Tiny Tim*	0.08-0.10 ppm; 5 hours/day, 5 days/week, 5 weeks (greenhouse)	86% reduction in fruit number, 91% reduction in fruit weight.
Soybean	0.02 to 0.097 ppm; 341 hours, intermittent over 113 days; greenhouse	Yield was +15%, -34%, and -40% at 0.046, 0.070 and 0.097 ppm in comparison to 0.02 ppm. The increase at 0.046 ppm was not significant (Endress and Greenwald, 1985).
Cotton, *Acala SJ-2*	0.20 ppm; 6 hours, 2 times/ week, 1 group started at 8 day, 1 at 42 days from seed (greenhouse)	Vegetative biomass and boll production reduced; greatest reduction in boll and root weights; 48% reduction in boll number.
Clover, *Ladino*	0.03, 0.05, 0.08 ppm; 7 hours/ day, 6 months (field: pots)	Total forage and forage regrowth reduced for clover and clover-fescue mixture in relation to O_3 concentration; fescue unaffected.
Rice (3 cvs)	0.05 to 0.20 ppm, 5 hr per day, compared to 5 da per wk, 15 wks; in pots in open top chambers	Yield at highest O_3 lowest was reduced from 12 to 29% in the three cultivars (Kats et al., 1985).

SOURCE: Heck et al., 1986. Individual references, not included here, are found in the original table.

Growth, Biomass, and Yield Effects

Greenhouse and controlled-environment research were used in early assessments of O_3 effects. These studies underestimated the effects of O_3 found in later field studies, but the estimates were reasonable. The effects of O_3 on growth and yield are briefly summarized in Table 4. Research results have generally shown that:

- O_3 affects crop growth and productivity;
- cultivar differences are usually observed;
- root growth is affected more than shoot growth;

- changes in growth rate occur when exposures are during early vegetative growth;
- recovery occurs after the exposure terminates;
- changes in the quality of usable product are often found; and
- growth and yield reductions occur at ambient O_3 concentrations.

CROP YIELD RESPONSES TO OZONE UNDER FIELD CONDITIONS: A CASE STUDY OF THE NCLAN PROGRAM

The National Crop Loss Assessment Network (NCLAN) developed O_3 dose/crop yield response functions for use in assessing the economic effects of O_3 on crop production. The NCLAN program started in 1980 and ended in 1988 after seven years of intensive field research and two years of data analysis and writing to complete the documentation of results from all research sites (Heck et al., 1982b, 1983a, 1984b, 1984c, 1988a, 1988b). The program was designed to define the relationships between yields of major agricultural crops and varying exposure to O_3 and to assess the primary economic consequences resulting from the exposure of agricultural crops to O_3.

The NCLAN program developed O_3 dose/yield response data for 14 species (39 cultivars) over the seven years of research (Table 5). Experiments involving O_3 interactions, cultivars, soil moisture, SO_3, and exposure dynamics were completed during this time. A complete summary of yield data reported by the various investigators with predicted losses at different seasonal (7-hr and 12-hr) mean O_3 concentrations is found in Heagle et al. (1988).

Data from the early experiments were analyzed using linear equations (Heck et al., 1982b). However, it was clear that nonlinear models would better reflect yield losses associated with O_3. Thus, a number of three-parameter models were tested and the Weibull model was chosen for several reasons:

- its flexible form covers the range of responses observed;
- its form is biologically realistic;
- its parameters are easily interpreted;
- it provides direct estimates of proportional yields; and
- it tests for homogeneity of proportional yield responses over data sets are easy to accomplish (Rawlings and Cure, 1985).

The Weibull model is given as

$$y = \alpha\ exp\{-(x/\sigma)^c\} + \epsilon$$

where y is the observed yield and x is the O_3 concentration in ppm. The three estimated parameters are: α, the hypothetical maximum yield at zero

TABLE 5 Summary of crop studies in the NCLAN Program.[a]

Crop	Number of Studies[b]	Number of Cultivars	Other Factors (# of Studies)
Alfalfa	2 (1-2yr)	2	Moisture (1), SO_2 (1)
Barley	2	2	Moisture (1)
Corn	2	5	SO_2 (1)
Cotton	5	3	Moisture (4), SO_2 (1)
Forages	3 (1-2 yr)	2/2	Moisture (1), SO_2 (1)
Bean	2	1	---
Lettuce	1	1	---
Peanut	1	1	---
Sorghum	1	1	---
Soybean	14	9	Moisture (7), SO_2 (4)
Tobacco	1	1	---
Tomato	2	1	SO_2 (2)
Turnip	1	4	---
Wheat	4	4	SO_2 (1)

[a] Ozone was studied at 4 to 6 concentrations in all studies.
[b] One alfalfa and one forage study ran for two years.

O_3 concentration; σ, the O_3 concentration when y is 0.37 α; and c, a dimensionless shape parameter that gives the model flexibility. The term ϵ is the random variation associated with each experimental unit. Factors that affect yield will affect α. Predicted relative yield losses (percent) at four seasonal 7-hr/day or 12-hr/day mean O_3 concentrations are shown for the major field crops tested in Table 6.

Figures 2 and 3 show results drawn from summaries of NCLAN data through 1983 (Heck et al., 1986): Figure 2 shows proportional yield loss for five crop species as predicted by the Weibull model, and Figure 3 summarizes proportional yield losses for soybean across years and research sites. Figures 4 and 5 show results drawn from summaries of NCLAN data by Heagle et al. (1988): Figure 4 shows that SO_2 can affect the yield response in an additive way to O_3; although several studies suggested an interaction, no interaction was found in this study. Figure 5 shows that soil-moisture stress can affect the yield response to O_3. Additional graphic results are shown for nine species in the paper by Adams et al. (1988) and in Heagle et al. (1988). The results are representative of data from the NCLAN program collected from 1980 through 1986.

METHODOLOGY FOR ASSESSMENT OF THE IMPACT OF AIR POLLUTANTS SUCH AS OZONE ON CROPS

An assessment of impact, by implication, includes an economic aspect as a final step; however, an in-depth assessment of effects can be done without economics. The NCLAN program included an economic analysis that utilized an assessment of crop losses as a basis for the assessment.

TABLE 6 Predicted relative yield losses (percent) at four seasonal (7-hr or 12-hr/day) mean ozone concentrations using the Weibull function.

Crop (# of Studies)	CV[a]	Ozone Concentration (ppm)[b,c] 0.04	0.05	0.06	0.08
7 hr/day Seasonal Means					
Bean, Kidney (2)	15.5	4.3	8.9	14.9	30.8
Peanut (1)	7.3	6.5	12.5	19.8	36.6
Sorghum (1)	5.1	0.8	1.7	2.6	5.3
Wheat (5)	10.9	9.1	16.0	23.4	38.4
12 hr/day Seasonal Means					
Alfalfa (2)	8.0	3.9	7.1	10.6	18.8
Corn (6)	9.9	1.2	3.3	7.3	23.6
Cotton (7)	6-18	6.0	14.0	26.0	57.5
Forage Mix (2)	8.9	3.8	7.7	12.5	24.6
Soybean (22)	4-20	11.0	17.6	23.7	34.6
Tobacco (1)	5.3	6.2	11.1	16.4	27.4

[a] The coefficient of variation came from Heagle et al., 1988 and is shown for the study or studies from which the modeled data were obtained.

[b] The predicted relative yield losses came from Lesser et al., 1989 (Tables 2 and 4). Table 2 shows the homogeneous models identified for various crops (where more than one study was conducted there were often several models). The losses shown in Table 4 give the 95% confidence limits, the mean values are shown in the above table. Yield losses are calculated relative to a seasonal O_3 mean of 0.025 ppm.

[c] The following models were used from Table 2 of Lesser et al., 1989:
-- Peanut, sorghum, alfalfa and tobacco had only one model;
-- Bean, kidney: used model 2 with the longer exposure period;
-- Wheat: used model 1 (one of intermediate sensitivity);
-- Corn: used model 1 (one of intermediate sensitivity);
-- Cotton: used model 3 (included 3 studies);
-- Forage: Mix used model 2 (ladino clover/fescue, 2 yr study);
-- Soybean: used model 2 (one of intermediate sensitivity, included 8 studies).

While the economic aspects of a crop loss assessment are not included in this paper, these issues are well covered by Adams et al. (1984, 1988).

The Needs of an Assessment Program

Local, regional, national, or international assessments of crop losses require three basic types of information:

- a *response function* relating crop yield to an exposure statistic;
- an *air quality database* that can be used to estimate crop exposure *on a county level* using the same exposure statistic as used in the response function; and
- a *crop census*, i.e., what crops are grown and their yield within a county (Heck et al., 1984b; Shriner et al., 1984).

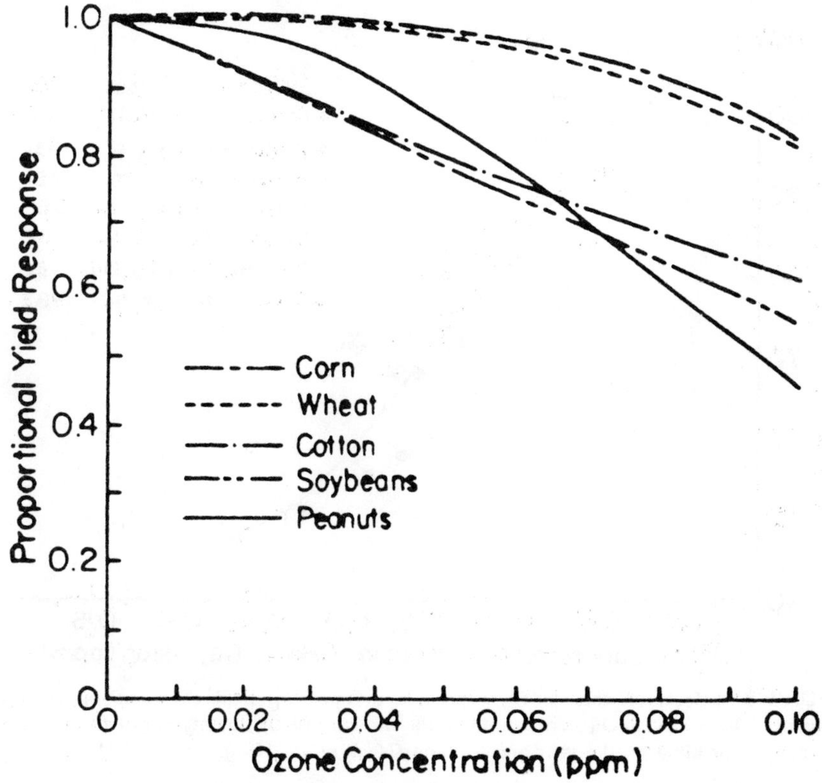

FIGURE 2 Effects of chronic O_3 exposures on the proportional yield loss of five crop species as predicted by the Weibull model, using results from open-top chambers. The O_3 concentration is the seasonal 7-hr/day mean (Heck et al., 1986).

Ozone Functions for Use in Crop Loss Assessments

The NCLAN program determined that the Weibull function had a number of redeeming features and accepted its relative response portion for use in assessment efforts. The Weibull permitted the assessment either to utilize a homogeneous response function for a species (i.e., a single or several cultivars) developed from experimental designs across years or sites, or to use a heterogeneous response function if such a function showed no apparent inconsistencies. In the assessment of yield effects (Table 6, Figure 2), many of the combined data sets were homogeneous; in the 1988 economic assessment, however, heterogeneous data sets were used where necessary to describe the effects on a single species (Adams et al., 1988).

Ozone dose/crop yield response functions can take on a variety of forms depending upon the dose statistic used. There is no preferred function,

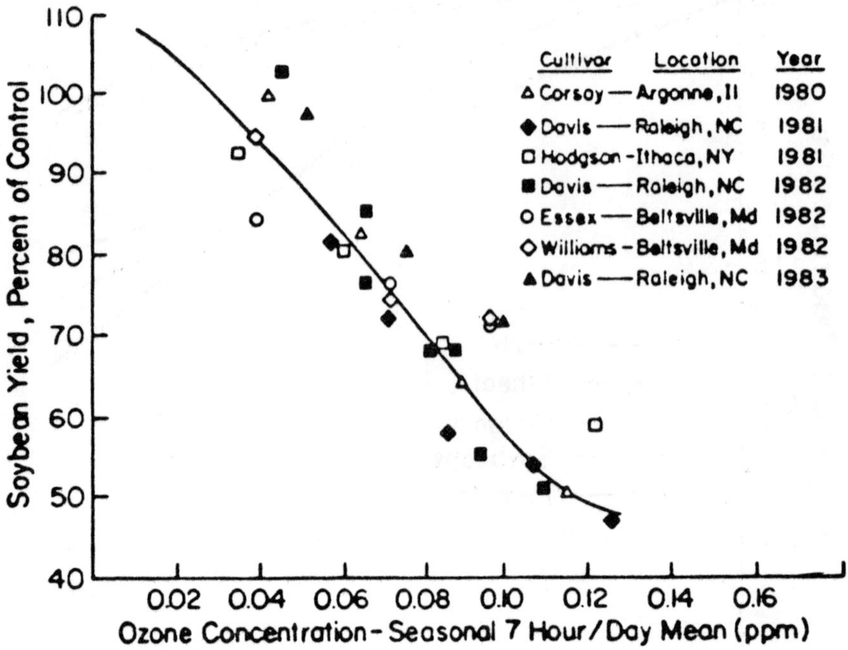

FIGURE 3 Effects of chronic O_3 exposures in open-top field chambers on yield of soybeans. The seasonal O_3 concentrations for the control treatment ranged from 0.02 to 0.03 ppm, depending on the year and location (Heck et al., 1986).

but it should utilize an exposure statistic that adequately describes the biological response to O_3. The NCLAN program used 7-hr/day and 12-hr/day seasonal mean O_3 values because they adequately described the yield responses of the crops tested. These functions were used by Adams et al. (1988) in their final economic assessment for NCLAN.

Interpolations of Ozone Data to a County Level

The NCLAN program utilized data from the U.S. Environmental Protection Agency (EPA) Storage and Retrieval of Aerometric Data (SAROAD) system for interpolation processes. In the SAROAD system most monitoring sites are urban, with few in rural crop-growing areas, and many areas of the United States have only a few monitoring sites. Several factors enhanced our ability to interpolate O_3 data across broad areas. First, O_3 precursors are transported over great distances. Second, O_3 is more stable as air masses move into rural areas because concentrations of reactive chemical species are reduced (Heck et al., 1984b). NCLAN used the kriging spatial interpolation process to develop county-level seasonal (7-hr/day,

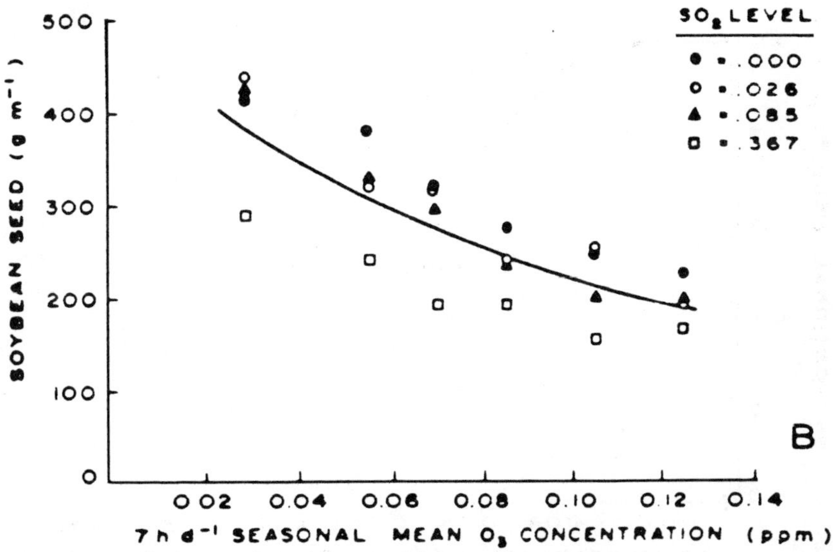

FIGURE 4 Yield response of 'Davis' soybean to combinations of O_3 and SO_2 at Raleigh, NC in 1982. Curve was estimated by polynomial analysis using combined SO_2 data (Heagle et al., 1988).

12-hr/day) mean O_3 concentrations for use in crop-loss assessment (Heck et al., 1983b, 1984b). This technique was fully reviewed by an outside group and found to be a reasonable approach to estimate county-level O_3 concentrations (Heck et al., 1985; Lefohn et al., 1987). However, it should be noted that this technique does not handle mountainous terrain well, and it is weak where monitoring points are too far apart, as in much of the western United States.

Crop Census

The Census of Agriculture, conducted by the U.S. Department of Agriculture, provides a county-level yield statistic for crops of interest (Heck et al., 1982b; Shriner et al., 1984; Adams et al., 1988). It involves an extensive national inventory of crops (not broken down by cultivars) and acreage grown. Data are obtained by analyzing responses to questionnaires mailed out approximately every five years. County estimates are adjusted for nonrespondents.

Analysis of Crop Yield Reductions

Ozone dose/crop yield response functions, crop yields at the county level, and seasonal 7-hr/day mean O_3 concentrations at the county level are

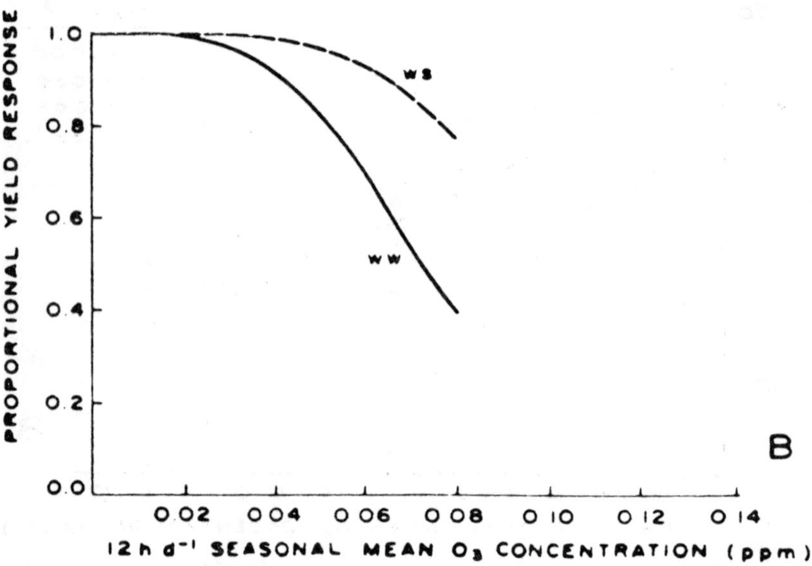

FIGURE 5 Proportional yield response to O_3 of cotton grown with intermittent periods of soil-moisture stress (WS) or grown with well-watered conditions (WW) for 'McNair 235' at Raleigh, NC in 1985. Curves were derived using Weibull analyses with the *a* value set at 100 (Heagle et al., 1988).

used to calculate crop losses related to reductions in yield. The impacts of O_3 are reflected in the yield data found in the Census of Agriculture. Using the county-level O_3 values and the response function for the crop of interest, the expected percent yield reduction of the crop in the county is calculated. The yield reduction is based on comparing the yield found at the observed county-level O_3 values with the expected yield at an 0.025 ppm O_3 concentration (seasonal 7-hr/day mean O_3 concentration in clean air). Increases in yield with different percentage improvements in air quality can then be calculated. Values for each crop and each county where a given crop is grown are then used to calculate national yield losses for each crop of interest.

This approach was first used for an analysis with four species (i.e., corn, peanut, soybean, and wheat) by Heck et al. (1982b) as part of a larger assessment by Shriner et al. (1984). The assessment was done to show the value of the selected approach in documenting regional and national losses of crops due to O_3. Nine NCLAN type data sets (i.e., 1 corn, 1 peanut, 3 soybean, and 4 wheat) obtained from two NCLAN and seven pre-NCLAN studies at North Carolina State University were utilized in the assessment effort. County O_3 and crop inventory data for 1978 were used. Kriging of the 1978 O_3 data was first attempted and used on a national basis by James Reagan (EPA) to predict county-wide yield losses (Heck et al., 1983a). The data sets were used with the 7-hr/day seasonal mean O_3 statistic in calculating yield reductions using a linear response function. Results were calculated for county units, and tables and maps were developed to summarize and show patterns of the O_3 effects on soybean, corn, wheat, and peanuts. The assessment estimated that an approximate $3 billion of productivity in the four crops would be gained if current maximum 7-hr/day seasonal O_3 concentrations were reduced to 0.025 ppm. Soybean represented 64% of the impact, corn 17%, wheat 12%, and peanut 7%. The methodology is sound but the assessment of dollar losses cannot be considered an economic analysis.

The same basic approach was utilized by Adams et al. in developing yield reduction/increase estimates for their interim (1984) and final (1988) NCLAN economic assessment efforts. In the interim assessment, the Weibull model was used for the response function. Two O_3 data sets were utilized: the kriged 1980 O_3 data on a county basis and the 1978-82 five-year average O_3 data base kriged to the county level. Crop inventory data were used and losses were based on 1980 dollars. Six crops (i.e., barley, corn, cotton, sorghum, soybean, and wheat) were used in this assessment. The final assessment utilized a similar approach, incorporated three additional crops (i.e., alfalfa, hay, and rice) and used 1982 as the base year for O_3.

The Economic Effects of Ozone on Agriculture

Based on crop response data from many experimental designs, a number of regional and national economic assessments have been made since 1980 (Table 7). Regional estimates ranged from about $30 million (Minnesota) to about $670 million (Corn Belt). National estimates were based on different groups of crops but ranged from $1.2 to 3.0 billion; corn and soybean were included in all estimates. The national economic assessments by NCLAN (Adams et al., 1988) was in good agreement, showing about $2.7 billion surplus with a 40% reduction in seasonal O_3 concentrations (Table 8). A summary of economic estimates suggests that current seasonal O_3 concentrations are causing in excess of $3 billion annual loss in crop productivity (Adams et al., 1988).

CONCLUSIONS

The information presented here supports the thesis that O_3 has a major impact on crop production in North America. Assessment methodologies are developed along with results of field research that are critical to the prediction of O_3 effects on crop productivity.

TABLE 7 Summary results of several regional and national economic assessment using NCLAN data.[a]

Region	Assumptions of O_3 Concentration	Crops	Benefits ($\$ \times 10^6$)
California	Reduce to 0.04 ppm (seasonal)	18	$ 45
Corn Belt	Reduce NAAQS from 0.12 to 0.08 ppm	Corn, Soybean Wheat	$ 668
Illinois	10% reduction	Corn, Soybean	$ 226
U.S.	Reduce to 0.04 ppm	Corn, Soybean Cotton, Wheat	$ 2,400
U.S.	Reduce to 0.04 ppm	Same + Peanut	$ 1,300

[a] Reports from 1984/1985; results in 1980 U.S. dollars.

SOURCE: Adams et al., 1988.

TABLE 8 National economic assessment from NCLAN.[a]

	O_3 Assumption	Total Surplus (millions in 1982 dollars)
1984 Model	25% increase	-2,165
	10% reduction	699
	25% reduction	1,828
	40% reduction	2,637
1988 Model	25% increase	-2,053
	10% reduction	808
	25% reduction	1,890
	40% reduction	2,780

[a] Reports from 1984 and 1988; results in 1982 dollars.

SOURCE: Adams et al., 1988

Summary of Current Knowledge on the Effects of Ozone (and Other Air Pollutants)

- Ozone is responsible for most of the crop-yield losses from air pollutants on both a regional and national scale within North America.
- Losses from other pollutants are minimal, relative to O_3, and primarily source related or related to joint effects with O_3.
- Pollutant dose/crop yield response functions are essential for predicting yield losses; non-linear models give the best fit to available field data for O_3.
- Based on available technology, the open-top chamber system is the best approach for the development of predictive models for O_3, but open release systems may function well for other pollutants.

- The extrapolation of O_3 data on a regional basis, using the interpolation technique of kriging, is a useful and necessary part of an assessment effort. However, the technique is not suitable with SO_2 or other point source related pollutants.
- Foliar symptoms on crops under field conditions often appear as early senescence and may be difficult to assess.
- Although the mechanism of plant response to air pollutants is not understood, the cell membrane is probably the site of initial impact for O_3.
- Ozone and SO_2 affect photosynthesis and carbon allocation in plants; reduced allocation to roots and reproductive structures is usually found. Similar effects may be true for other pollutants.
- Differences in both species and cultivars within species response to air pollutants is found in all crops.
- Interactions between air pollutants (O_2, SO_2, and HF) and both biotic and abiotic factors on plant responses are documented.
- The response of many crop species to O_3 is affected by the presence of other pollutants (specifically SO_2 and NO_2).
- Most crops show growth, biomass and yield reduction when grown under current ambient air concentrations of O_3. These responses may be true for SO_2 and acidic precipitation, but the data are equivocal.
- The NCLAN data base has permitted a reasonable first estimate of crop yield losses associated with O_3 as an air pollutant of national and international importance.

Areas of Uncertainty in Assessing the Effects of Air Pollutants on Crop Production

- The available data base is small and thus is not fully representative of North America. For O_3, only 10 field crops were studied in the NCLAN program; five had four or more experimental designs. Data for other pollutants is not as strong.
- Potential effects of field methodology have not been fully addressed for any pollutant.
- Ozone dose/crop yield response models are empirical and not based on mechanistic considerations. Models for other pollutants are weak.
- The effect of soil moisture on crop yield response to air pollutants has been studied for O_3 and SO_2, but the results are not definitive.
- The effects of other biotic or abiotic stresses on crop response to air pollutants are not understood and are not included in predictive models.
- The importance of exposure dynamics (i.e., peak O_3 values) occurring throughout the growing season is not well understood.

- There are insufficient rural monitoring sites to corroborate the kriging interpolative process for O_3; interpolation for other pollutants is not feasible.
- The estimates of economic loss from O_3 would probably range from $1-7 billion or more if all crops were considered; losses will vary from year to year depending on both O_3 concentrations and meteorological conditions.

REFERENCES

Adams, R.M., J.D. Glyer, and B.A. McCarl. 1988. The NCLAN economic assessment: Approach, findings and implications. In Assessment of Crop Loss from Air Pollutants, W.W. Heck, D.T. Tingey, and O.C. Taylor, eds. London: Elsevier Science Publishers.

Adams, R.M., S.A. Hamilton, and B.A. McCarl. 1984. The Economic Effects of Ozone on Agriculture. EPA. Environ. Res. Lab., Corvallis, OR. EPA-600/3-84-090. September 1984.

Altshuller, A.P., and R.A. Linthurst, eds. 1984. The acidic deposition phenomenon and its effects: Critical assessment review papers, Vol. II: Effects Sciences, EPA-60018-83-016B, EPA, Office of Research and Development, Washington, DC.

Ashenden. 1987. Effects of ambient levels of air pollution on grass swards subjected to different defoliation regimes. Environmental Pollution 45:29-47.

Baker, C.K., J.J. Colls, A.E. Fullwood, and G.G.R. Seaton. 1986. Depression of growth and yield in winter barley exposed to sulphur dioxide in the field. New Phytologist 104:233-241.

Banwart, W.L., P.M. Porter, J.J. Hassett, and W.M. Walker. 1987. Simulated acid rain effects on yield response of two corn cultivars. Agricultural Journal 79:497-501.

Bressan, R.A., L.G. Wilson, and P. Filner. 1978. Mechanisms of resistance to SO_2 in the cucurbitaceae. Plant Physiolology 61:761-767.

Dochinger, L.S. and T.A. Seliga, eds. 1976. Proceedings of the First International Symposium on Acid Precipitation and the Forest Ecosystem. U.S.D.A. Forest Service, General Technical Report NE-23. Upper Darby, PA.

DuBay, D.T., and A.S. Heagle. 1987. The effects of simulated acid rain with and without ambient rain on the growth and yield of field-grown soybeans. Environmental and Experimental Botany 27:395-401.

Endress, A.G., and C. Grunwald. 1985. Impact of chronic ozone on soybean growth and biomass partitioning. Agriculture, Ecosystems and the Environment 13:9-23.

Evans, L.S., K.F. Lewin, and G.R. Hendry. 1986. Yields of field-grown soybean exposed to simulated acidic rainfalls. BNL 52009. Dept. Applied Science. Brookhaven National Laboratory, Upton, NY.

Grimes, H.D., K.K. Perkins, and W.F. Boss. 1983. Ozone degrades into hydroxyl radical under physiological conditions. Plant Physiology 72:1016-1020.

Guderian, R., ed. 1985. Photochemical oxidants-formation, distribution, control and effects on vegetation. Berlin and New York: Springer-Verlag.

Heagle, A.S. 1982. Interactions between air pollutants and parasitic diseases. Pp. 333-348 in Effects of Gaseous Air Pollution in Agriculture and Horticulture, M.H. Unsworth and D.P. Ormrod, eds. London: Buttersworth Scientific.

Heagle, A.S., W.W. Heck, J.O. Rawlings, and R.B. Philbeck. 1983. Effects of chronic doses of ozone and sulfur dioxide on injury and yield of soybeans in open-top chambers. Crop Science 23:1184-1191.

Heagle, S.S., L.W. Kress, P.J. Temple, R.J. Kohut, J.E. Miller, and H.E. Heggestad. 1988. Factors influencing ozone dose/yield response relationships in open-top field chamber studies. In Assessment of Crop Loss from Air Pollutants, W.W. Heck, D.T. Tingey, and O.C. Taylor, eds. London: Elsevier Science Publishers.

Heath, R.L. 1980. Initial events in injury to plants by air pollutants. Annual Review of Plant Physiology 31:395-431.

Heck, W.W. 1984. Defining gaseous pollution problems in North America. Pp. 35-48 in Gaseous Air Pollutants and Plant Metabolism, M.J. Koziol, and F.R. Whatley, eds. London: Butterworth.

Heck, W.W., and C.S. Brandt. 1977. Effects on vegetation: Native crops, Forest. Pp. 157-229 in Air Pollution, A.C. Stern, eds. 3rd Edition, Vol. 2b. New York: Academic Press.

Heck, W.W., A.S. Heagle, and D.S. Shriner. 1986. Effects on vegetation: Native crops, forest. Pp. 247-350 in Air Pollution, A.C. Stern, ed. 3rd Edition, Vol. 6. New York: Academic Press.

Heck, W.W., J.B. Mudd, and P.R. Miller. 1977. Plants and microorganisms. Chapter 11 in Ozone and Other Photochemical Oxidants. National Academy of Sciences, Washington, DC.

Heck, W.W., D.T. Tingey, and O.C. Taylor, eds. 1988a. Assessment of crop loss from air pollutants. London: Elsevier Science Publishers.

_____. 1988b. Assessment of crop loss from air pollutants. Environmental Pollution. Special Issue 53.

Heck, W.W., U. Blum, R.A. Reinert, and A.S. Heagle. 1982a. Effects of air pollution on crop production. Pp. 333-350 in Strategies of Plant Reproduction, W.J. Meudt, ed. BARC Symposium No. 6. Totowa, NJ: Allanheld, Osman & Co. Publisher, Inc.

Heck, W.W., J.A. Dunning, R.A. Reinert, S.A. Prior, M. Rangappa, and P.S. Benepal. 1988c. Differential response of four bean cultivars to chronic doses of ozone. Journal of the American Society of Horticultural Sciences 113:46-51.

Heck, W.W., R.M. Adams, W.W. Cure, A.S. Heagle, H.E. Heggestad, R.J. Kohut, L.W. Kress, J.O. Rawlings, and O.C. Taylor. 1983a. A reassessment of crop loss from ozone. Environmental Science and Technology 17:573-580A.

Heck, W.W., O.C. Taylor, R. Adams, G. Bingham, J. Miller, E. Preston, and L. Weinstein. 1982b. Assessment of crop loss from ozone. Journal of the Air Pollution Control Association 32:353-362.

Heck, W.W., U. Blum, W.F. Boss, A.S. Heagle, R.A. Linthurst, R.A. Reinert, J.F. Reynolds, and H.H. Rogers. 1984a. Perspectives of air pollution research on plants. Pp. 173-249 in Reviews in Environmental Toxicology I, Ernest Hodgson, ed. Amsterdam: Elsevier Scientific Publishing Co.

Heck, W.W., W.W. Cure, J.O. Rawlings, L.J. Zaragoza, A.S. Heagle, H.E. Heggestad, R.J. Kohut, L.W. Kress, and P.J. Temple. 1984b. Assessing impacts of ozone on agricultural crops: I. Overview. Journal of the Air Pollution Control Association 34:729-735.

_____. 1984c. Assessing impacts of ozone on agricultural crops: II. Crop yield functions and alternative exposure statistics. Journal of the Air Pollution Control Association 34:810-817.

Heck, W.W., O.C. Taylor, R.M. Adams, G. Bingham, J.E. Miller, E.M. Preston, L.H. Weinstein, R.G. Amundson, R.J. Kohut, J.A. Laurence, W.C. Cure, A.S. Heagle, J.T. Gish, H.E. Heggestad, L.W. Kress, G.E. Neely, J.O. Rawlings, and P. Temple. 1983b. National Crop Loss Assessment Network (NCLAN) 1982 Annual Report. EPA-600/3-84-049. EPA, Corvallis, OR.

Heck, W.W., O.C. Taylor, R.M. Adams, J.E. Miller, D.T. Tingey, L.H. Weinstein, R.G. Admundson, A.S. Heagle, D.A. King, R.G. Kohut, L.W. Kress, J.A. Laurence, A.S. Lefohn, V.M. Lesser, J.R. Miller, G.E. Neely, P.J. Temple, and J.O. Rawlings. 1985. National Crop Loss Assessment Network (NCLAN) 1984 Annual Report. EPA/600/3-86/041, Environ. Res. Lab., Office of Res. and Develop., EPA, Corvallis, OR.

Heggestad, H.E., J.H. Bennett, E.H. Lee, and L.W. Douglass. 1986. Effects of increasing doses of sulfur dioxide and ambient ozone on tomatoes: plant growth, leaf injury, elemental composition, fruit yield and quality. Phytopathology 76:1338-1344.

Heggestad, H.E., T.J. Gish, E.H. Lee, J.H. Bennett, and L.W. Douglass. 1985. Interaction of soil moisture stress and ambient ozone on growth and yields of soybeans. Phytopathology 75:472-477.

Hughes, P.R., A.I. Dickie, and M.A. Penton. 1983. Increased success of the Mexican bean beetle on field grown soybeans exposed to sulfur dioxide. Journal of Environmental Quality 12:565-568.

Irving, P.M. 1987. Effects on agricultural crops. Chapter 6 in Interim Assessment: The Causes and Effects of Acidic Deposition. Vol. IV. The National Acid Precipitation Assessment Program. Washington, D.C.

Jager, H.J., and H. Klein. 1980. Biochemical and physiological effects of sulfur dioxide on plants. Angewandte Botanik 54:337-348.

Kats, G., P.J. Dawson, A. Bytnerowicz, J.W. Wolf, C.R. Thompson, and D.M. Olszyk. 1985. Effects of ozone or sulfur dioxide on growth and yield of rice. Agriculture, Ecosystems and the Environment 14:103-117.

Laurence, J.A. 1981. Effects of air pollutants on plant pathogen interactions. Zeitschrift fur Pflanzenkrankheiten und Pflanzenschutz 88:156-172.

Lefohn, A.S. and D.P. Ormrod, eds. 1984. A review and assessment of the effects of pollutant mixtures on vegetation- Research recommendations. EPA-600/3-84-037. Corvallis Environmental Research Lab., Office of Research and Development, EPA, Corvallis, OR.

Lefohn, A.S., H.P. Knudsen, J. Logan, J. Simpson, and C. Bhumralkar. 1987. An evaluation of the Kriging method, as applied by NCLAN, to predict 7-hr seasonal ozone concentrations. Journal of the Air Pollution Control Association 37:595-602.

Mansfield, T.A., M.E. Whitmore, and R.M. Law. 1982. Effects of nitrogen oxides on plants: two case studies. Pp. 511-520 in Air Pollution by Nitrogen Oxides, T. Schneider and L. Grant, eds. Amsterdam: Elsevier Scientific Publ. Co.

McLeod, A.R., K. Alexander, and D.M. Cribb. 1986. Effects of open-air fumigation with sulfur dioxide on the growth of cereals. I. Grain Yield of Winter Barley (Hordeum vulgare) cv Sonja 1982-83. Report No. TPRD/L/3071/R86. Central Electricity Research Laboratories, Leatherhead, Surry, U.K.

Meredith, F.I., C.A. Thomas, and H.E. Heggestad. 1986. Effect of the pollutant ozone in ambient air on lima bean. Journal of Agriculture and Food Chemistry 34:179-185.

Mudd, J.B., S.K. Banerjee, M.M. Pooley, and K.L. Knight. 1984. Pollutants and plant cells: effects on membranes. Chapter 8 in Gaseous Air Pollutants and Plant Metabolism. M.J. Koziol, and F.R. Whatley, eds. London: Butterworth.

National Academy of Sciences. 1977. Nitrogen Oxides. Chapter 9. National Academy Press, Washington, D.C.

Olszyk, D.M., and D.T. Tingey. 1984. Phytotoxicity of air pollutants. Plant Physiolology 74:999-1005.

Ormrod, D.P., 1978. Pollution in Horticulture. Amsterdam and New York: Elsevier Scientific Publishing Company.

Pell, E.J., C.J. Arny, and N.S. Pearson. 1987. Impact of simulated acidic precipitation on quantity and quality of a field grown potato crop. Environmental and Experimental Botany 27:7-14.

Rawlings, J.O., and W.W. Cure. 1985. The Weibull function as a dose/response model for studying air pollution effects on crop yields. Crop Science 25:807-814.

Reich, P.B., and R.G. Amundson. 1984. Low level O_3 and/or SO_2 exposure causes a linear decline in soybean yield. Environmental Pollution (Series A) 34:345-355.

Reinert, R.A. 1984. Plant response to air pollutant mixtures. Annual Review of Phytopathology 22:421-442.

Roberts, T.M. 1984. Effects of air pollution on agriculture and forestry. Atmospheric Environment 18:629-652.

Rogers, H.H., J.C. Campbell, and R.J. Volk. 1979. Nitrogen-15 dioxide uptake and incorporation by *Phaseolus vulgaris L.* Science 206:333-335.

Sekiya, J., L.G. Wilson, and P. Filner. 1982. Resistance to injury by sulfur dioxide. Plant Physiology 70:437-441.

Shriner, D.S., C.R. Richmond, and S.E. Lindberg, eds. 1980. Atmospheric sulfur deposition, environmental impact and health effects. Ann Arbor: Ann Arbor Science Publishers, Inc.

Shriner, D.S., W.W. Cure, A.S. Heagle, W.W. Heck, D.W. Johnson, R.J.Olson, and J.M. Skelly. 1984. An analysis of potential agriculture and forest impacts of long range transport air pollutants. ORNL-5910. Oak Ridge National Laboratory, Oak Ridge, TN.

Shupe, J.L., H.B. Peterson, and N.C. Leone, eds. 1983. Fluorides: Effects on vegetation, animals, and humans. Proceedings of the International Fluoride Symposium, Paragon Press, Inc., Salt Lake City, UT.

Sprugel, D.G., J.E. Miller, R.N. Muller, H.J. Smith, and P.B. Xerikos. 1980. Sulfur dioxide effects on yield and seed quality in field-grown soybeans. Phytopathology 70:1129-1133.

Temple, P.J., and O.C. Taylor. 1983. Worldwide ambient measurements of peroxyacetyl nitrate (PAN) and implications for plant injury. Atmospheric Environment 17:1583-1587.

Tingey, D.T., and W.E. Hogsett. 1985. Water stress reduces ozone injury via a stomatal mechanism. Plant Physiology 77:944-947.

Tingey, D.T., G.L. Thutt, M.L. Gumpertz, and W.E. Hogsett. 1982. Plant water status influences ozone sensitivity of bean plants. Agriculture and the Environment 7:243-254.

Treshow, M., ed. 1984. Air Pollution and Plant Life. Chichester, England: John Wiley & Sons Limited.

Unsworth, M.H., and D.P. Ormrod, eds. 1982. Effects of gaseous air pollution in agriculture and horticulture. Butterworth, London.

U.S. Environmental Protection Agency (EPA). 1978. Air Quality Criteria for Ozone and Other Photochemical Oxidants. Chapters 10/11, EPA-60018-79-004. Office of Research and Development, Washington, D.C.

U.S. EPA. 1982a. Air Quality Criteria for Oxides of Nitrogen. Chapter 12, EPA-600/8-82-026. Office of Research and Development, Environmental Criteria and Assessment Office, Research Triangle Park, NC.

U.S. EPA. 1982b. Air Quality Criteria for Particulate Matter and Sulfur Oxides. Vol. III, Chapter 8, EPA-600/8-82-029C. Office of Research and Development, Environmental Criteria and Assessment Office. Research Triangle Park, NC.

U.S. EPA. 1986. Air Quality Criteria for Ozone and Other Photochemical Oxidants. Vol III, Chapter 6, EPA-600/8-84-020CF. Environmental Criteria and Assessment Office, Research Triangle Park, NC.

Wellburn, A.R., C. Higginson, D. Robinson, and C. Walmsley. 1981. Biochemical explanations of more than additive inhibitory effects of low atmospheric levels of sulfur dioxide plus nitrogen dioxide upon plants. New Phytologist 88:223-237.

Winner, W.E., H.A. Mooney, and R.A. Goldstein, eds. 1986. Sulfur dioxide and vegetation: Ecology, physiology and policy Issues. Stanford Univ. Press, Pala Alto, CA.

Yoneyama, T., H. Sasakawa, S. Ishizuka, and T. Tatsuka. 1979. Absorption of atmospheric NO_2 by plants and soils. Soil Science and Plant Nutrition 25:267-275.

Impacts of Air Pollution on Agriculture and Horticulture in Poland

STEFAN GODZIK
Institute of Environmental Engineering
Polish Academy of Sciences

In contrast to the condition of forestry in Poland (Chapter 10, this volume), a country-wide survey of the impact of air pollution on the quantity and quality of crops does not exist. Investigations carried out near sources of air pollution and in the Upper Silesian industrial region have shown a significant reduction in crop yield. It has also been shown that the quality of plants is reduced in areas of high heavy-metal concentration. However, a direct cause-and-effect relationship has not been established in all cases. In addition, the problem of air pollution generally is not considered to be important for agricultural sciences or practices (Rutkowski, 1986).

The purpose of this chapter is to confirm the importance of air pollutant effects on agricultural and horticultural crops, in terms of their quantity and quality, relying on research results from the Upper Silesian industrial region as the major source of information.

ENVIRONMENTAL POLLUTION

The greatest threat to the terrestrial environment is air pollution, which comes primarily from industrial sources. Sulfur dioxide and nitrogen oxides are the most important gaseous pollutants in Poland. Emission of these gases from Polish sources ranges from 2.4 to 4.3 TG per year, with emission of nitrogen oxides calculated at about 1.1 TG per year. Differences in the emission of air pollutants given are due to calculation method (GUS, 1986; Juda et al., 1980; Cofala and Bojarski, 1988; National Program, 1988).

Fluorine compounds, hydrocarbons, and particulates, including metals, are of local importance, and may have a significant impact on the local environment. The processing and use of coal for energy production is limited to fairly small areas of the country; for this reason, levels of

TABLE 1 Emission of sulfur dioxide in Poland in 1982.

Region or Area	Area (km^2)	Population	SO_2 emission Total (tons)	tons/km^2
Poland	312,683	35,163,505	2,433,000	7.78
AEH	35,237	12,335,862	1,986,700	56.4
AEH-SO_2*	23,413	8,551,428	1,805,900	77.1
AEH-Upper Silesia**	3,134	2,753,366	604,600	192.9
AEH-Rybnik**	1,038	554,731	177,400	170.9

* 12 areas with largest SO_2 emission
** SO_2 amounts were taken into account in AEH and AEH-SO_2

SOURCE: GUS, 1984.

environmental pollution differ significantly within Poland (Figure 1, Table 1). Similarly, areas of forest and other natural vegetation occur almost as islands within the agricultural and industrial regions. This has led to the designation of 27 locations as Areas of Ecological Hazard (AEH) (Kassenberg, 1986; Chapter 22, this volume). Their distribution is shown in Figure 1 along with the index of agricultural land quality (GUS, 1984). Table 1 contains data on sulfur dioxide emissions for 12 of the 27 AEHs. Data for the Upper Silesian and Rybnik AEHs were separated, as detailed air pollution measurements are available for these two areas (Karczmarz and Cimander, 1988).

Air pollutants from industrial regions of Czechoslovakia (e.g., the Ostrava region) are transported to these areas as well. Transboundary transport is based on modeling of distribution of sulfur dioxide (Figure 2), but has been confirmed by measurements carried out for the Katowice district (Figures 4 and 5). This is a significant contribution to the major sources located in Upper Silesian and Rybnik areas, where the highest concentrations of all pollutants measured have been determined (Figures 3-8).

Measurements carried out for several years in the Katowice district indicate that the concentration of sulfur dioxide is increasing (Karczmarz and Cimander, 1988). Annual mean concentrations higher than the national standard (64 $\mu g\ m^{-3}$) exists mainly in the central part of Upper Silesia (Karczmarz and Cimander, 1988). Other parts of the region typically have concentrations below the national standard (Figure 3).

A similar pattern exists for concentrations of nitrogen oxides and some other pollutants, e.g., aerosols (Karczmarz and Cimander, 1986). The national standard for nitrogen oxides (32 $\mu g\ m^{-3}$) is exceeded in the entire area of Upper Silesia, and especially in the central part (Figure 4).

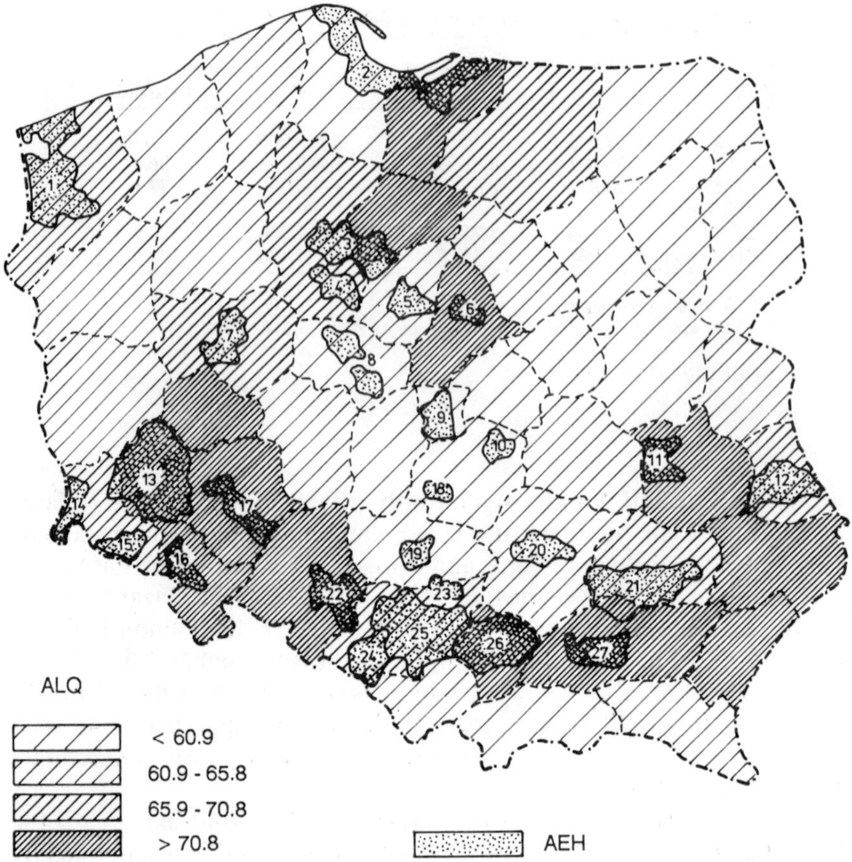

FIGURE 1 Twenty-seven Areas of Ecological Hazard (AEH) and index of agricultural land quality (ALQ) (GUS, 1984).

On the basis of calculations presented earlier, the total amount of nitrogen oxides is about one-third that of sulfur dioxide.

The national standard for dust fall (250 tons km^{-2} year) is exceeded in a large part of the Katowice district, with the highest amount deposited in the central part of Upper Silesia (Figure 5). There has been some improvement in controlling the emission of particulates from industrial sources, and dust fall is declining. However, the concentration of aerosols is increasing simultaneously (Figure 6). This is of concern not only because of possible impact on plant quality but, more significantly, on human health. In contrast to the pattern for dust fall, for a decade there has been a trend of increasing aerosol concentrations in the Katowice district (Karczmarz and Cimander, 1988). The direct impact of particulate matter on crop

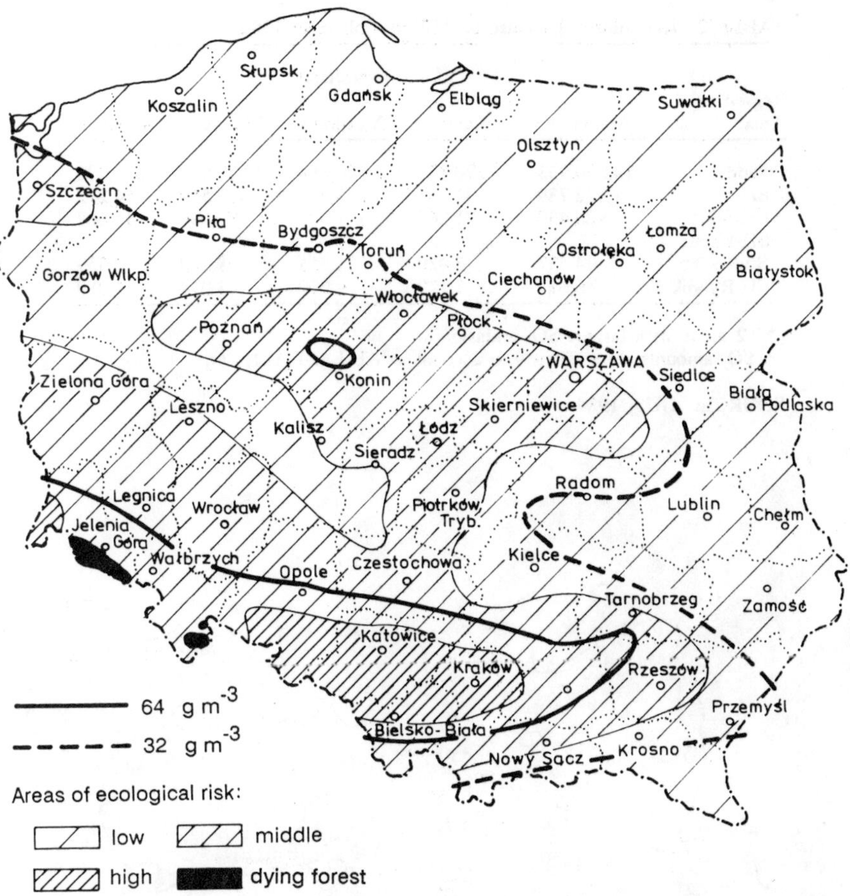

FIGURE 2 Distribution of sulfur dioxide in Poland. Yearly mean values of sulfur dioxide concentration for the year 1985 (National Program, 1988).

yield seems to be of limited importance, and only in locations where the deposited amounts are very high due to the decline in crop quality from heavy metal content. This effect has been found for areas in which the national standard for dust fall has not been exceeded.

AGRICULTURE AND PLANT PRODUCTION

While the Katowice district, like other AEHs, is not a significant source of agricultural production on a national scale (Table 2), most agricultural and horticultural crops produced on farmland and in alloted garden plots

TABLE 2 Agricultural land use in AEH in Poland in 1982.

Region or Area	Land in hectares				
	Arable	Orchards	Meadows	Pasture	Vegetables
Poland	14,550,953	270,151	2,520,365	1,550,063	232,090
AEH	1,422,789	31,945	288,991	139,519	54,212
AEH-SO_2*	926,430	21,267	197,302	89,630	35,662
AEH-Upper Silesia **	94,637	3,430	21,855	9,269	9,559
AEH-Rybnik **	46,617	1,129	9,090	3,274	1,989

* 12 areas with largest SO_2 emission
** SO_2 amounts were taken into account in AEH and AEH-SO_2

SOURCE: GUS, 1984.

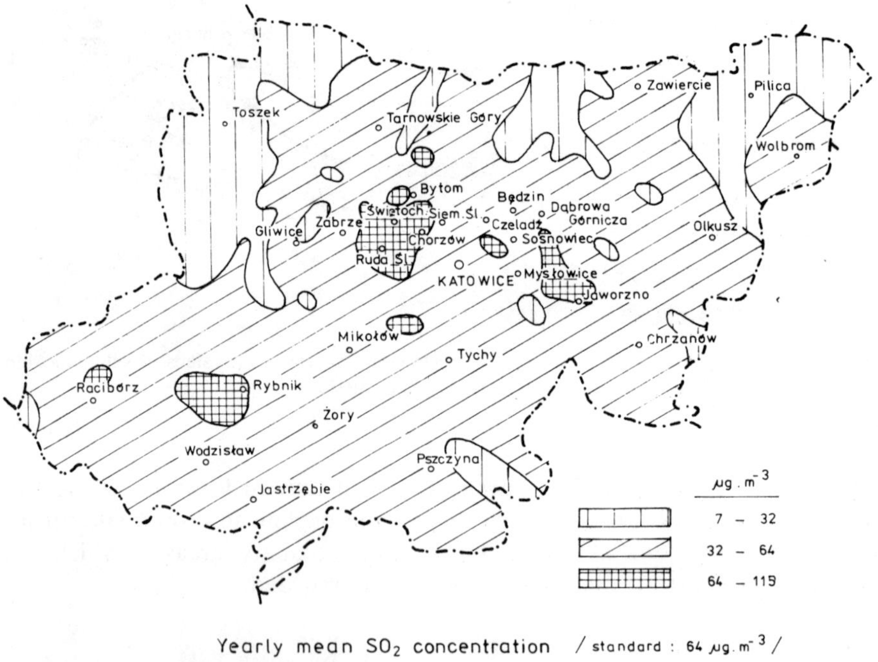

FIGURE 3 Concentration of sulfur dioxide in Katowice district with yearly mean values (in μg m^{-3}). The permissible concentration (national standard) is 64 μg m^{-3} (District Office for Public Health and Epidemiology, Katowice).

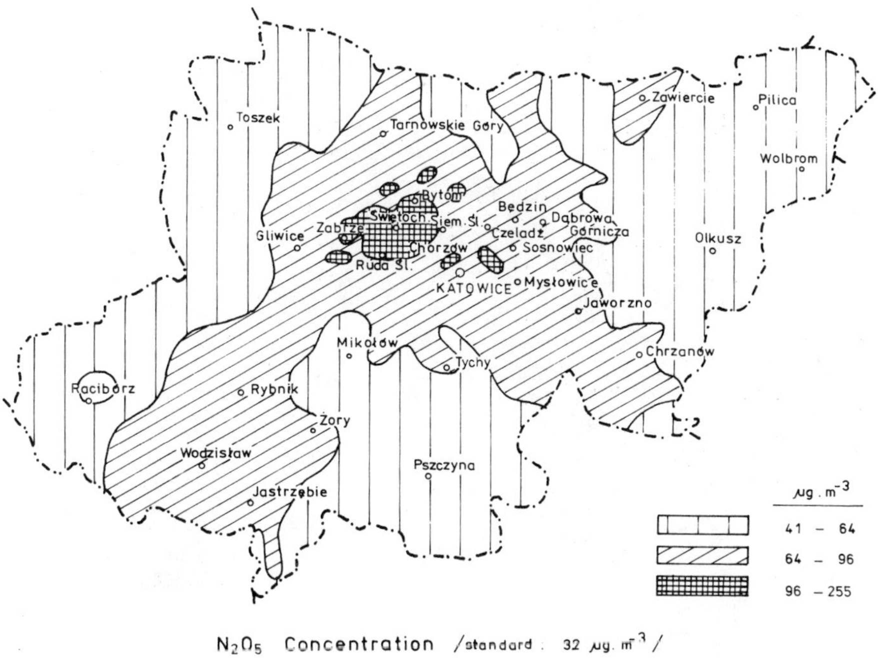

FIGURE 4 Concentration of nitrogen oxides in the Katowice district with yearly mean values (in $\mu g\ m^{-3}$). The permissible concentration (national standard) is 32 $\mu g\ m^{-3}$ (District Office for Public Health and Epidemiology, Katowice).

are used by the local population in the area. Thus, local contamination is of great concern.

A good database exists for heavy metal content in plants, air, and soil for the Katowice region. Data on the reduction of quality and quantity of agricultural and horticultural plant species can be used to predict effects in other polluted areas of Poland. Information is available on yields of the most important agricultural crops, the use of fertilizers, emissions of air pollutants, and the index of agriculture land quality (Table 3). These specific areas have been selected for discussion purpose based on yield of crops (highest and lowest) and emissions of air pollutants.

There is no clear correlation between emission of air pollutants and yield of crops. Yield data for the most polluted Katowice district are higher than or equal to the mean value for the whole country. Except for cereals, the correlation coefficient is also low between yield and agricultural land quality. However, a strong correlation exists between use of fertilizers and yield: 0.9991 for cereals, 0.8356 for potatoes, and 0.7570 for beets. These correlations suggest that the application of fertilizers increases the yield of

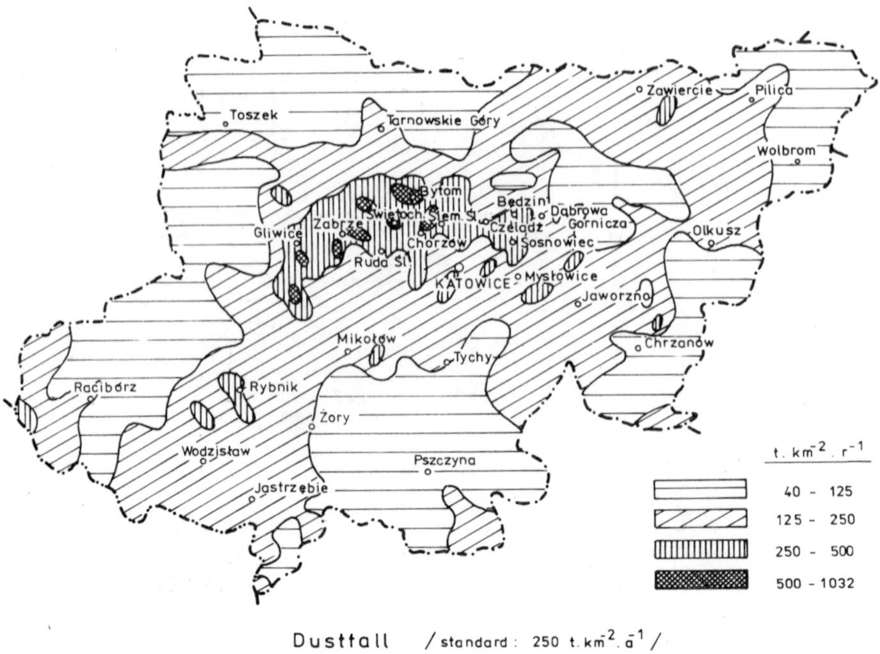

FIGURE 5 Dust fall in the Katowice district with yearly mean values. The permissible concentration (national standard) is 250 t/km^2 (District Office for Public Health and Epidemiology, Katowice).

crops in large areas of Poland, including polluted areas. However, if yield is considered alone, the actual impact of air pollution will not be understood. Of equal importance is the quality of agricultural and horticultural crops as related to contamination by heavy metals and organic compounds, e.g., polycyclic hydrocarbons. Air pollution impacts on the quality, more than the quantity, of any food product from areas of high environmental pollution. This should be the crucial criterion for further public concern.

In addition to application of fertilizers, proper pest management is another possible means of increasing crop yield. According to some estimates, weeds, insects, and plant diseases reduce the output of agriculture on a national scale by about 15% (Committee on Plant Protection, 1986). This may be the reason why the effects of environmental pollution on agricultural production is of little interest to agricultural scientists (Rutkowski, 1986; Czembor et al., 1986). As will be shown below, however, there is evidence of adverse effects of air pollutants on agricultural crops.

In the Silesian region alone, 400,000 people consume vegetables and fruit from garden plots located in areas of high pollution deposition

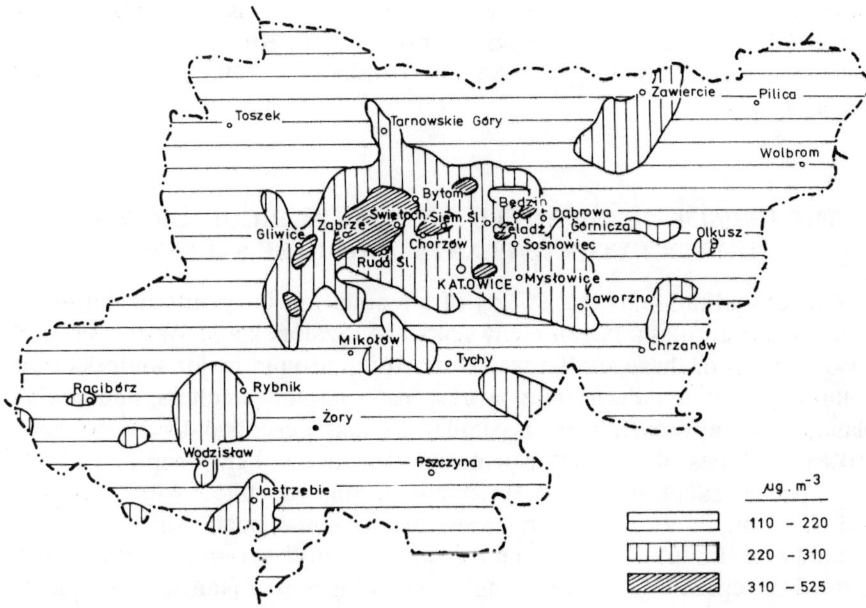

FIGURE 6 Concentration of aerosols in the Katowice district with yearly mean values (in $\mu g\ m^{-3}$). The permissible concentration (national standard) is 22 $\mu g\ m^{-3}$ (District Office for Public Health and Epidemiology, Katowice).

TABLE 3 Emission of pollutants, use of fertilizers, yield of crops, and index of agricultural land quality.

Area or District	Emission of pollutants (1,000 kg/km^2)	NPK kg/h	Yield cereal	Potatoes (100 kg/ha)	Beets (100 kg/ha)	Index
Poland	21.5	175.2	27.6	168	331	65.8
Katowice	300.0	201.9	29.8	180	331	68.1
Krakow	215.7	173.6	28.0	159	298	85.7
Opole	29.1	276.0	36.2	176	364	81.7
Ostroleka	16.2	101.0	21.5	147	315	51.0

SOURCE: GUS, 1986; GUS, 1984.

(Kucharski et al., 1984). This figure would be much higher for all the AEHs (Tables 1 and 2). The concentration of several heavy metals in vegetables and fruits from these areas contributes to the overall intake of these metals by a large portion of the population, and is therefore of great concern for public health reasons (Karweta, 1980; Kucharski et al., 1984; Marchwinska and Kucharski, 1986; Grodzinska et al., 1987; Niklinska and Maryanski, 1988).

THE IMPACT OF AIR POLLUTANTS ON AGRICULTURAL AND HORTICULTURAL PLANTS: TWO CASE STUDIES

Investigations into the effects of air pollution on plants in industrial parts of Upper Silesia began many years ago (Szalonek and Warteresiewicz, 1966a). Most of these studies were carried out around major sources of air pollution, such as metallurgical works, lead and zinc smelters, and coking plants. The approach used is similar to that described by Schoenbeck (1968). A series of holes 35 cm in diameter and 110 cm deep and sealed by a plastic sheet from the surrounding soil, are filled with a unified soil. Test plants are grown in these "experimental pots" which could be located at different distances and directions from the sources of pollutants. Instead of separate holes for a single or small group of plants, "microplots" of 1 m^2 with unified soil are used (Warteresiewicz and Szalonek, 1972). For statistical reasons multiple plots are used, usually four on each site (location) and for each crop. To investigate the effect of a given source of pollutants, microplots are located at different distances and directions around the source.

Assessment of losses is based on comparisons of harvest parameters for the same species at the location under investigation with that of those reference (control) locations. To avoid any major influence of variables other than air pollutants, the reference point must be located very close to the area under investigation. In areas with several or many air pollutant sources, it is difficult or even impossible to have an unpolluted locality. For this reason, crop losses may be underestimated when compared to clean air responses. Concentrations of air pollutants in locations accepted as reference points are much lower than in the areas investigated, but still higher than in areas characterized by background concentrations of pollutants.

Effects of air pollutants on horticultural species were investigated both in experimental orchards of the Institute of Pomology in Skierniewice and in containers with uniform soil (Blidy et al., 1983; Kulawik, 1985). Containers were located in areas of both low and high pollution; and growth and yield parameters of several plant species, or cultivars, were measured.

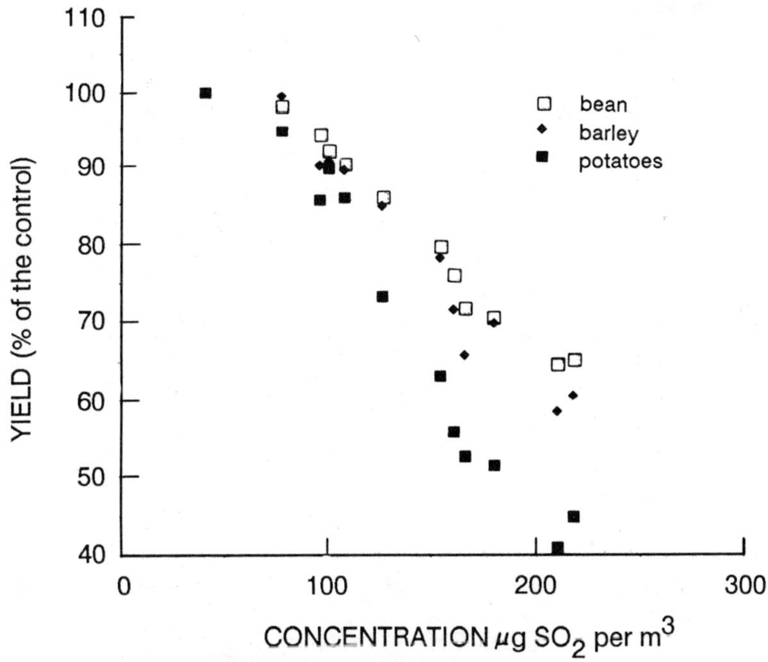

FIGURE 7 Effects of sulfur dioxide on yield of plants in the field; means of three years of experiments. Sulfur dioxide concentration (in $\mu g\ m^{-3}$) is calculated from sulfation rate (Warteresiewicz, 1979).

Impacts on the yield of two apple cultivars (McIntosh and Jonathan), pear (c.v. *Konferencja*), and cherry (c.v. *Lutowka*) were evaluated, at the experimental orchards of the Institute of Pomology. At weekly intervals, dust—a mixture of coal powder and feldspar—was suspended in water and spread in amounts equal to and double the national standard. The application of dust to plants started after the flowering period (Blidy et al., 1983).

RESULTS OF THE CASE STUDIES

Agriculture

Depending on the concentration of sulfur dioxide on the plant species studied in the microplots described above, decreases in yield have been found from less than 10% up to 55% for potatoes, 40% for bean, and 35% for barley (Warteresiewicz, 1979; Figure 7). Comparison of data on air pollution measurements obtained by Warteresiewicz (1979, 1987) with data obtained using other methods for sulfur dioxide concentration in the same

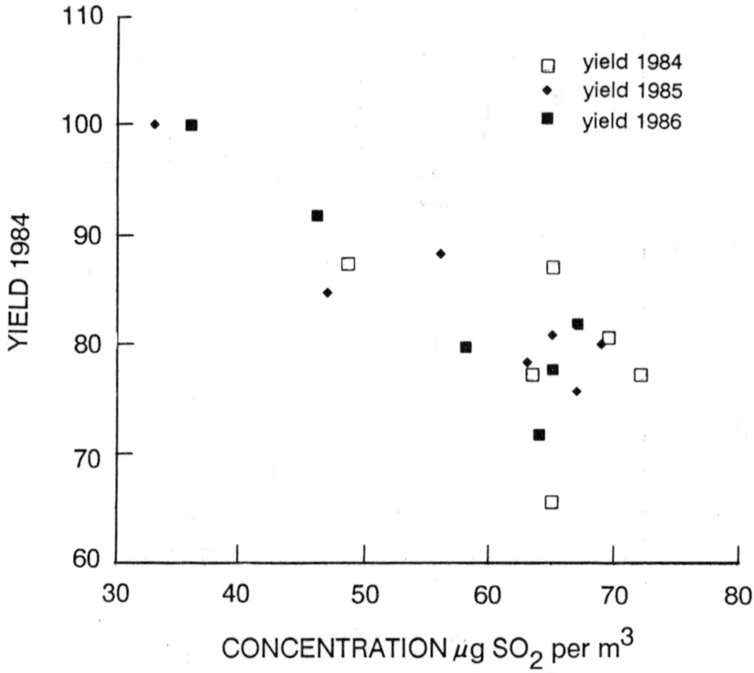

FIGURE 8 Yield of barley (grains) in the vicinity of a metallurgical plant, as percent of the reference point. Data from three years of experiments (Warteresiewicz, 1987).

area is difficult (Karczmarz and Cimander, 1988; Figure 3). However, if the coefficient for calculating sulfur dioxide concentration from sulfatation rates is accepted, data presented in Figure 7 would agree with Roberts (1984) that a decrease in yield can be expected if the sulfur dioxide concentration is higher than 60 $\mu g\ m^{-3}$.

More recent data obtained by Warteresiewicz (1987) in experiments still in progress are presented in Figures 8-10. These data are from areas surrounding a steel mill in Upper Silesia, where there is high background air pollution independent of emissions from the source under study. Independent measurement of concentrations of some air pollutants were also carried out in the same region; the results are shown in Figures 3-6. For the area where the reference point was located, the concentration of SO_2 was 29 $\mu g\ m^{-3}$, and NO_x was 51 $\mu g\ m^{-3}$. No major differences in pollutant concentration within the area under investigation (i.e., surrounding the steel mill) have been found. The data collected are as follows: for SO_2, from 50 to 70 $\mu g\ m^{-3}$, and for NO_x, from 65 to 85 $\mu g\ m^{-3}$ (Karczmarz and Cimander, 1988). However, sulfur dioxide concentrations measured using sulfatation methods and presented in Figures 8-10 are higher.

TABLE 4 Some growth and harvest parameters of black currant (*c.v. Roodknop*) cultivated at Brzezna and Zabrze. Mean of three-year experiment, calculated for one plant.

Location	Sum of shoot growth (cm)	Harvest number	weight (g)	Parameters, fruit size distribution (%) < 8	8-10	> 10
Brzezna	12,949	7,011	3,227	16.3	36.5	47.2
Zabrze	1,092	2,233	879	49.2	47.4	3.4

SOURCE: Kulawik, 1985.

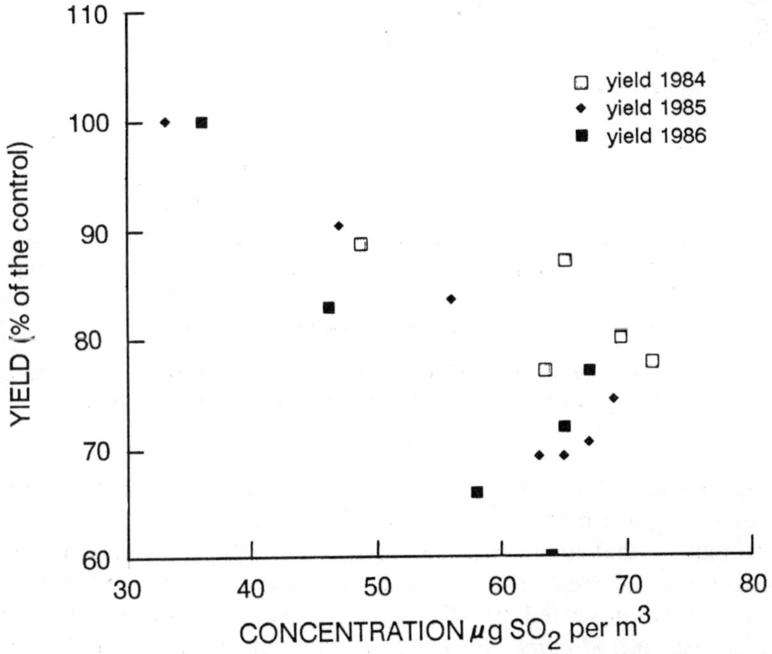

FIGURE 9 Yield of bean (*Vicia faba* f. *minor*) in the vicinity of a metallurgical plant, as percent of the reference point. Data from three years of experiments (Warteresiewicz, 1987).

The data for yield loss and pollution level of sulfur dioxide are rather poorly correlated; however, the threshold concentration is lower than has been demonstrated by similar field measurements obtained earlier (Warteresiewicz, 1979). Data published by Warteresiewicz (1979) indicate that crop losses between 3 and 4% occurred when the sulfur dioxide concentration increased by 10 $\mu g\ m^{-3}$ in a concentration range of 70-210 $\mu g\ m^{-3}$ (Figure 7). According to the most recent field experiments (Warteresiewicz, 1987), a crop loss of 4.5 to 7.1% takes place for every 10 $\mu g\ m^{-3}$

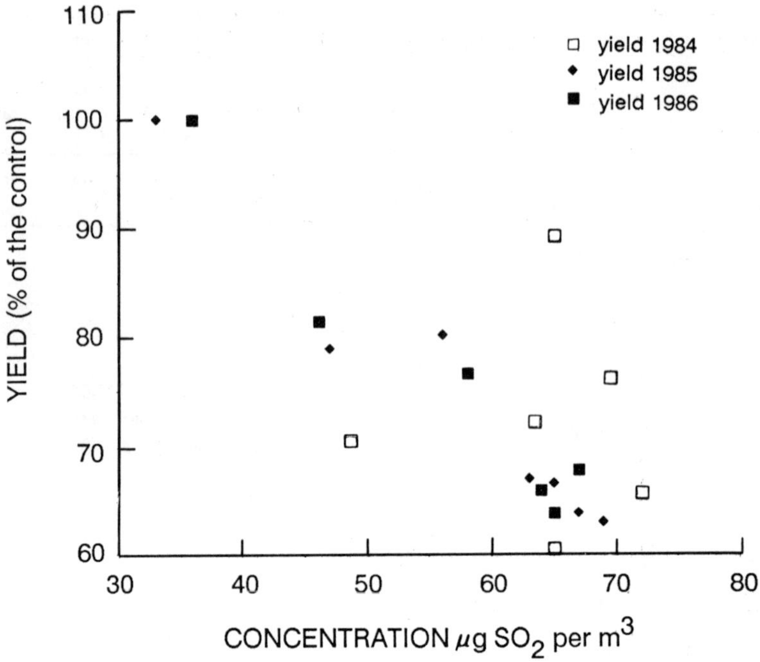

FIGURE 10 Yield of potatoes in the vicinity of a metallurgical plant, as percent of the reference point. Data from three years of experiments (Warteresiewicz, 1987).

increase (Figures 8-10). Data for higher concentrations are disparate and any numbers given must be interpreted with care.

It may be that nitrogen oxides play some role in reduction of yields at concentrations measured near the industrial activity. These data are similar to that presented by Whitmore (1985) for long-term exposure of *Poa pratensis* cv. *Monoply* to a mixture of sulfur dioxide and nitrogen dioxide at weekly mean concentrations of 0.062 ppm. Concentrations of both pollutants in this range are very likely to occur in most AEHs (Table 1, Figures 1 and 2).

Horticulture

Data for horticultural plants are fewer than for agricultural crops. Experiments were carried out in two locations: Zabrze as a polluted site, and Brzezna as a reference point. As in the previous discussion, data on sulfur dioxide data were collected. No data for nitrogen oxide concentrations are available for the reference point. Data for the growing season are in the range of 20 μg SO_2 m^{-3} for Brzezna, while concentrations for the polluted location (Zabrze) are 50 $\mu g\ m^{-3}$ for SO_2 and 150 $\mu g\ m^{-3}$ for nitrogen oxides (Karczmarz and Cimander, 1988). The differences between years are not large. Results from another study using similar SO_2 methods show a level of 60 $\mu g\ m^{-3}$ for the growing season in the polluted site (Kulawik, 1987).

Dust applied to apple trees in the reference area caused a drop in yield up to 40% for McIntosh and 80% for Jonathan as compared to untreated trees in the same orchard. The impact of similar treatments in the polluted orchard was significantly less pronounced, with losses of about 8% and 18% for the two cultivars, respectively (Blidy et al., 1983). Deposition of particulate matter in amounts above 250 metric tons/km^2 per year is not typical in areas of large orchards. However, horticultural plants in garden plots are very commonly located in areas of high environmental impact (Table 2).

A substantial reduction of plant growth and yield of fruits has been found for apple trees, raspberries, strawberries, and black currants grown in containers; the data for black currants are given in Table 4. All parameters, both qualitative (i.e., size of berries) and quantitative, are significantly better in the less polluted area at Brzezna. It is difficult to say which of the parameters measured provides a better example of the effect of air pollutants. The total yield, size, number, and weight of fruits examined by Kulawik (1985) for other species were reduced in polluted areas up to two-thirds when compared to the reference point. In addition, elevated concentrations of heavy metals like lead (Pb) and/or cadmium (Cd) were found in some fruits: 2.2-5.5 ppm and 0.9 ppm for Pb and Cd, respectively.

For Upper Silesia and some other AEHs, the reduction of crop quality because of elevated concentrations of some heavy metals may be of greater importance for consumers than the reduction of yield. However, data are scarce (Karweta, 1980; Marchwinska and Kucharski, 1984; Grodzinska et al., 1987; Niklinska and Maryanski, 1988). The highest concentration of heavy metals in soils and plants are found surrounding zinc and lead smelters, all of which are located in Upper Silesia. The deposition of Pb and Cd for this region is shown in Figures 11 and 12.

Heavy metal concentrations which exceed the recommendations of the World Health Organization (WHO) or the Council for Mutual Economic Assistance (COMECON) also have been found in Krakow (Figures 13 and 14) (Grodzinska et al., 1987; Niklinska and Maryanski, 1988). However, the concentration of those pollutants in vegetables like carrots or potatoes, which are consumed in larger quantities, seems to be more dangerous for the population. No data are available on yield reductions for vegetables.

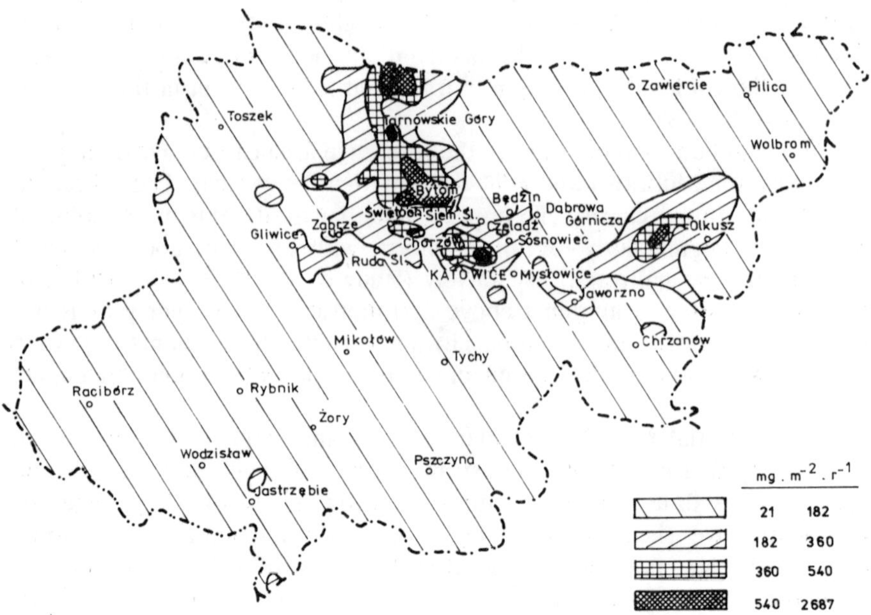

Pb deposition

FIGURE 11 Deposition of lead in the Katowice district in mg m^{-2} per year (District Office for Public Health and Epidemiology, Katowice).

CONCLUSION

Existing databases on the impact of air pollutants on plants in Poland are limited to coverage of only a few locations; therefore, countrywide assessments are not possible except by estimation. The yield reduction of all agricultural and horticultural crops may be greater than that shown by data presented so far, due to the assessment method used. Air pollution by sulfur dioxide in reference locations was much higher than background. In addition to reduction of yield, the quality of agricultural products is reduced. This is an effect of greater importance for consumers.

On the basis of data available so far, the reduction of yield in all AEHs is not greater than 10%, except in Upper Silesia where it is significantly higher. On a national scale, this figure is lower because agricultural land in AEHs accounts for only 10% of the total for the country. Reductions in crop yield due to other factors—including insufficient application of fertilizers, poor management, and inadequate use of pesticides—seem to be higher than those caused by air pollutants. On a regional scale, however, the effects of air pollution are acute in both quantitative and qualitative

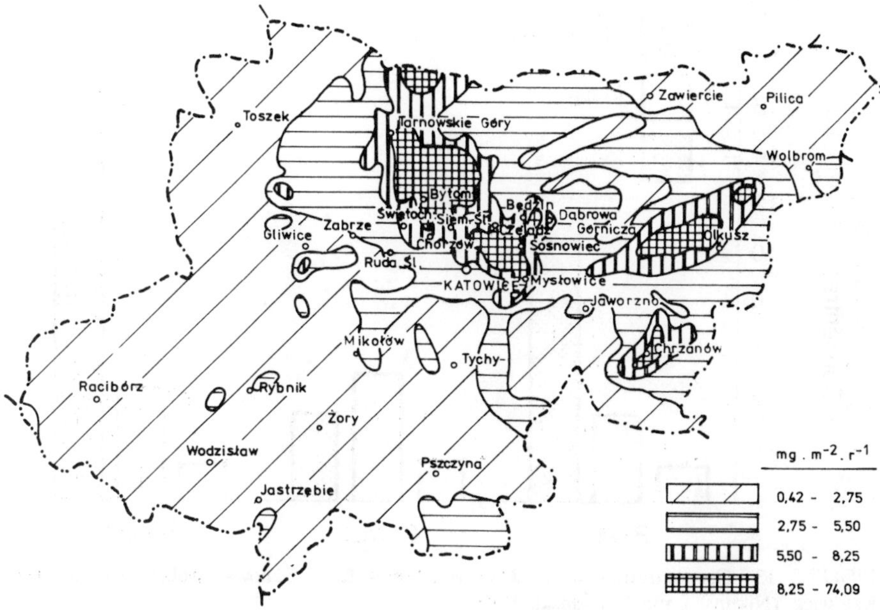

Cd deposition

FIGURE 12 Deposition of cadmium in the Katowice district in mg m^{-2} per year (District Office for Public Health and Epidemiology, Katowice).

terms. Of particular importance for the population is the condition of fruit and vegetables grown in highly polluted areas, both in terms of their aesthetic quality and their possible effect on human health.

As stated previously, more accurate determination of crop losses due to air pollution on a countrywide scale is not possible yet. The three main reasons for this are:

- limitation of methods actually used for determination of yield reduction in polluted areas;
- lack of data on air pollution concentrations on larger than local- and, at best, regional-scale;
- lack of experimental results concerning yield decreases in regions other than Upper Silesia.

Rough estimates of losses are possible for the AEHs in terms of SO_2 only. Data presented in Figures 7-10 seem to confirm that some losses due to air pollution do occur in locations used as reference points. If this is correct, the threshold concentration will be well below that measured actually on relative large areas. Therefore, the results from microplots

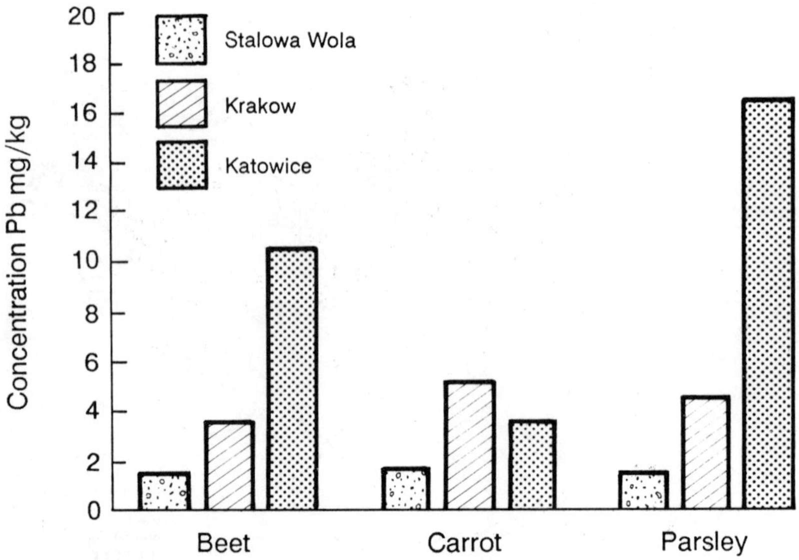

FIGURE 13 Concentration of lead in vegetables from Stalowa Wola, Krakow, and Katowice (Niklinska and Maryanski, 1988).

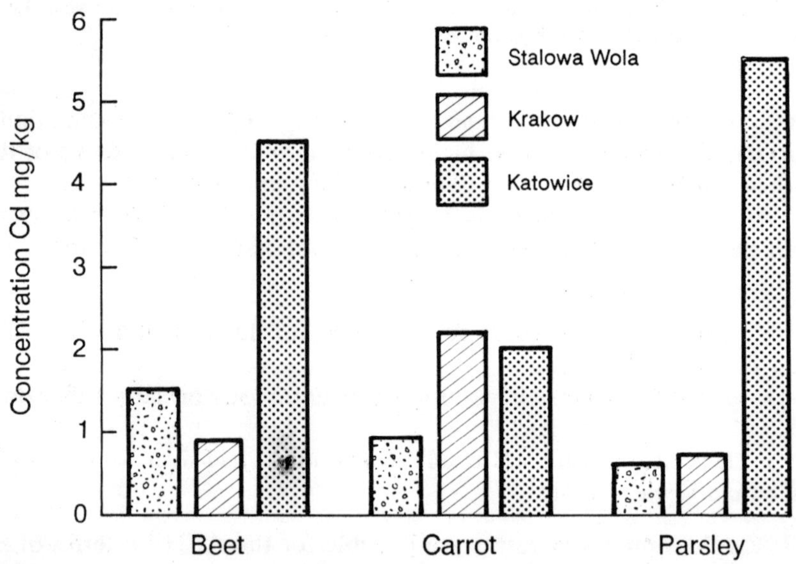

FIGURE 14 Concentration of cadmium in vegetables from Stalowa Wola, Krakow, and Katowice (Niklinska and Maryanski, 1988).

located in Upper Silesia are underestimated. Clearly, additional research is needed to obtain data which will be more accurate on regional and countrywide scales.

REFERENCES

Blidy, J., P. Kolodziejczyk, S. Godzik, and B. Cisek. 1983. Yield of apple crops from trees treated with water suspension of non-toxic dusts. Prace Instytutu Sadownictwa i Kwiaciarstwa Ser C Biul Inf 1-3:155-156.

Cofala, J., and W. Bojarski. 1988. Sulfur and nitrogen oxides emissions resulting of energetic utilization of fuels—the case of Poland. Archiwum Ochrony Srodowiska 3/4 (in press).

Commitee for Plant Protection. 1986. Assumption concerning developments of science in the field of plant protection till the year 2000 and in the subsequent years. Postepy Nauk Rolniczych 2-3/86:293-300 (in Polish).

Czembor, H., J. Jakubiec, B. Kubicki, E. Nalborczyk, S. Starzycki, and K. Swierzynski. 1986. The main trends in development of plant physiology, genetics and breeding till the year 2000 and in subsequent years. Postepy Nauk Rolniczych 2-3/86: 179-196 (in Polish).

Godzik, S., and S.V. Krupa. 1982. Effects of sulfur dioxide on the growth and yield of agricultural and horticultural crops. Pp. 246-265 in Effects of Gaseous Air Pollution in Agriculture and Horticulture, M.H. Unsworth and D.P.Ormrod, eds. Butterworth, London.

Grodzinska, K., B. Godzik, and S.Szarek. 1987. Vegetables and soil contamination by heavy metals in allotment gardens in Krakow agglomeration (S.Poland). Bull. Pol. Acad. Sci. Biol. Sci. 35:111-122.

GUS (Glowny Urzad Statystyczny [Main Statistical Office]). 1984. Regional Documents. Areas of Ecological Hazard (AEH) in Poland. Warszawa.

GUS. 1986. Statistical yearbook 1986. Warszawa.

Juda, J., M. Nowicki, and W. Jaworski. 1980. Expert opinion on air pollution in Poland in the year 1978, as cited in GUS Environmental Protection and Water Management. Statistical Data. Warszawa (in Polish).

Karczmarz, T., and B.Cimander. 1988. State and trends in air pollution in the Katowice region. Presentation at 2nd International Conference on Health and Environmental Pollution. Zabrze. To be published in Archiwum Ochrony Srodowiska.

Karweta, S. 1980. Effects of air pollution on soil and plant contamination by heavy metals in Upper Silesian industrial region. Pp. 314-325 in the Proceedings of the Conference on Effects of Air Pollution on the Environment. Naczelna Organizacja Techniczna, Warszawa.

Kassenberg, A. 1982. Areas of ecological hazards—a new planning category. Kosmos 35:153-160 (in Polish).

Kucharski, R., E. Marchwinska, and J. Gzyl. 1984. A method for evaluation of vegetable crops in areas of increased emission of pollutants. Roczniki PZH 35:221-232 (in Polish).

Marchwinska, E., and R. Kucharski 1986. Thresholds for agriculture and horticulture in industrial areas. Kosmos 35:125-132 (in Polish).

National Program of Natural Environment Protection till the Year 2010 Project. Ministry for Environmental Protection and Natural Resources. Warsaw 1988 (in Polish).

Niklinska, M., and Marianski. 1988. Heavy metals in vegetable crops. Aura 4/88:20-22 (in Polish).

Roberts, T.M. 1984. Long-term effects of sulfur dioxide on crops: Analysis of dose/response relations. Phil. Trans. R. Soc. London B 305:299-316.

Rutkowski, A. 1986. Trends in development of agriculture and forest science. Postepy Nauk Rolniczych 2-3/86:35-56 (in Polish).

Schoenbeck, H. 1968. Die Anwendung von Testpflanzenmethode, eine Modifikation des Sorauschen Fangpflanzenverfahrens zum Nachweis von Pflanzenschaedigenden Immissionen. Pp. 313-325 in the Proceedings of the VI International Conference on Effects of Air Pollution on Forests,“ Katowice 1968.

Szalonek, I., and M. Warteresiewicz. 1966a. Effects of industrial pollutants on some vegetables. Biuletyn ZBN GOP PAN 8:59-74 (in Polish).

Szalonek, I., and M. Warteresiewicz. 1966b. Growth and yield of potatos and industrial air pollution. Biuletyn ZBN GOP PAN 8:59-74,75-84 (in Polish).

Warteresiewicz, M. 1979. Effects of air pollution by SO_2 and same plant species in Upper Silesian industrial region. Archiwum Ochrony Srodowiska 7:95-166 (in Polish).

Warteresiewicz, M. 1987. Growth and yield of some agricultural crops around large sources of air pollutants. Polish Academy of Sciences, Institute of Environmental Engineering, internal report. Zabrze.

Warteresiewicz, M., and I. Szalonek. 1972. Untersuchungen ueber die Schaeden an Kulturpflanzen in der Naehe eines metalurgischen Kombinates. Pp. 201-218 in Effects of Air Pollution on Forests. VII International Symposium of Forest Fume Damage Experts.

Whitmore, M.E. 1985. Effects of SO_2 and NO_x on plant growth. Pp. 281-295 in Sulfur Dioxide and Vegetation, W.E. Winner, H.A. Mooney, and R.A. Goldstein, eds. Stanford: Stanford University Press.

Distribution and Movement of Selected Elements in Poland Using Pine Needle Analysis

BOGUSLAW A. MOLSKI
WOJCIECH DMUCHOWSKI
Botanical Garden
Polish Academy of Sciences

Poland is under strong air pollution stress from industrial souces in both Western and Eastern Europe. In addition, since Poland's primary source of energy is coal, pollution generated within the country is also a serious problem. Because of this, the Botanical Garden of the Polish Academy of Sciences (PAN) began research in 1975 on problems of air pollution impacts on vegetation. The first stage, from 1975 to 1980, was in the area of methods development research. Scots pine needles (*Pinus sylvestris* L.) were chosen as the best bioindicator of pollutant impact on vegetation because Scots pine grows everywhere in the country (except in high mountains) and pollutant accumulation in needles occurs.

The method development research was conducted in the Bialowieza Forest District, representing the cleanest place in Poland, and in Panewnik Forest District near Katowice, a location with the highest level of pollution in the country. Studies have been performed in these areas for several years.

Pine needles were collected every second month and analyzed for all major elements, including such pollutants as sulfur, fluorine, lead, cadmium, arsenic, chromium, copper, iron, nickel, and zinc. It has been found that the best time to collect samples is the winter months. In addition, the range of basic elements in clean and heavily polluted areas was determined (Figure 1). The greatest changes were in the content of polluting elements such as lead, zinc, sulfur, fluorine, cadmium, chromium, and iron (300% to 600% increases). Only manganese concentration decreased by 50% in comparison to the control area (Bytnerowicz et al., 1980, 1981/1982, 1983/1984; Dmuchowski et al., 1981/1982; Molski and Dmuchowski, 1985, 1986; Dmuchowski and Molski, 1986).

In 1981, a survey of the entire territory of Poland was initiated. The

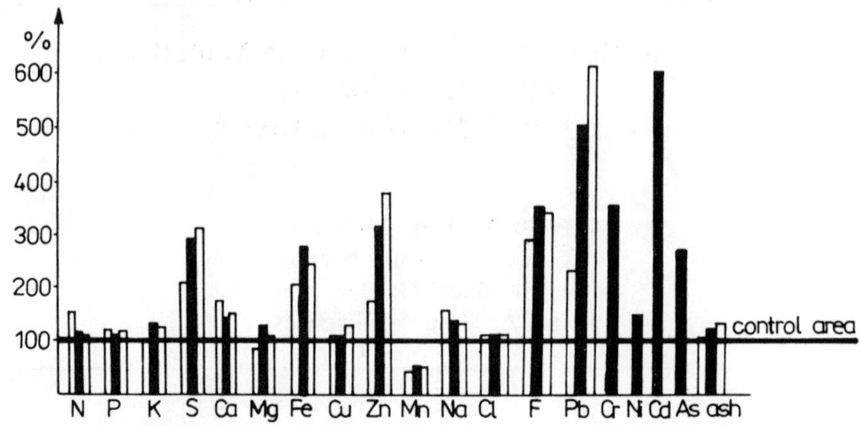

FIGURE 1 Chemical composition of pine needles in a fairly clean area (Bialowieza Forest) shown as a 100% line and in a heavily polluted area as percent of the content of the clean area. The content of each element is shown in three columns: the first represents current growth; the second represents previous year's growth, and the third represents a third year's growth.

country was divided into 8 × 8 km grids, and about 300 squares were selected by random sampling for sample collection. A single pine needle sample was representative of an area of about 1,200 km^2. The Warsaw *voivodship* (district) was surveyed in more detail, i.e., one sample represented only 70 km^2. After elemental analyses were completed, maps of Poland and the Warsaw district were developed using a computer program (Molski et al., 1987). Maps have been completed showing distribution of nine pollutant elements in Poland and in the Warsaw district: sulfur, chromium, arsenic, iron, copper, zinc, cadmium, lead, and nickel. Maps for an additional six elements are in preparation: fluoride, phosphorus, nitrogen, manganese, calcium, and potassium. The following sections discuss the importance and distribution of selected elements.

SAMPLING RESULTS AND DISCUSSION

Sulfur

Sulfur is an essential macronutrient which is required by plants in relatively large amounts. Sulfur is normally absorbed from soils in the form of sulfate as well as from the air as SO_2. Sulfur as a macronutrient has an optimum concentration, and deficiencies or excesses can be deleterious for plant growth.

Extensive studies conducted by the PAN Botanical Garden have identified the normal, physiological requirement of sulfur for pine needles (*Pinus*

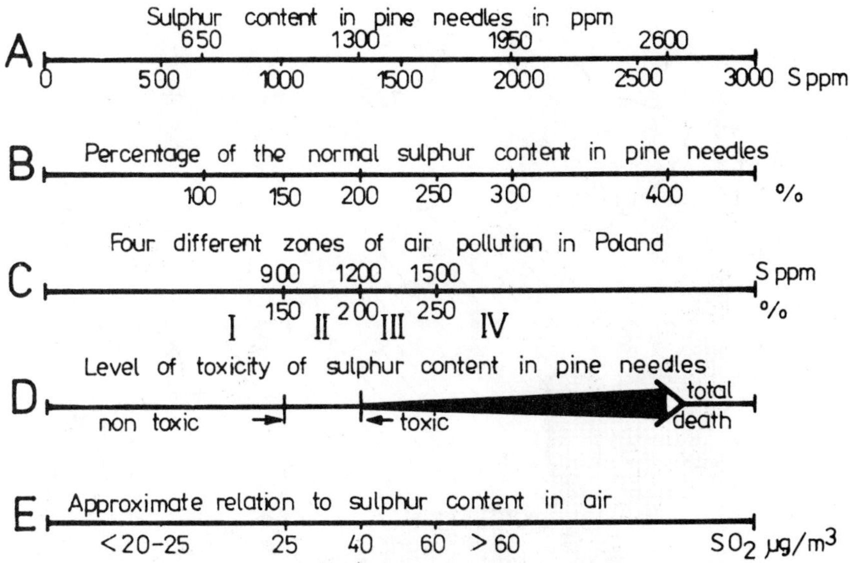

FIGURE 2 Relationship between total sulfur content in pine needles in ppm and as normal or exceeded levels in percentage, as well as possible levels related to its toxicity to trees. Levels of total sulfur content in pine needles can be used to distinguish zones of air pollution impact on vegetation. In Poland, four different zones were differentiated and approximate SO_2 content in air in micrograms per cubic meter are indicated.

sylvestris L.). The relationship between sulfur content in pine needles in their second year of growth is shown at normal or excessive levels in Figure 2. A comparison is made of four distinct zones of air pollution in Poland and approximate SO_2 content in air in micrograms per cubic meter. These data indicate a relationship between the expected and experimentally determined level of toxic effects and the sulfur content in pine needles.

Sulfur concentration of 650 ppm (0.065%) in second-year pine needles is a minimum for growth; below that amount, there can be a deficiency. Therefore, 650 ppm can be considered a "normal" level of sulfur, and presented as 100% of its content. Elevated content up to 1,300 ppm (or 0.13%) can be considered a "luxury" concentration, i.e., not needed but also not toxic. Levels greater than 1,300 ppm can be considered phyto-toxic (Linzon et al., 1979). Investigations published by Gasch and Wentzel (1981) on sulfur fractions in spruce needles showed that organic sulfur occurs at levels of 500-800 ppm, so that the excess sulfur is in inorganic fractions. A sulfur level of 3,000 ppm (0.3%) for pine needles in the second year of growth is critical, as above that level the tissue dies. All pine trees (*P. sylvestris*) with sulfur content greater than 1,300 ppm may exhibit symptoms of toxicity, but 3,000 ppm generally results

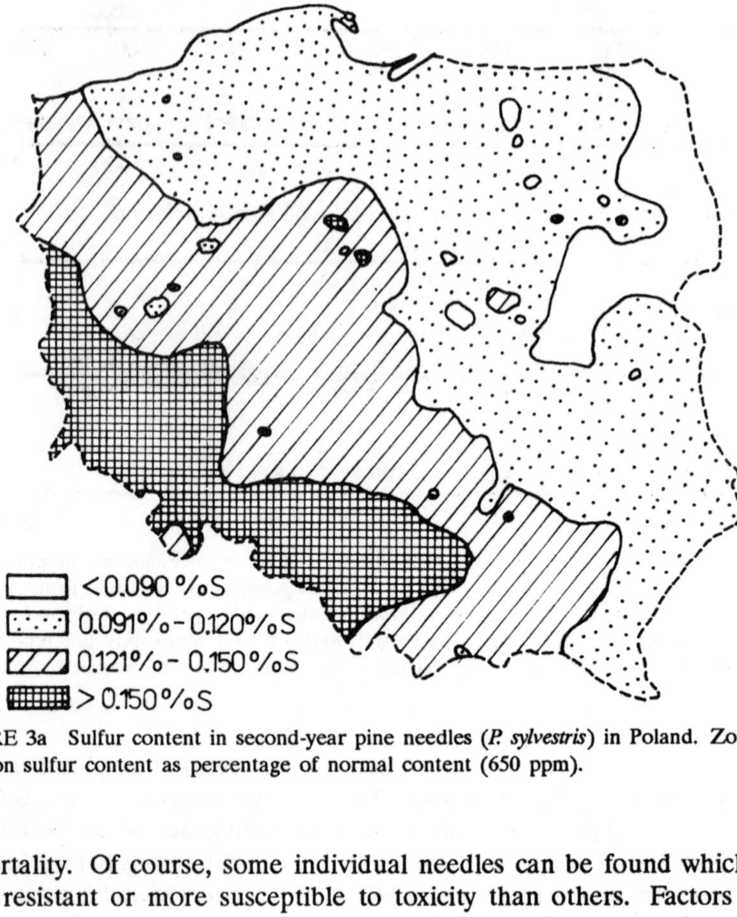

FIGURE 3a Sulfur content in second-year pine needles (*P. sylvestris*) in Poland. Zonation based on sulfur content as percentage of normal content (650 ppm).

in mortality. Of course, some individual needles can be found which are more resistant or more susceptible to toxicity than others. Factors such as ecological conditions, water supply, availability of other nutrients, and extreme temperatures can influence greatly the survival or death of trees.

The distribution of different zones of pollution presented in terms of sulfur content in pine needles is shown in Figure 3a. Only a small part of Poland has air relatively low in sulfur. About 50% of Poland's territory has pine forest where sulfur content exceeds 200% of the normal level, e.g., above 1,200 ppm. A similar situation occurs in the Warsaw district (Figure 3b) and around a steel mill in Warsaw (Figure 3c) where one sample represents about 5 km^2.

The highest sulfur content in pine needles in Poland, as well as in the Warsaw district, is about 2,200 ppm (0.22%), while the majority of samples showed a content between 900 ppm and 1,300 ppm. Table 1 presents sulfur content in different plant materials collected in contaminated and presumed uncontaminated areas in Europe and Canada, according to

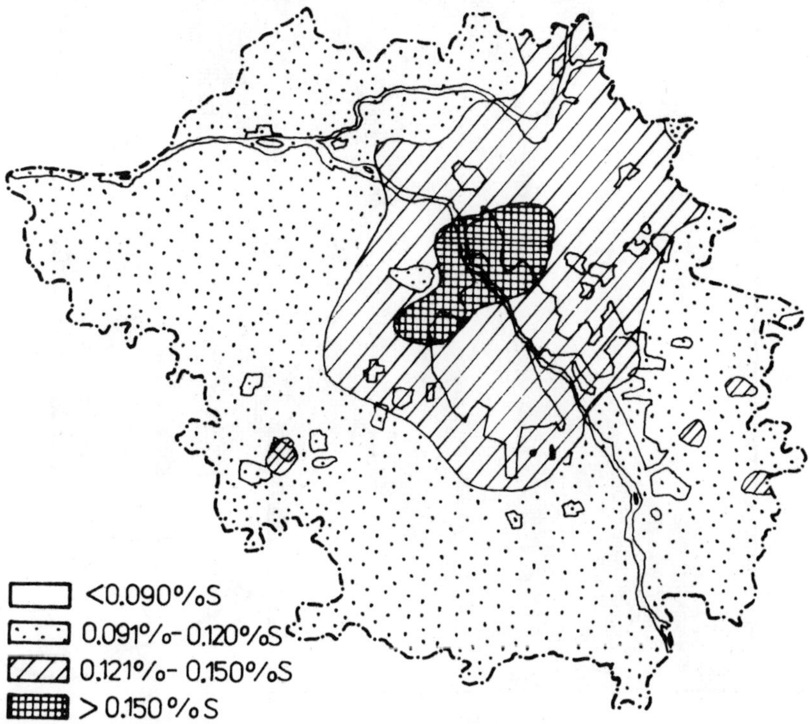

FIGURE 3b Sulfur content in second-year pine needles in Warsaw district (% in dry matter).

different authors. All pine needles collected throughout Europe, regardless of the intensity or time, show that in areas of extensive forest damage, sulfur content is above 1,500 ppm, while in areas remote from industrial centers, sulfur content is about 650-900 ppm. In areas near large industrial centers, sulfur content is about 1,200 ppm (Molski and Dmuchowski, 1986).

Table 2 presents sulfur deposition in accordance with the European Monitoring and Evaluation Programme (EMEP) and with the estimations of the PAN Botanical Garden. Sulfur deposition in different zones is consistent with the estimates of the authors, although higher than the estimates of EMEP (Table 2). The number of zones presented by EMEP and in research conducted at the Botanical Garden are the same, i.e., four. There is also agreement with regard to the distribution of these zones in Poland, although the size of the zones are different.

According to EMEP, sulfur deposition in Poland ranges from 15 to 120 kg per hectare, or 0.15-1.2 kg per km^2. These calculations would put total deposition in Poland at approximately 1,632 million tons of sulfur (or

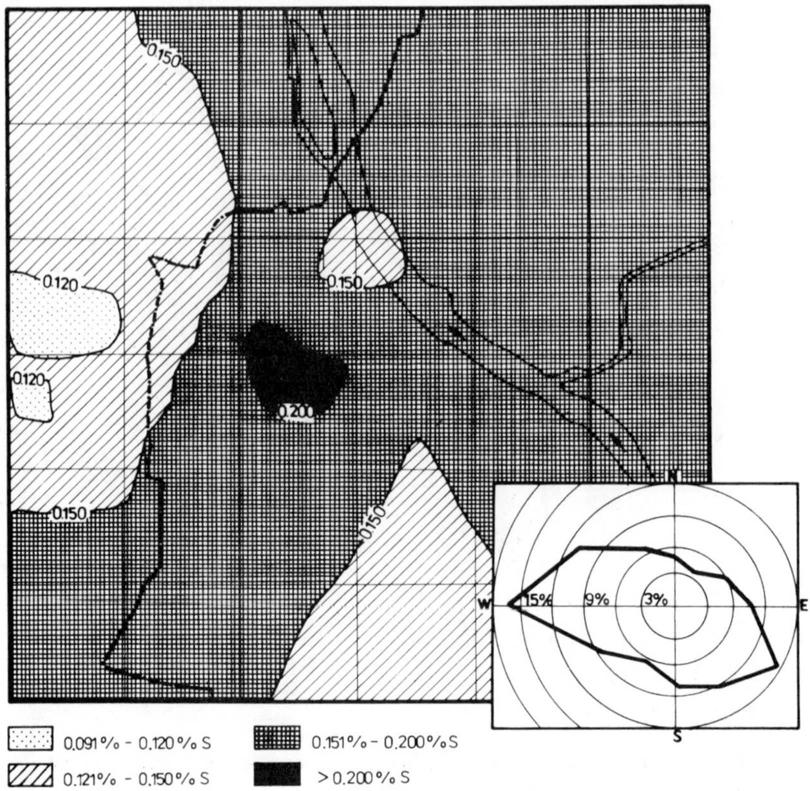

FIGURE 3c Sulfur content in second-year pine needles around a steel mill in Warsaw (as percentage). In this area, each sample represents about 5 km^2.

3.2 million tons of SO_2). However, according to estimates of the PAN Botanical Garden, sulfur deposition is higher—from 20 kg per hectare in the cleanest zone to 200 kg in the most polluted zone (Table 2). Estimates suggest that total deposition of sulfur in Poland would be 2,562 million tons (or 5.1 million tons of SO_2).

Figure 4 represents a schematic illustration of dry (SO_2) and wet (H_2SO_4) deposition of sulfur in different areas of Poland. In the Katowice and Krakow districts, dry deposition of sulfur is much higher than wet deposition with precipitation, primarily due to heavy pollution transported from the German Democratic Republic and from Czechoslovakia, where large industrial centers are located near the Polish border. In the Warsaw district, which is located in the center of Poland, dry deposition is less than in southern Poland (expressed in total amount as well as in percentage).

TABLE 1 Content of sulfur (in %) in leaves or needles of trees collected in contaminated and uncontaminated areas according to different authors.

PLANT SPECIES OR TYPES	PLACE OF COLLECTION	PRESUMED UNCONTAMINATED AREA	CONTAMINATED AREA	REFERENCES
Pinus sylvestris	industrial region	0.06-0.09	0.10-0.17	Huttunen et al. 1979/1980
Pinus sylvestris	industrial region	0.12-0.13	0.13-0.24	Grodzinska 1977
Pinus sylvestris	industrial region	0.03-0.09	0.35	Palvik 1965, cit. Bengton et al. 1977
Pinus sylvestris	uncontaminated	0.08-0.17	----	Materna 1978
Pinus sylvestris	industrial region	0.11-0.18	0.29	Themlitz 1960
Pinus sylvestris	uncontaminated	0.14	----	Linzon et al. 1979
Pinus sylvestris	power station	0.07	0.08-0.19	Boratynski 1983
Pinus sylvestris	fertilizer plant	----	0.12-0.26	Borowiec 1983
Pinus sylvestris	power plant	----	0.19-0.34	Karwata et al. 1987
Pinus strobus	industrial region	0.09-0.13	0.11-0.22	Linzon et al. 1979
Pinus nigra	uncontaminated	0.17	----	Linzon et al. 1979
Picea abies	industrial region	0.08-0.18	0.16-0.36	Materna 1978
Picea abies	industrial region	0.02-0.06	0.33	Palvik 1965, cit. Bengton et al. 1977
Picea abies	uncontaminated	0.03-0.08	----	Stefan 1968
Picea abies	uncontaminated	0.04-0.25	----	Thomas et al. 1965
Picea abies	uncontaminated	0.11	----	Guderian 1970
Picea abies	uncontaminated	0.13	----	Linzon et al. 1979
Picea glauca	uncontaminated	0.14	----	Linzon et al. 1979
Picea pungens	uncontaminated	0.15	----	Linzon et al. 1979
Populus tremuloides	industrial region	0.16-0.22	0.17-0.64	Linzon et al. 1979
Betula papyrifera	industrial region	0.10-0.24	0.20-0.58	Linzon et al. 1979
Tilia cordata	industrial region	----	0.23	Karkanis 1976
Tilia cordata	urban	----	0.29	Chmielewski et al. 1985
Platanus acerifolia	urban	----	0.42	Chmielewski et al. 1985
Quercus robur	industrial region	----	0.13	Karkanis 1976
Quercus robur	chemical plant	0.13	0.18-0.72	Bytnerowicz et al. 1980
Betula verrucosa	chemical plant	0.16	0.34-1.06	Bytnerowicz et al. 1980
Populus nigra	chemical plant	0.22	0.76-1.01	Bytnerowicz et al. 1980
Prunus serotina	chemical plant	0.07	0.10-0.37	Bytnerowicz et al. 1980

Wet deposition is less in total amount but is similar in percentage. In the Suwalki district, which is considered the cleanest part of Poland, the fraction of wet deposition from long-range transport is larger; however, the total amount of sulfur is much smaller. Dry deposition, mainly from local sources, is much smaller in this region than in other parts of Poland.

Cadmium

Cadmium is one of the most dangerous pollutants for humans and all mammals. It is dangerous in any quantity above background in foodstuffs, which is considered to be approximately 0.05 to 0.5 ppm. However, there is very often a much higher cadmium content in plants, and 5 ppm is

TABLE 2 Sulfur deposition from air pollution in Poland according to EMEP data and estimates from the Botanical Garden (BG) of the Polish Academy of Sciences according to zones of pollution.

ZONE	SULFUR CONTENT OF PINE NEEDLES	SULFUR DEPOSITION ($kg/km^2/year$) according to EMEP	BG	AREA OF ZONE (thousands of km^2)	SULFUR DEPOSITION (x 1,000 tons/year) according to EMEP	BG
I	< 900	1,500	2,000	18	27	36
II	901-1,200	3,000	4,000	149	447	596
III	1,201-1,500	6000	10,000	101	606	1,010
IV	> 1500	12,000	20,000	46	552	920
TOTAL	----	----	----	314	1,632	2,562

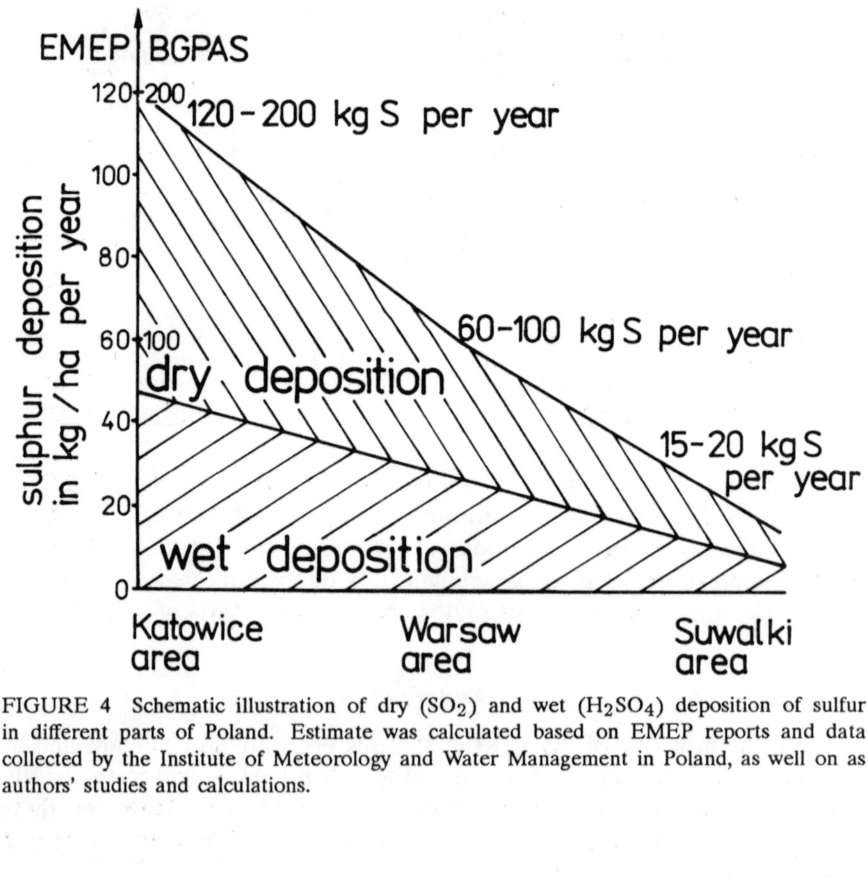

FIGURE 4 Schematic illustration of dry (SO_2) and wet (H_2SO_4) deposition of sulfur in different parts of Poland. Estimate was calculated based on EMEP reports and data collected by the Institute of Meteorology and Water Management in Poland, as well on as authors' studies and calculations.

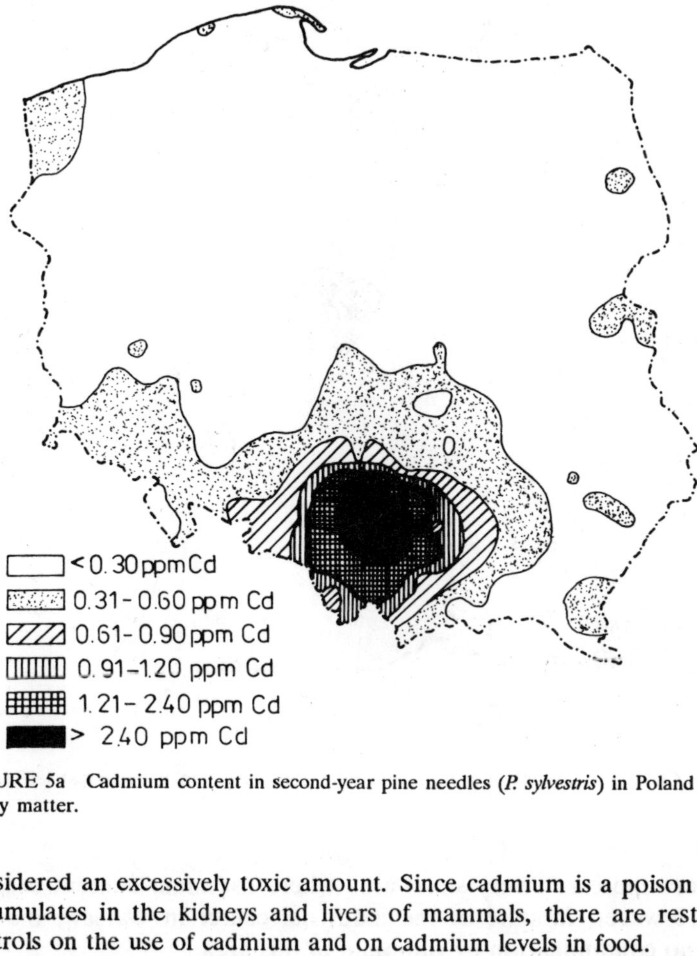

FIGURE 5a Cadmium content in second-year pine needles (*P. sylvestris*) in Poland in ppm in dry matter.

considered an excessively toxic amount. Since cadmium is a poison which accumulates in the kidneys and livers of mammals, there are restrictive controls on the use of cadmium and on cadmium levels in food.

Cadmium primarily enters the terrestrial environment through use of phosphate fertilizers in agricultural areas and from emissions of the mining and metal industries, especially as a by-product of zinc refinement. Cadmium content in pine needles in Poland is shown in Figure 5a. There are very few samples where cadmium content exceeds 2 ppm, and most samples have cadmium contents below 0.5 ppm. The only area where the cadmium level in pine needles is higher is the traditional metal mining and refining area of Poland, where such activities have occurred for several hundred years. Excess cadmium content in pine needles follows the area of zinc contamination, where zinc content is about 100-200 ppm.

Figure 5b represents cadmium content in pine needles in the Warsaw

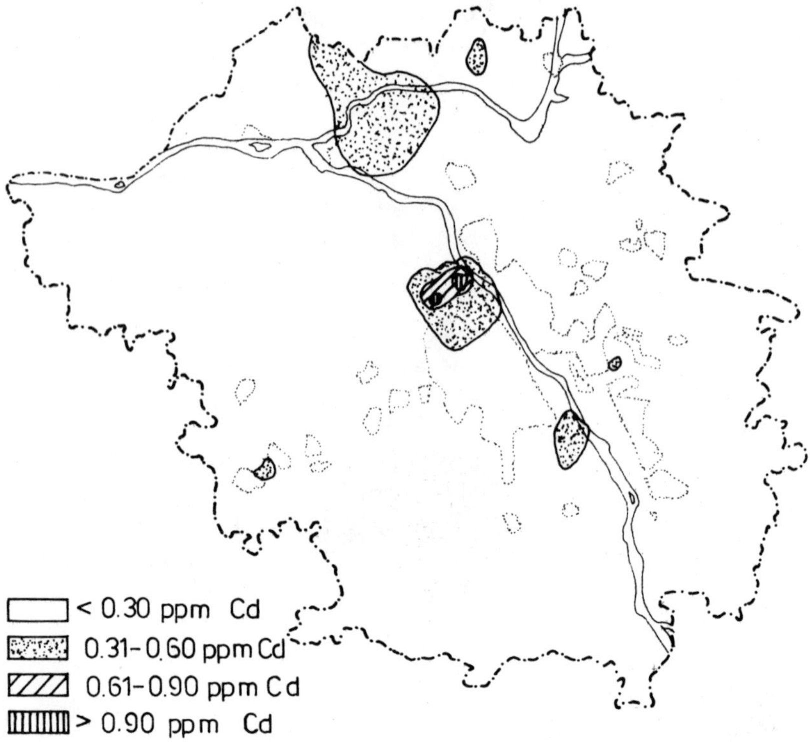

FIGURE 5b Cadmium content in second-year pine needles in the Warsaw district in ppm in dry matter.

district. Here there is a very small area where cadmium content exceeds 0.5 ppm. However, in spite of the industrial center in Warsaw, there is practically no contamination by cadmium in this area.

Lead

Lead is a common pollutant and is usually found locally near mines and metal industries, as well as along roads and highways from the combustion of leaded fuel. Unlike cadmium, lead is not mobile in soils or in plants; therefore, it is not generally toxic to plants, except under certain conditions. Even in cases of very large concentrations of lead in localized plant environments, or even associated in or on plants, there are few reports of lead-induced toxic effects on plants grown in natural ecosystems that have been severely impacted with lead (Koeppe, 1981).

Studies performed in 1974 by Bazzaz et al. (cited in Koeppe, 1981) reported that leaf lead concentrations of 193 micrograms per gram of dry weight reduced photosynthesis in sunflowers by 50%. Rolfe and Bazzaz (1975) found no effect of lead on photosynthesis or transpiration of loblolly pine (*Pinus taeda*) or autumn olive (*Elaeagnus umbellata*) at tissue concentrations below 60 and 72 micrograms per gram of dry weight. At these concentrations (60 and 72 μg/g), photosynthesis was reduced in *Pinus taeda* by 11% and in *Elaeagnus umbellata* by 17%. In other studies carried out by Bazzaz et al., there was a strong correlation between lead-effected decreases in photosynthesis and decreases in rates of transpiration. The decrease in rates of whole-plant photosynthesis may be due to induced closure of stomata rather than to a direct effect on the process of photosynthesis residing directly within the chloroplasts (Koeppe, 1981). While lead may not be very toxic to the plant itself, its concentrations may be very deleterious to human health.

Under certain soil conditions with low pH, low organic matter levels, and low phosphorus levels, large quantities of lead can be taken up by roots of higher plants. However, lead absorbed by roots generally has no toxic effect on plants, except at extremely high root media concentrations that have little relevance to natural conditions. Movement of lead in flowering plants has been demonstrated through roots, but not from lead particles deposited on leaf surfaces. Lead deposits on leaves have little effect on gas exchange, but are of considerable importance to grazing herbivores. Quite possibly the most important effect of lead associated with plant leaves is in food chains where plants act as passive lead carriers (Koeppe, 1981).

The normal lead content of plants ranges from 2 to 10 ppm; concentrations of 30 to 300 ppm is considered toxic (Kabata-Pendias and Piotrowska, 1984). In Poland, lead content in pine needles is very low, with a background level below 10 ppm. Levels exceed 10 ppm in only a few locations, and only in one very small area do levels exceed 30 ppm. Figure 6a shows lead concentration in pine needles in Poland. A similar situation exists in the Warsaw district (Figure 6b).

Zinc

Zinc is an essential micronutrient needed by all organisms as a constituent of many metaloenzymes and of several proteins. Zinc appears to play a role in the synthesis of auxin. Plants with an inadequate supply of zinc display symptoms that derive mainly from a lack of cell elongation (Raven and Johnson, 1986). In addition, zinc is regarded as an essential element in human nutrition, and deficiency effects include growth failure and impairment in wound healing (Underwood, 1971). It would seem, therefore, that zinc deficiency is likely to be more significant than its excess (Bevan et al., 1975). For example, zinc at higher concentrations is moderately toxic to plants, but only slightly toxic to mammals.

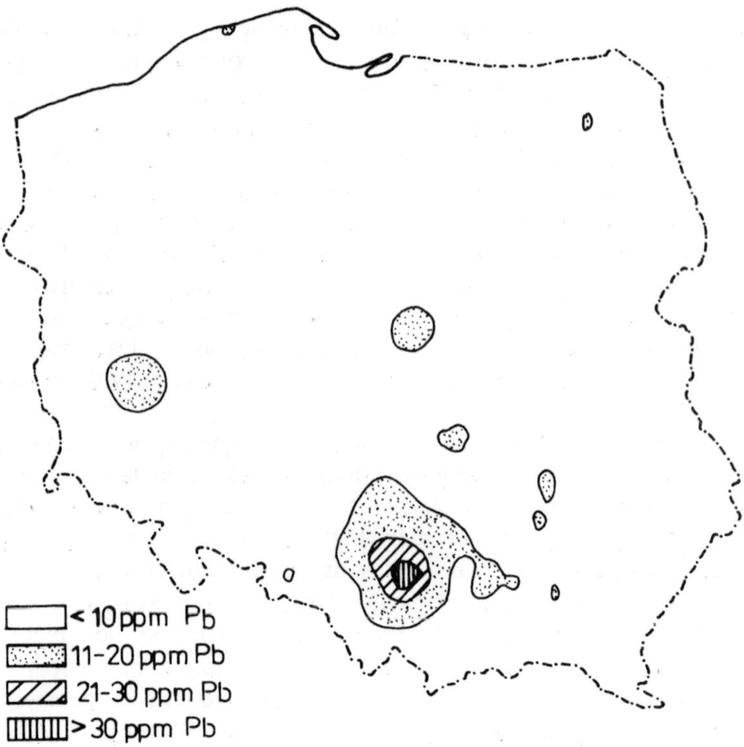

FIGURE 6a Lead content in second-year pine needles (*P. sylvestris*) in Poland in ppm in dry matter.

However, different authors are not consistent in establishing toxicity concentrations for plants. For example, Kabata-Pendias and Pendias (1979) consider zinc content in plants to be toxic between 100 and 400 ppm, with normal levels ranging from 20 to 100 ppm in uncontaminated areas.

In this study, the background concentration of zinc found in pine needles was below 70 ppm. A slightly higher concentration (from 70 to 100 ppm) was found in the southern, industrialized part of Poland, and around Szczecin, Gdansk, and other areas in the northern lake area (Figure 7a). High concentrations of zinc in pine needles (above 100 ppm) were found only in the area of Katowice and Krakow, which is an area traditionally characterized by the metal industry. Zinc appears to be a pollutant from local sources and is rarely associated with long-range transport. The maps of zinc content in pine needles of Poland (Figure 7a) and the district of Warsaw (Figure 7b) support this assumption.

In coniferous trees, zinc content ranges from 13 to 80 ppm (Ahrens, 1964). Materna (1978) achieved similar results with spruce. Angiosperm

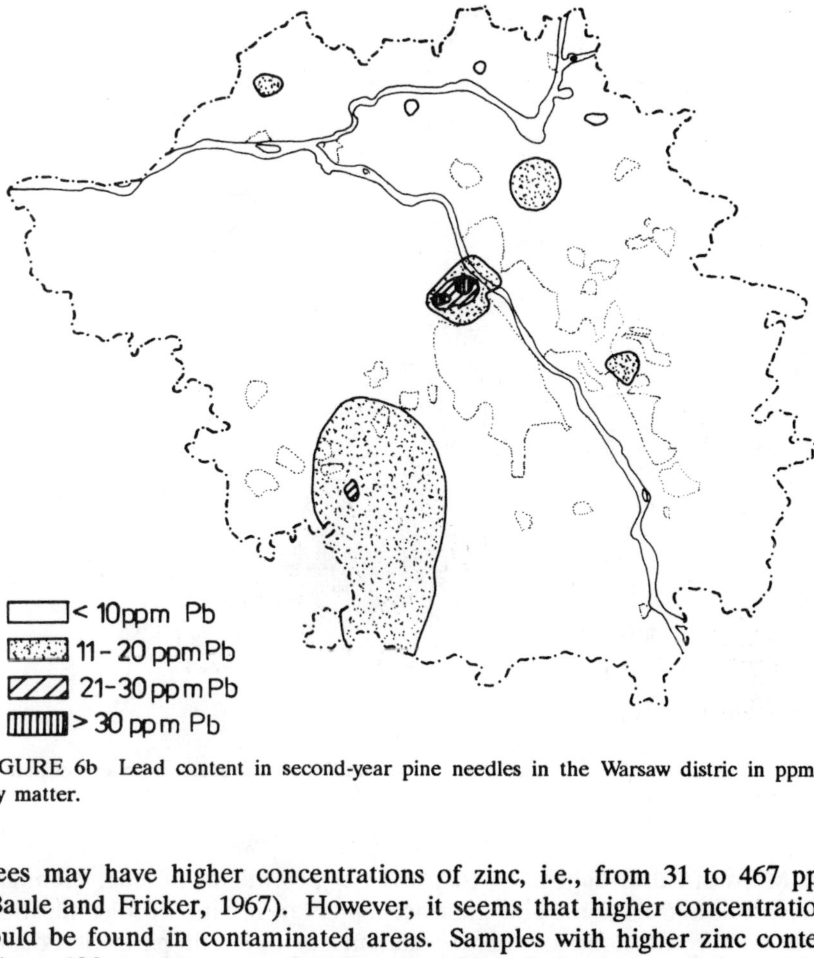

FIGURE 6b Lead content in second-year pine needles in the Warsaw distric in ppm in dry matter.

trees may have higher concentrations of zinc, i.e., from 31 to 467 ppm (Baule and Fricker, 1967). However, it seems that higher concentrations could be found in contaminated areas. Samples with higher zinc content (above 130 ppm) are very few, less than 30 in Poland as a whole and less than five in the Warsaw district.

CONCLUSION

Chemical analysis of pine needles and pollutant content transported by air allow the production of maps which document air pollution impact on vegetation. These maps can be produced on different levels: for a large region, e.g., Poland (312,000 km^2), where one pine needle sample was collected per 1,200 km^2; for a *voivodship* (district), e.g., Warsaw (approximately 3,800 km^2), where one sample was collected per 70 km^2; and for a specific site, e.g., the area around the steel mill in Warsaw (approximately 500 km^2), where one sample was collected per 5 km^2.

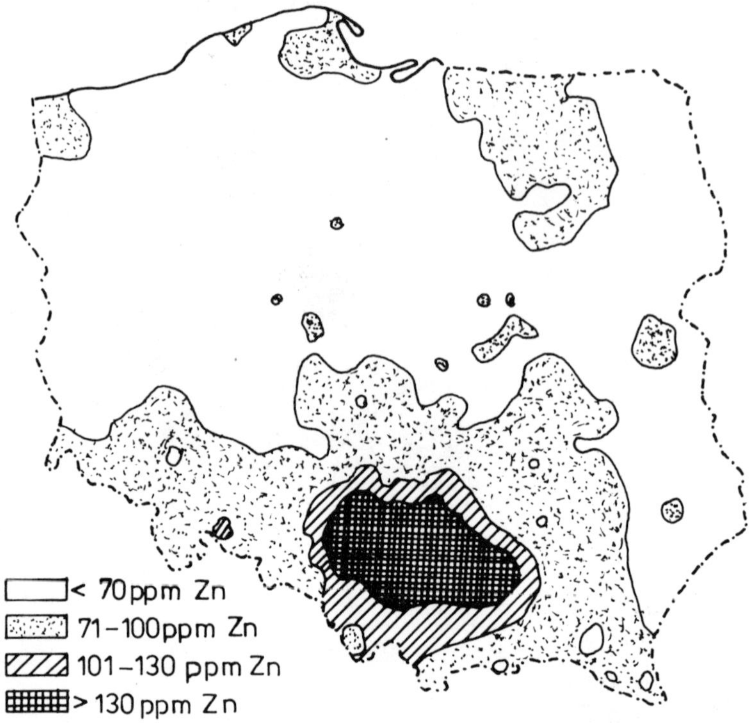

FIGURE 7a Zinc content in second-year pine needles (*P. sylvestris*) in Poland in ppm in dry matter.

Maps of sulfur deposition (dry and wet), as well as of lead, cadmium, zinc, and other elements and their accumulation in pine needles illustrate quantitatively the load of these pollutants in different parts of Poland. Sulfur deposition in Poland is very high, with more than 50% of the country above toxic levels for conifers. Cadmium content in pine needles is rather low (below 2 ppm). The only area in Poland where the cadmium level in pine needles is higher is the traditional area for metals mining and refining. Lead content is low (below 10 ppm); only a few localities exceed 10 ppm, and only one very small area exceeds 30 ppm. Zinc content is rather low; only in the area of an old zinc industry does the content of zinc in pine needles exceed 100 ppm.

The results presented in this chapter provide a basis for further research in the area of air pollution impacts on vegetation. There is also a need for further development of the foliar chemical analysis method. First, it would be very important to compare the pollutant content in pine needles with the direct measurement of these pollutants in the air. Second, it would

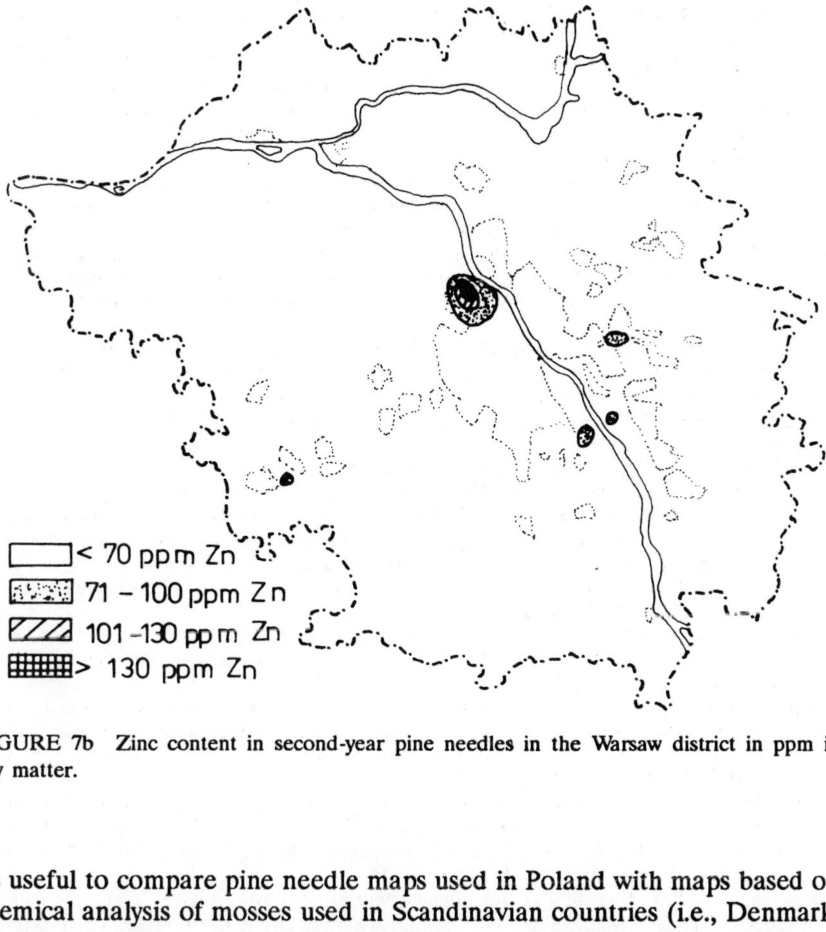

FIGURE 7b Zinc content in second-year pine needles in the Warsaw district in ppm in dry matter.

be useful to compare pine needle maps used in Poland with maps based on chemical analysis of mosses used in Scandinavian countries (i.e., Denmark, Norway, Sweden, and Finland). Development of this method would detect direct relationships between pine needle pollutant content and the content of these elements in agricultural crops and other tree species. All of these approaches have been initiated at the PAN Botanical Garden.

REFERENCES

Ahrens, E. 1964. Untersuchungen uber den Gehalt von Blattern und Nadaln verschiedener Baumarten an Kupfer, Zinc, Bor, Molybdan und Mangan. AFJZ 135:1,8-16.

Baule, H., and C. Fricker, 1973. Nawozenie drzew lesnych. PWRiL, Warszawa, p. 315.

Bevan, R.T., et al. 1975. Report of a collaborative study on certain elements in air, soil, plants, animals, and humans in the Swansea/Neath/Port Talbot area, together with a report on a Moss Bag study of atmospheric pollution across South Wales, pp. 1-264.

Borowiec, S., and Z. Zablocki. 1983. Zawartosc fluoru i siarki w mchach i szpilkach sosny pospolitej jako wskaznik skazenia srodowiska lesnego przez Zaklady Chemiczne "Police". Pp. 351-354 in Materialy II Krajowego Sympozjum "Reakcje biologiczne drzew na zanieczyszczenia przemyslowe," R. Siwecki, ed. Kornik 16-19 maja 1984. Wydawnictwo Naukowe Uniwersytetu im. Adama Mickiewicza, Poznan.

Bytnerowicz, A., W. Dmuchowski, and B. Molski. 1980. The air pollution accumulation capabilities of some tree species in the vicinity of the chemical plant in Torun. Rocznik Dendrologiczny 33:15-28.

Bytnerowicz, A., W. Dmuchowski, and B. Molski. 1981/1982. The effect of needle harvest time, age of needles, and age of Scots pine (*Pinus silvestris* L.) trees on the accumulation of total sulfur. Rocznik Dendrologiczny 34:51-68.

Bytnerowicz, A., W. Dmuchowski, and B. Molski. 1983/1984. Fluorine content of Scots pine needles in the presence of long-term low fluorine air pollution. Influence of forest site, time of sample collection, trees age, and needle age. Rocznik Dendrologiczny 35:9-14.

Chmielewski, W., W. Dmuchowski, and B. Molski. 1985. Trees in the city as sinks for air pollution: Field study with the use of portable lysimeters conducted in Warsaw. Pp. 103-109 in the Proceedings of the Symposium on Creation and Protection of Verdure in the Urbanized Landscape, J. Supuka, ed. Nitra, July 2-4, 1984. Veda, Bratislava, Czechoslovakia.

Dmuchowski, W., A. Bytnerowicz, and B. Molski. 1981/1982. The influence of the boreal site on the accumulation of total sulfur in pine needles. Rocznik Dendrologiczny 34:69-77.

Dmuchowski, W., B. Molski. 1985. The influence of the forest site and needle age on the processes of sulfur, fluorine, and some chosen metals accumulation in Scots pine needles on highly polluted and control areas. Proceedings of the IUFRO Air Pollution Symposium, XIII International Meeting of Specialists in Air Pollution Damages in Forests, Most (CSRS), August 27-September 1, 1984.

Grodzinska, K. 1977. Changes in the forest environment in southern Poland as a result of steel mill emissions. Vegetation Science and Environmental Protection. Tokyo: Maruzen Co. Ltd.

Grodzinska, K. 1980. Mchy i kora drzew jako czule wskazniki skazenia srodowiska gazami i pylami przemyslowymi. Referat przedstawiony na sesji naukowej Ocena przydatnosci i perspektywy ujednolicenia testow i reakcji stosowanych w szacowaniu skazenia srodowiska, PAN, Wroclaw, November 14-15, 1980.

Grodzinska, K., and B. Godzik. 1985. Skazenie metalami ciezkimi gleb i warzyw w ogrodach dzialkowych krakowskiej aglomeracji miejskiej. Pp. 41-44 in Materialy z III Krajowej Konferencji "Wplyw zanieczyszczen pierwiastkami sladowymi na przyrodnicze warunki rolnictwa," Pulawy May 28-30, 1985, cz. II.

Guderian, R. 1970. Quantitative relations between the sulfur content of plants and the sulfur dioxide content of the air. Z. Pflanzenkr. Pflanzenschutz 77:200-220 (abstract).

Huttunen, S., K. Laine, and H. Tormalehto. 1979. The stress effects of air pollution on conifers. Symposium on the Effects of Air-Borne Pollution on Vegetation. Komisja Ekonomiczna dla Europy, ONZ, Warszawa, pp. 216-218.

Huttunen, S., L. Karenlampi, K. Laine, S. Soikkeli, T. Pakonen, and M. Karhu. 1980. Dispersion and effects of air pollutants in forest environments. Symposium on the Effects of Air Pollution on Terrestrial Ecosystems. Oulu, Finland.

Kabata-Pendias, A., and H. Pendias. 1979. Pierwiastki sladowe w srorowisku naturalnym. Wydwanictwa Geologiczne, Warszawa.

Kabata-Pendias, A., and M. Piotrowska. 1984. Zanieczyszczenie gleb i roslin uprawnych pierwiastkami sladowymi, CBR - opracowanie problemowe, Warszawa, p. 28.

Karakanis, M. 1976. Obieg siarki w ekosystemie lesnym Tilio-Carpinetum (gradu niskiego) w polnocnej czesci Puszczy Niepolomickiej kolo Ispiny. Fragmenta Floristica et Geobotanica 22:351-363.

Karweta, S., A. Kawalec, and Z. Harabin. 1987. Badania porownawcze szpilek *Pinus silvestris* L. w rejonie elektrowni Rybnik przed i po jej uruchomieniu. Pp. 361-366 in Materialy II Krajowego Sympozjum "Reakcje Biologiczne Drzew na Zanieczyszczenia Przemyslowe," R. Siwecki, ed. Kornik May 16-19, 1984. Wydawnictwo Naukowe Uniwersytetu im. Adama Mickiewicza, Poznan.

Koeppe, D.E. 1981. Lead: Understanding the minimal toxicity of lead in plants. Pp. 55-76 in Effects of Heavy Metal Pollution on Plants, Lepp, ed. Vol. 1. London and New Jersey: Applied Science Publisher.

Linzon, S.N., P.J. Temple, and R.G. Pearson. 1979. Sulfur concentrations in plant foliage and related effects. J. Air Pollut. Contr. Assoc., 29:520-525.

Materna, J. 1978. Pouziti listove analysy k prukazu vlivu kyslicniku siriciteho. Prace VULHM 52:101-114.

Molski, B., and W. Dmuchowski. 1985. Evaluation of air pollution pressure of forest and agricultural areas (mapping on macro-regions scale) on the basis of sulfur and fluorine accumulation by the needles of Scots pine. Proceedings of IUFRO S 2.09. Air Pollution Symposium, XIII International Meeting of Specialists in Air Pollution Damages in Forests. Most (CSRS), August 27-September 1, 1984, pp. 110-123.

Molski, B.A., and W. Dmuchowski. 1986. Effects of acidification on forests and natural vegetation. Wild animals and insects, acidification, and its policy implications. Pp. 29-51 in the proceedings of an international conference, T. Schneider, ed. Amsterdam, May 5-9, 1986. UN-ECE. Amsterdam: Elsevier.

Molski, B., K. Glebicki, and W. Dmuchowski. 1987. Data management computer system of air pollution impact on forest used in the Botanical Garden of the Polish Academy of Sciences and its relation to existing systems in Poland. Pp. 45-52 in the Proceedings of the Workshop on Forest Decline and Reproduction: Regional and Global Consequences, L. Kairiukstis, S. Nilsson, and A. Straszak, eds. Krakow, Poland, March 23-28, 1987. International Institute for Applied Systems Analysis (IIASA). Luxenburg, Austria.

Raven, P.H., and G.B. Johnson. 1986. Biology. St. Louis, Toronto, and Santa Clara: Times Mirror/Mosby College Publishing. pp. 1198.

Stefan, K. 1968. Uber den naturlichen Schwefelgeheld von Fichten nadeln und seine Bedeutung fur die Rauchschadendiagnose. Materialy VI Miedzynarodowej Konferencji "Wplyw zanieczyszczen powietrza na lasy," Katowice, pp. 297-312.

Themlitz, R. 1960. Die inviduelle Schwankung des Schwefelgehaltes gesunder und rauchgeschadigter Kiefern und seine Beziehung zum gehalt an den uberingen Hauptnakstoffen. Allg. Forst. und Jagdzt. 131:261-264 (abstract).

Long-term Ecological Monitoring in the National Parks of Poland

KRYSTYNA GRODZINSKA
Institute of Botany
Polish Academy of Sciences

Pollution of the environment is usually determined by means of physico-chemical methods, i.e., by recording concentrations of toxic elements and their compounds in the air, water, and soil. It can also be determined by means of biological methods with the help of bioindicators (Lepp, 1981a and b; Manning and Feder, 1980; Martin and Coughtrey, 1982). Biological methods are relatively simple, quick, and inexpensive. They also have a great advantage in that organisms themselves record toxic effect of the pollution. The intent of this chapter is to discuss the use of bioindicators as a tool in the long-term monitoring of ecosystems for pollution.

The level of environmental pollution can be estimated according to changes in the geographical distribution of various groups of organisms, as well as their morphological, cytological, physiological, biochemical, and chemical changes.

Three types of bioindicators are usually distinguished:

- scales of indicator species, noting the presence or absence of each species;
- true indicators, e.g., individual species that exhibit damage proportional to dose; and
- accumulators or collectors of potentially toxic materials with or without internal damage (Grodzinski and Yorks, 1981).

In assessments of environmental contamination by pollutants produced by industry and motor vehicles, bioindicators of the accumulator type are most frequently used (Martin and Coughtrey, 1982). Various groups of organisms, both plant and animal, can be accumulators (Grodzinska, 1982). Mosses are particularly effective accumulators of heavy metal (Folkeson,

1979; Grodzinska, 1978; Groet, 1976; Hvatum et al., 1983; Ruhling et al., 1987; Steiness, 1980; and Tyler, 1971 and 1972). Mosses also have several advantages as indicator organisms:

- Many species have a vast geographical distribution, and they grow abundantly in various natural habitats, even in industrial and urban agglomerations.
- Mosses have no epidermis or cuticle; therefore, their cell walls are easily penetrable for metal ions.
- Mosses have no organs for uptake of minerals from the substrate; they obtain minerals mainly from precipitation and dry deposition.
- Some species have layer structure, and annually produced organic matter forms distinct segments.
- Transport of minerals between segments is very poor because of the lack of vascular tissues.
- Mosses accumulate metals in a passive way, acting as ion exchangers.
- Mosses show the concentrations of metals as a function of the amount of atmospheric deposition.

For these reasons, mosses are used to determine the current degree of contamination of the environment, in both very extensive and quite small areas. They are also used for monitoring levels of contamination over a certain period of time—years, decades, or even centuries (Grodzinska, 1982).

The use of mosses for long-term monitoring of environmental contamination by heavy metals is widespread in the Scandinavian countries (Gydeson et al., 1983; Ruhling and Tyler, 1969, 1971, 1973; Ruhling et al., 1987) and also in Poland (Grodzinska, 1978; Grodzinska et al., in prep.). In this chapter, long-term monitoring of levels of contamination in Poland's national parks during the decade 1976-1986 is discussed.

MATERIALS AND METHODS

Poland has 14 national parks and several hundred nature reserves which occupy less then 1% of the area of the country (Figure 1). They cover fairly small areas, from 1,600 to 22,000 hectares, and are located from the Baltic coast through the central lowlands and uplands to the mountains (Table 1). Virtually all of these areas are under stress from both air and water pollution and tourism.

The mosses *Pleurozuim schreberi* Mitt. and *Hylocomium splendens* (Hedw.) Br.eur. were chosen as test species. The reasons for this choice were:

TABLE 1 General characterization of Polish national parks.

No.	Name of park	Altitude above sea level (m)	Annual sum of precipitation (mm)	Area (ha)
1	Slowinski	0-56	619	18,247
2	Wolinski	0-115	581	4,844
3	Wielkopolski	55-132	501	5,198
4	Kampinos	60-106	500	35,482
5	Bialowieza	147-170	585	5,317
6	Roztocze	300-390	710	6,843
7	Swietokrzyski	260-611	560, 660	5,897
8	Ojcow	305-478	791	1,592
9	Karkonosze	410-1,605	1,158	5,562
10	Babia Gora	800-1,725	960-1,400	1,741
11	Tatry	800-2,499	1,112-1,810	21,164
12	Gorce	600-1,310	900-1,220	5,945
13	Pieniny	420-982	805	2,328
14	Bieszczady	630-1,346	1,035	5,587

- both *Pleurozium* and *Hylocomium* are common species and occur in abundance in forest ecosystems through Poland;
- *Hylocomium* has bi- to tripinnate stems, with a very distinct separation between each year's shoot, which makes it easy to estimate the age of the sampled segments; *Pleurozium* has unipinnate stems without any clear separation between the annual shoots. The green parts of this species represent usually two- to five-year increments, whereas the brown parts are older;
- these species are more effective accumulators than others;
- these species have been repeatedly used for estimates of environmental pollution in many countries (e.g., Rinne and Barclay-Estrup, 1980; Ruhling and Tyler, 1969, 1984; Ruhling et al., 1987).

Pleurozium schreberi and *Hylocomium splendens* were collected from the same plots in 14 national parks in Poland twice in the autumns of 1976 and 1986. In each park, the material was collected at several points in both central and border areas. Unwashed mosses were separated into young, green parts and old, brown parts, dried at 85°C, and wet digested in a 4:1 mixture of concentrated nitric and perchloric acid. Seven heavy metals (i.e., cadmium [Cd], chromium [Cr], nickel [Ni], lead [Pb], copper [Cu], zinc [Zn], and iron [Fe]) were identified spectro-photometrically using the Perkin-Elmer and Varian Techtron Atomic Absorbtion Spectrophotometer (AAS).

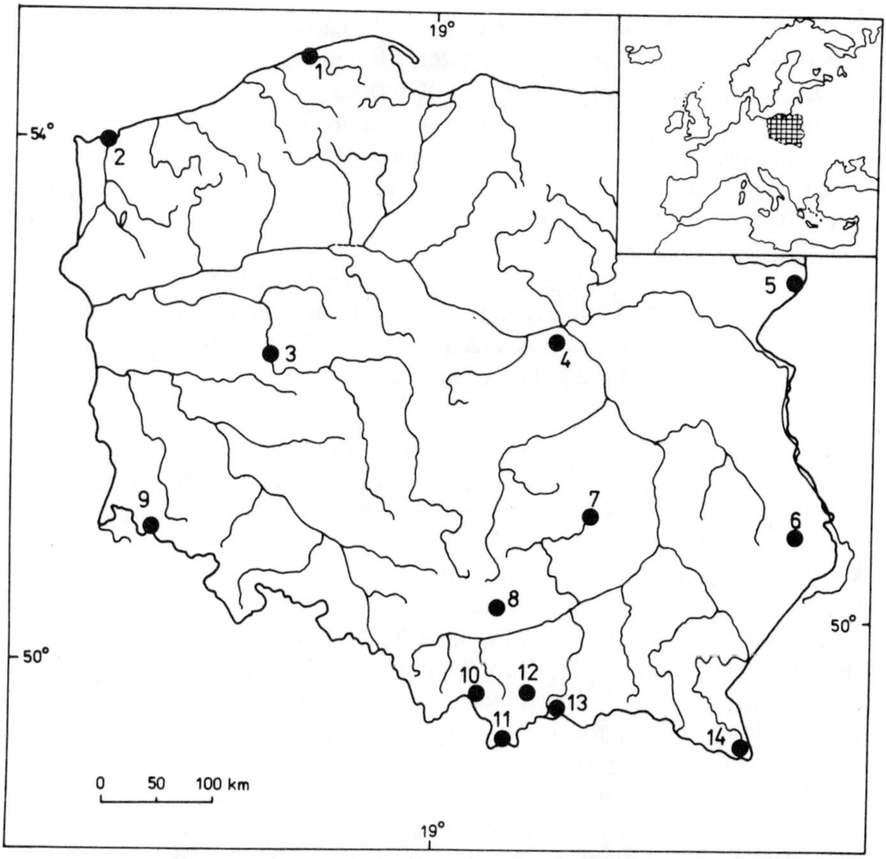

FIGURE 1 Location of Polish national parks. 1-Slowinski, 2-Wolinski, 3-Wielkopolski, 4-Kampinos, 5-Bialowieza, 6-Roztocze, 7-Swietokrzyski, 8-Ojcow, 9-Karkonosze, 10-Babia Gora, 11-Tatry, 12-Gorce, 13-Pieniny, 14-Bieszczady.

RESULTS

Significant differences were found in the levels of heavy metals in mosses representing particular parks (Figure 2). The greatest difference among parks was found in the cases of lead, cadmium, and zinc (about 4-5 times), with smaller differences in the case of chromium and nickel (2-3 times). Most of the heavy metals occurred in lowest concentrations in the mosses of the national parks in northern and eastern Poland; their highest concentrations were found in parks in the southern part of the country. In 1976, in the first group of parks, the concentration of cadmium was approximately 1 $\mu g\ g^{-1}$, whereas in the second one it reached 6 $\mu g\ g^{-1}$. The respective accumulations of chromium were approximately 4 and 8 μg

g^{-1}; nickel, approximately 3.5 and 7 $\mu g\ g^{-1}$; lead, approximately 60 and up to 270 $\mu g\ g^{-1}$; and zinc, approximately 70 and up to 300 $\mu g\ g^{-1}$ (Figure 2). In 1986, the cadmium content of mosses in the least contaminated national parks was about 0.5, and up to 2 $\mu g\ g^{-1}$ in the mosses of the most contaminated parks. Perspective accumulations of chromium were approximately 3 and 8 $\mu g\ g^{-1}$; nickel, approximately 3 and 5 $\mu g\ g^{-1}$; lead, approximately 40 and up to 100 $\mu g\ g^{-1}$; and zinc, approximately 50 and 100 $\mu g\ g^{-1}$ (Figure 2).

In order to assess the general situation regarding the contamination of Polish national parks by heavy metals, a synthetic pollution index was used. It represents the mean, standardized contents of six elements (Cd, Cr, Ni, Cu, Pb, and Zn) in mosses. For example:

$$\text{Standardized value of Cd} = \frac{Cd\ conc.\ in\ Ojcow\ National\ Parks}{\text{mean Cd conc. in 14 National Parks}}$$
$$\text{index} = \text{(Ojcow NP)} = \text{sum of standardized values of 6 heavy metals}$$

According to this index, parks were classified as being relatively clean, moderately contaminated, or heavily contaminated. The maritime parks and those in the eastern part of Poland were classified in the first category. The moderately polluted group is represented by parks in the central, lowland part of the country as well as some mountain parks. Parks located in the uplands and in the higher mountain ranges in southern Poland are the most polluted (Figure 3).

A change was found in the contamination of national parks by heavy metals over the last ten years. The present concentration in mosses of toxic metals such as cadmium, lead, and zinc has fallen twofold, while that of nickel, chromium, and iron is 20-26% lower than in 1976 (Table 2). Except for iron and nickel, these differences are significant, and highly significant statistically (Table 3).

However, the changes in the contamination of mosses by heavy metals are different in particular parks. In those parks which were very heavily contaminated 10 years ago (e.g., Ojcow, Karkonosze, and Swietokrzyski), the mosses at present contain smaller quantities of heavy metals. In the three parks that were slightly and moderately contaminated in 1976 (e.g., Bialowieza, Bieszczady, and Kampinos), the level of concentration of heavy metals is similar to current levels, while in three other parks (Tatry, Pieniny, and Wielkopolski), it has increased considerably (Figure 4).

The differences in the concentration of heavy metals across Poland were much more pronounced in 1976 than in 1986. This is supported by the results of statistical analysis (Table 3). In 1976, the differences were highly significant statistically for all the heavy metals, while they were highly significant only for chromium, copper, and lead in 1986 (Table 3).

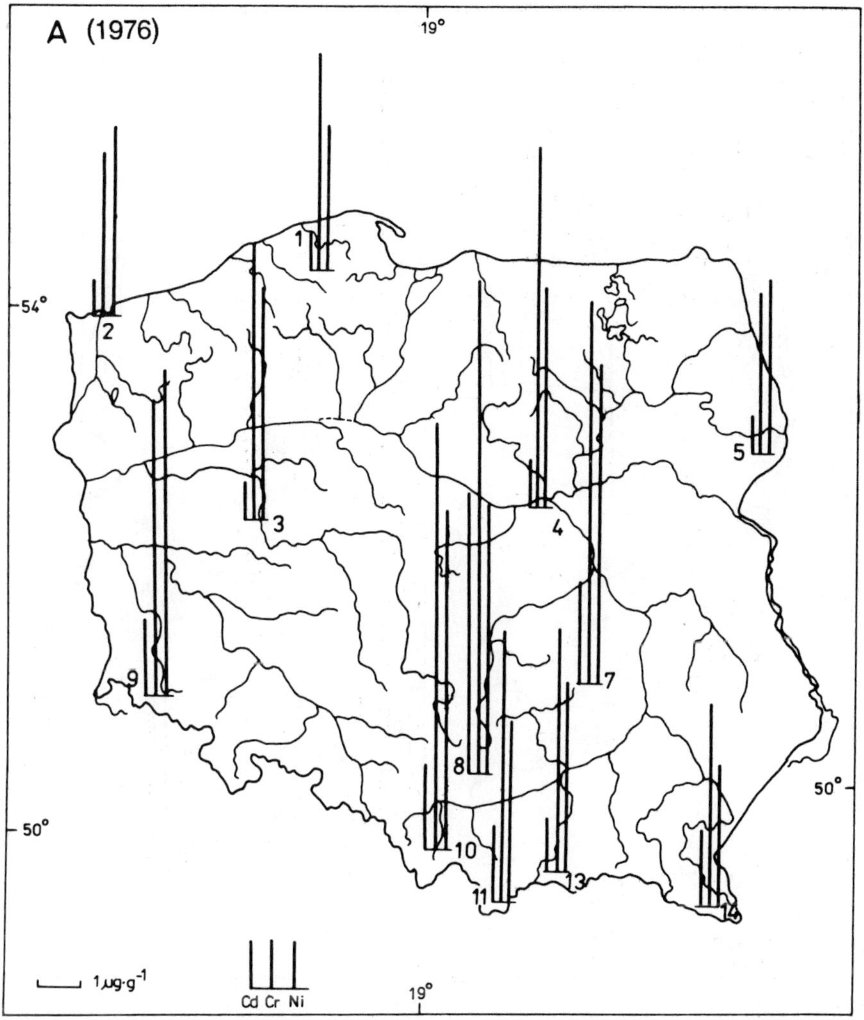

FIGURE 2a Cd, Cr, Ni, Pb, Zn, (μg g^{-1} dw) concentrations in *Pleurozium schreberi* in the Polish national parks. The mean contents of metals in the green and brown parts of the mosses are indicated for both the central and the peripheral areas of the parks. A-1976, B-1986. Parks no. 1-14, as identified in Figure 1.

The differences in the metals content between mosses in the central and border parts of the national parks were also determined. These were highly significant in 1976 for all the heavy metals analyzed, except cadmium. The metals also occurred in higher concentrations in mosses growing in the outer parts of parks. This situation changed in 1986. The mosses

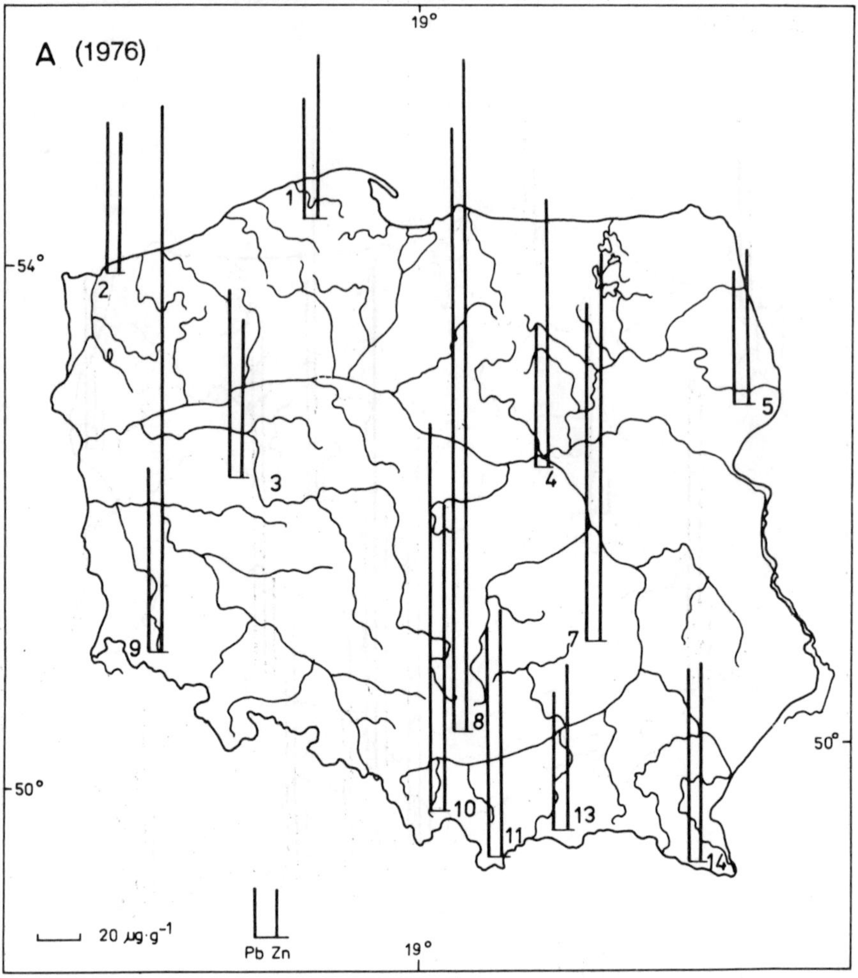

FIGURE 2b

accumulated metals, except nickel, in similar quantities, both in the center and at the edges of the parks (Table 3).

In addition, highly significant statistical differences were found in concentrations of heavy metals between the younger, green and older, brown parts of mosses, both in 1976 and at present (Table 3). These differences were much more pronounced in 1986.

Finally, it was shown that *Pleurozium schreberi* and *Hylocomium splendens* accumulated cadmium, nickel, and chromium in similar quantities in 1976, but differed statistically in the level of accumulation of copper, lead,

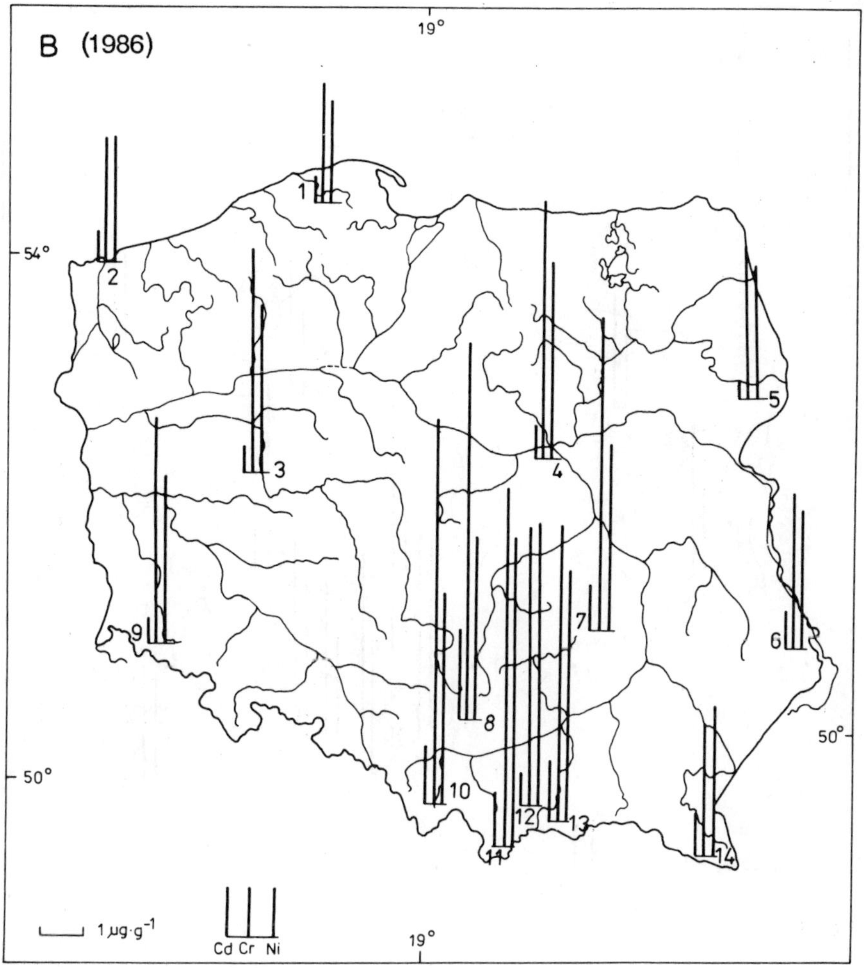

FIGURE 2c

zinc, and iron (Table 3). In 1986, four elements (Cd, Cu, Zn, and Fe) occurred in similar concentration in both mosses species, and three (Ni, Cr, and Pb) differed from them in concentration (Table 3).

DISCUSSION

The contamination of Polish national parks by heavy metals, determined by analyzing mosses, correlates well with the distribution of sources of industrial emissions in Poland (Kassenberg and Marek, 1986; Kassenberg and Rolewicz, 1985). The parks in the southern parts of Poland

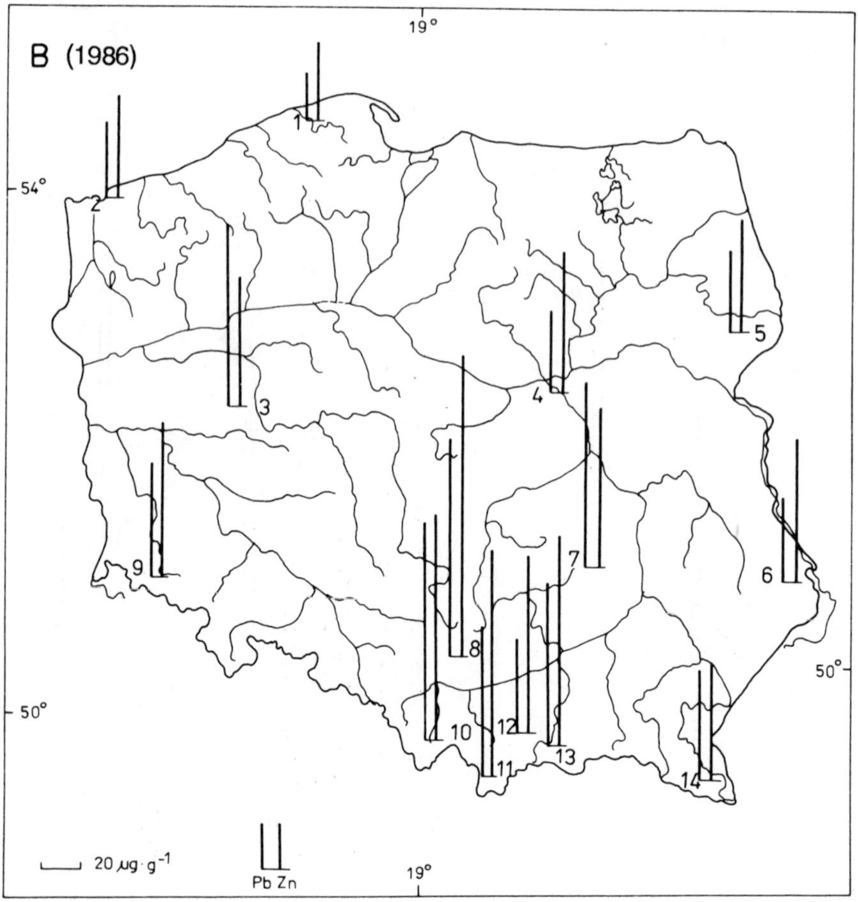

FIGURE 2d

are most polluted (e.g., Ojcow, Babia Gora, Tatry, Pieniny, Karkonosze, and Swietokrzyski). These parks lie within the range of emissions from the Silesian-Krakow, Legnica-Glogow, and Central Industrial regions. The moderately contaminated parks (i.e., Wielkopolski and Kampinos) are located at a considerable distance from the great industrial centers, but close to large cites such as Warsaw and Poznan. The cleanest parks (i.e., Slowinski, Wolinski, Bialowieza, Roztocze, and Bieszczady) lie in the least industrialized parts of Poland.

The results show that the contamination of mosses with heavy metals has decreased noticeably during the last decade in the Ojcow, Swietokrzyski, and Karkonosze parks, but only slightly in three others, i.e., Kampinos, Bieszczady, and Bialowieza; contamination has increased in the mountain

TABLE 2 Heavy metal concentration (ug g^1) in mosses in 14 Polish national parks (mean concentration over all parks).

	Pleurozium		*Hylocomium*	
Element	1976	1986	1976	1986
Cd	1.80	0.94	2.27	0.92
Cr	6.58	5.51	7.19	4.63
Ni	5.10	4.21	4.62	3.25
Pb	108	59.0	92.3	45.4
Zn	132	73.5	122	65.1
Fe	1,870	1,598	1,768	1,352

TABLE 3 Statistical analysis of the difference between heavy metals concentrations in *Pleurozium* (P) and *Hylocomium* (H).

F-VALUES AND LEVEL OF SIGNIFICANCE

	Species (P & H)		Parts of mosses		Central & border parts of parks		Parks		Years
	1976	1986	1976	1986	1976	1986	1976	1986	76/86
Cd	131.68^b	4.40	3.63 P	$8.56\ P^a$	3.30	0.01	144.8^b	3.72^a	5.73^a
			0.08 H	$9.58\ H^a$					
Ni	1.26	6.25^a	$48.93\ P^b$	$43.81\ P^c$	18.17^b	6.87^a	33.51^b	5.33^b	2.93
			$19.99\ H^b$	$22.16\ H^b$					
Cr	0.81	7.36^a	$35.56\ P^b$	$37.21\ P^c$	31.91^b	1.88	27.72^b	2.15	5.59^a
			$10.96\ H^b$	$22.94\ H^b$					
Cu	166.66^b	1.35	$17.87\ P^b$	0.75 P	49.70^b	2.58	61.55^b	2.51	5.14^a
			$13.91\ H^b$	4.48 H					
Pb	9.89^a	8.58^a	$14.41\ P^b$	$41.90\ P^c$	101.98^b	0.18	200.77^b	1.88	9.64^b
			$45.04\ H^b$	$54.23\ H^c$					
Zn	8.67^a	1.29	3.98 P	$43.23\ P^c$	202.49^b	0.45	384.96^b	3.95^a	9.37^a
			$6.48\ H^a$	$14.49\ H^b$					
Fe	7.37^a	1.04	$38.93\ P^b$	$56.15\ P^c$	46.00^b	2.89	62.00^b	4.48^a	1.95
			$41.15\ H^b$	$8.20\ H^a$					

a = $P < 0.05$ *b* = $P < 0.01$ *c* = $P < 0.001$

parks at Tatry, Pieniny, and Babia Gora. The deposition of metallurgical dusts in 1976 was 173,000 tons (0.6 t/km^{-2}) over the entire area of Poland, and under 140,000 tons per year (0.4 t/km^{-2}) in the period from 1980-1984, but rose again in 1985 and 1986 to nearly 170,000 tons per year (GUS Statistical Yearbook, 1976-1986). The decrease in the deposition of metallurgical dust was most pronounced in the large industrial regions (i.e., Silesia-Krakow, Legnica-Glogow, and the Central Region) and adjacent areas. In central and southern Poland, however, this change was quite small (GUS Statistical Yearbook, 1976-1986).

Changes in the amounts of heavy metals, as recorded by the analysis of

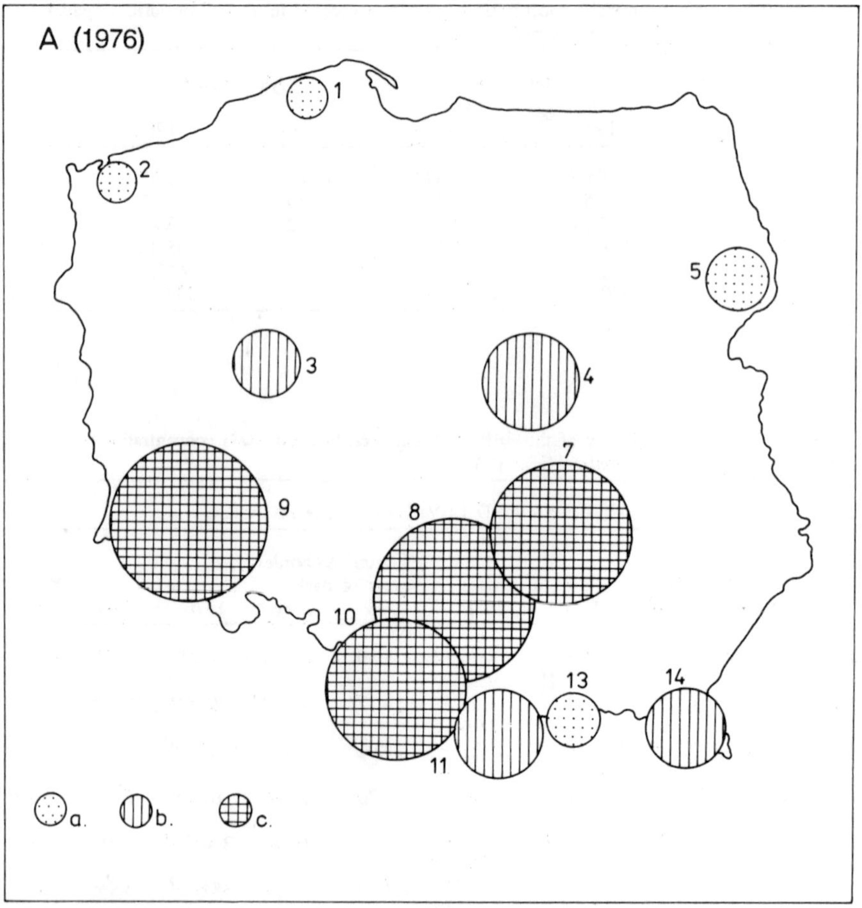

FIGURE 3a Pollution index values for the Polish national parks defined as a sum of standardized contents of heavy metals in mosses. a-relatively clean parks, b-moderately polluted parks, c-heavily polluted parks. A-1976, B-1986. Parks no. 1-14 as identified in Figure 1.

mosses in national parks, correlate well with changes in the total emission of dusts in the period from 1976-1986 in Poland. The mountain parks (Tatry, Babia Gora, and Pieniny) are the exception. These lie at Poland's southern borders, and consequently are exposed to additional heavy metals from Czechoslovakia.

Gydesen et al. (1983), Ruhling and Tyler (1984), and Ruhling et al., (1987) found an analogous decrease in the level of heavy metals in mosses in Scandinavia over the period 1968-1985. They cite data demonstrating a gradual decrease in emissions of metallic dusts in various countries of

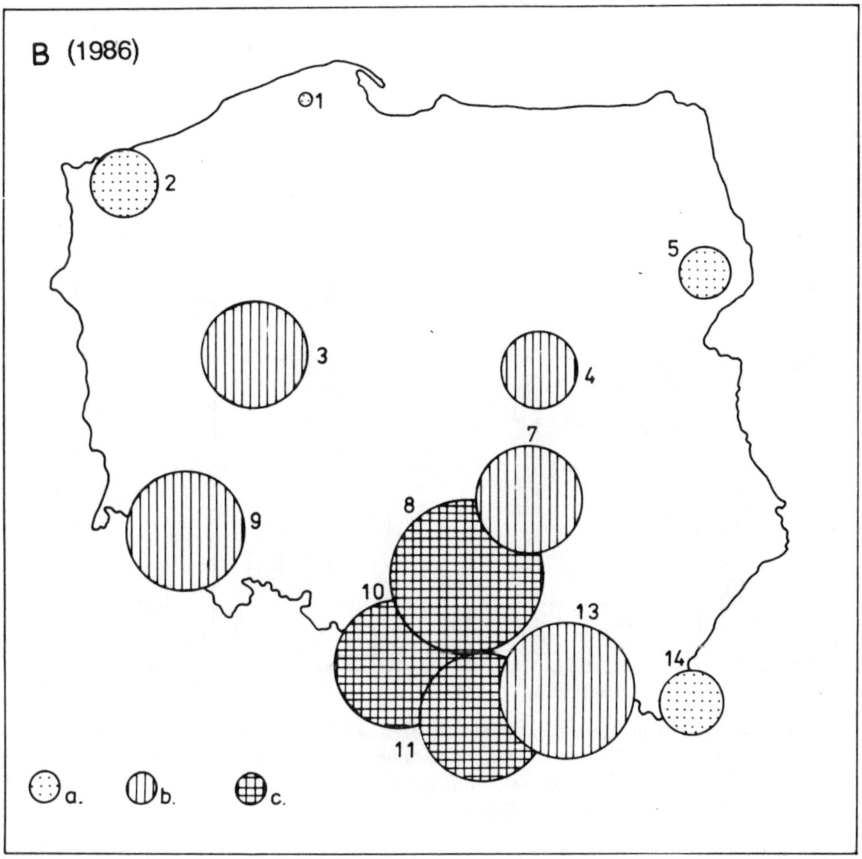

FIGURE 3b

western and central Europe. In Poland, there was also a decreasing trend in the emission of industrial dust during this period. This decrease was most pronounced in the early 1980s, due to a recession in heavy industry.

Mosses reflected this trend with precision because their biomass accumulates dust from the previous 3 or 4 years. It has been noted that the differences in concentration between the younger, green and older, brown parts of mosses were greater in 1986 than in 1976. This reflects the low deposition of metallic dusts in the early 1980s (presently the brown parts of mosses) and slightly greater deposition in the mid-1980s (presently the green parts of mosses). While in some mountain parks (e.g., Karkonosze) forest decline has been observed over the last decade, it should be attributed to other, gaseous, pollutants such as SO_2 and NO.

The differences in the contamination of mosses by heavy metals among

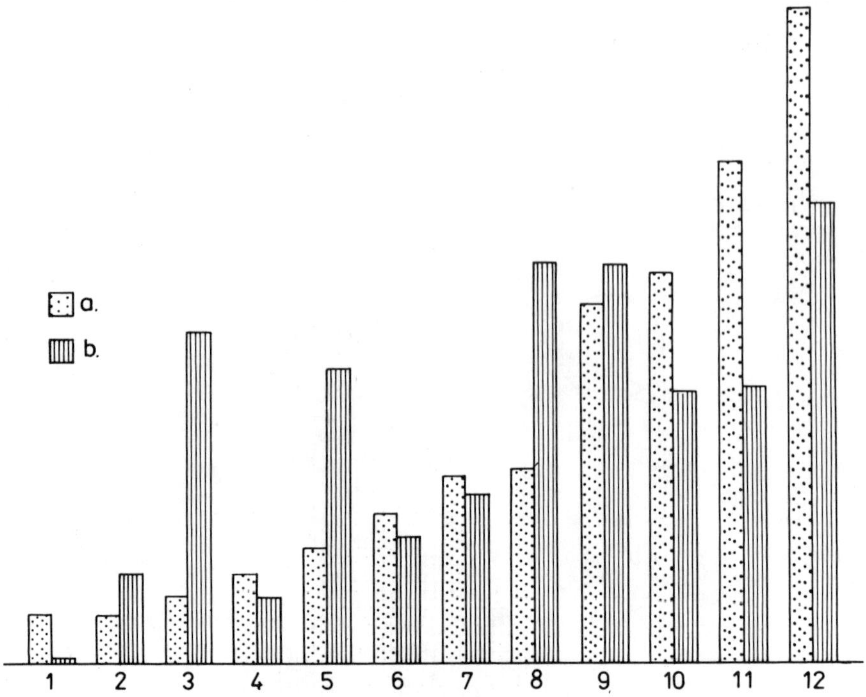

FIGURE 4 Changes in moss contamination by heavy metals in Polish national parks. a-pollution index value in 1976, b-pollution index value in 1986. Parks: 1-Slowinski, 2-Wolinski, 3-Pieniny, 4-Bialowieza, 5-Wielkopolski, 6-Bieszczady, 7-Kampinos, 8-Tatry, 9-Babia Gora, 10-Swietokrzyski, 11-Karkonosze, 12-Ojcow.

particular parks were very clear in 1976. At present they are less distinct. This can be explained by the creation of other, local sources of emission of metallic dusts in central and northeastern Poland. Current decreases in the concentration of heavy metals in mosses between the central and outer parts of parks demonstrates that entire park areas are now subjected to the pressure of emissions; hence, the functioning of the ecosystems of these parks is currently at greater risk.

The contamination of mosses by heavy metals in Poland is especially great when compared with Scandinavia (Ruhling and Tyler, 1973, 1984; Ruhling et al., 1987). The most endangered Polish parks have two to four times as much cadmium, nickel, lead, iron, zinc, and chromium than do the most heavily polluted parts of Sweden and Norway. As compared with the clean parts of northern Scandinavia, the heavily contaminated Polish national parks had three to five times as much nickel and zinc, over 70 times as much cadmium, and over 10 times as much lead and chromium in 1976. In 1976, the mosses from the cleanest parks in Poland accumulated about twice as much iron, zinc, chromium, and nickel; six times as much lead; and over 10 times as much cadmium as the mosses in northern Scandinavia. In 1986, the seriously contaminated Polish parks were three to five times more polluted, and the cleanest parks about twice as polluted, by these heavy metals than northern parts of Scandinavia.

MANAGEMENT RECOMMENDATIONS

Management of Polish national parks must take two special circumstances into consideration: the relatively small park areas and the heavy contamination of the air and water. The various types of protected areas in Poland such as national parks, nature reserves, and landscape parks presently cover 3.9 million hectares, which represents 10% of the total area of the country (GUS Statistical Yearbook, 1986). Most are grouped in the southern part of the country, where the natural landscape is most varied. At the same time, southern Poland is the most industrialized and under the greatest stress from industrial emissions. The value of the Polish national parks is reflected by the fact that three of them (Babia Gora, Bialowieza, and Slowinski) were declared UNESCO biosphere reserves.

National parks in Poland and central Europe are small, many times smaller then those of North America, the USSR, and even Scandinavia. The idea behind the establishment of such small national parks in the very heart of industrialized Europe is to protect the diversity of ecosystems and species rather than regenerate to the original climax.

Therefore, management of Polish national parks must include practical measures to allay the affects of air pollution until such time as the emission of gases and dusts can be controlled. These parks can, however, serve well for studies of long-term changes in the ecosystem. The network of national parks in Poland and other European countries can be utilized for monitoring long-term changes caused by air pollution. These parks are small and exposed to local and global air pollution. This study supports the use of mosses (*Pleurozium schreberi* and *Hylocomium splendens*) as very sensitive bioindicators (bioaccumulators) for air pollution by heavy metals. They can be used easily for ecological monitoring on various scales and areas. However, the procedures for sampling moss and subsequent laboratory analysis must be unified (Grodzinska, 1984).

REFERENCES

Folkeson, L. 1979. Interspecies calibration of heavy metal concentrations in nine mosses and lichens: Applicability to deposition measurements. Water, Air and Soil Pollution 11:153-260.

Grodzinska, K. 1978. Mosses as bioindicators of heavy metal pollution in Polish national parks. Water, Air, and Soil Pollution 9:83-97.

_____. 1982. Monitoring of air pollutants by mosses and tree bark. Pp. 33-42 in the Proceedings of the International Workshop, "Monitoring of Air Pollutants by Plants: Methods and Problems," L. Steubing and H.J. Jager, eds. Osnabruck (FRG), September 24-25, 1981. The Hague, Boston, and London: Dr. W. Junk Publishers.

_____. 1984. Bioindicators of environmental deterioration. Pp. 27-34 in Forest Ecosystems in Industrial Regions, W. Grodzinski, J. Weiner, and P.F. Maycock, eds. Ecological Studies 49:1-277. Berlin/Heidelberg/New York/Tokyo: Springer Verl.

Grodzinska, K., B. Godzik, and G. Szarek, in preparation. Heavy metal accumulation in Polish national parks: Changes over ten years.

Grodzinski, W., and T.P. Yorks. 1981. Species and ecosystem level bioindicators of airborne pollution: An analysis of two major studies. Water, Air, and Soil Pollution 16:33-53.

Groet, S.S. 1976. Regional and local variations in heavy metal concentrations of bryophytes in the northeastern United States. Oikos 27:445-456.

GUS (Statistical Yearbook of the General Statistical Department). 1976-1986. Warszawa.

Gydesen, H., K. Pilegaard, L. Rasmussen, and A. Ruhling. 1983. Mosses analyses used as a mean of surveying the atmospheric heavy metal deposition in Sweden, Denmark, and Greenland in 1980. National Swedish Environment Protection Board Bulletin 1670:1-44.

Hvatum, O.O., B. Bolviken, and E. Steinnes. 1983. Heavy metals in Norwegian ombrotrophic bogs. Pp. 351-356 in Environmental Biogeochemistry, R. Hallberg, ed. Ecol. Bull. (Stockholm) 35.

Kassenberg, A., and M.J. Marek. 1986. Ekologiczne aspekty przestrzennego zagospodarowania kraju (Ecological aspects of the spatial development of the country). Warszawa: PWN (State Scientific Publishers), pp. 1-174.

Kassenberg, A., and C. Rolewicz. 1985. Przestrzenna diagnoza ochrony srodowiska w Polsce (Spatial diagnosis of environmental protection in Poland). Komitet Przestrzennego Zagospodarowania Kraju. PAN. Studia 89:1-125. Warszawa: PW (State Economic Publishers), Ekon (Committee on Spatial Development of the Country of the Polish Academy of Sciences).

Lepp, N.W. 1981a, ed. Effect of heavy metal pollution on plants. Vol. 1. Effect of trace metals on plant functions. London: Applied Science Publishers, Ltd.

_____. 1981b, ed. Effect of heavy metal pollution on plants. Vol. 2. Metals in the environment. London: Applied Science Publishers, Ltd.

Manning, W.J., and W.A. Feder. 1980. Biomonitoring air pollutants with plants. London: Applied Science Publishers, Ltd.

Martin, M.H., and P.J. Coughtrey. 1982. Biological monitoring of heavy metal pollutions. London: Applied Science Publishers, Ltd.

Rinne, R.J.K., and P. Barclay-Estrup. 1980. Heavy metals in a feather moss—*Pleurozium schreberi*—and in soils in Northwest Ontario, Canada. Oikos 34:43-54.

Ruhling, A., L. Rasmussen, K. Pilegaard, A. Makinen, and E. Steinnes. 1987. Survey of atmospheric heavy metal deposition in the Nordic countries in 1985 monitored by moss analyses. Nord 21:1-44.

Ruhling, A., and G. Tyler. 1969. Ecology of heavy metals—a regional and historical pollution study. Bot. Notiser 122:248-259.

_____. 1971. Regional differences in the deposition of heavy metals over Scandinavia. J. Appl. Ecol. 8:497-507.

_____. 1973. Heavy metal deposition in Scandinavia. Water, Air, and Soil Pollution 2:445-455.

_____. 1984. Recent changes in the deposition of heavy metals in Northern Europe. Water, Air, and Soil Pollution 22:173-180.

Steinnes, E. 1980. Atmospheric deposition of heavy metals in Norway studied by analysis of moss samples using neutron activations analysis and atomic absorption spectrometry. J. Radioanal. Chem. 58:387-391.

Tyler, G. 1971. Moss analysis: A method for surveying heavy metal deposition. Pp. 129-132 in the Proceedings of the 2nd International Clean Air Congress, Washington, D.C.

_____. 1972. Heavy metals pollute nature, may reduce productivity. Ambio 1,2:29-52.

Agricultural Impacts

Ecological Guidelines for Management of Rural Areas in Poland

LECH RYSZKOWSKI
Institute of Agrobiology and Forestry
Polish Academy of Sciences

Increasing environmental degradation in various areas of Poland reflects neglect or disregard of environmental structures and processes in the pursuit of the economic development of the country. Rural areas comprise 60% of the total land area of Poland, and they are not excluded from widespread environmental degradation. From these failures one may conclude that some changes are needed in our attitudes about agricultural environments. We must stop treating each individual farm, industry, or landscape as though it were isolated from other aspects of the environment, economy, or social system. A larger, systemic approach is needed in decision-making regarding rural areas so that the long-term, functional effectiveness of the biological, economic, and social aspects of the system can be achieved and sustained.

Direct productive or social outcomes and the constraints involved in natural and managed environments should be recognized in formulating programs for agriculture, agricultural industries, and services for the agricultural sector as well as for recreation and other functions in rural areas. Realization of these goals requires:

- understanding the relationship between structures and levels of agricultural production with respect to conditions of the natural environment;
- changing the methods of planning for agricultural development to consider the costs of environmental degradation from traditional forms of economic development;
- stimulating the introduction of new technologies which are supportive of longer-term functioning of agroecosystems; and

• changing the rules of spatial planning to include greater concern for the ecological balances of natural and cultural environments.

ENVIRONMENTAL HAZARDS IN RURAL AREAS OF POLAND

The area in Poland used as arable land, grasslands, and orchards has decreased continuously since World War II. Agricultural area per capita decreased from 0.85 hectares in 1946 to 0.50 hectares in 1987 due to industrialization, urbanization, and growth of the country's population. At the same time, improvements in agriculture have led to increased yields. According to data from the Main Statistical Office (1986), yields per hectare of major crops increased as follows between the years 1946-1950 and 1981-1983: four cereals by 237%, potatoes by 132%, sugar beets by 171%, and rapeseed by 230%. Although agricultural productivity has increased, still larger yields are possible. For example, the average yield of the four cereals could be increased from the present level of three tons per hectare to about four tons per hectare; this is the average yield in East Germany and Czechoslovakia where soils and climatic conditions are similar to those in Poland.

Environmental hazards for rural areas give rise to several problems interfering with sustainable development of farming. The severity of the situation is reflected by the March 4, 1983, resolution of the Polish Cabinet in which 27 "areas of ecological hazard" were identified. Pollution and degradation of the environment threaten both the public health and the future economies of these areas.

Areas of ecological hazard cover 11.2 % of Poland's total area, according to the Main Statistical Office (1984). In 1983, these areas were inhabited by 35.5% of the nation's population. Agriculturally, 9.7% of these areas are devoted to cereals, 9.6% to potatoes, 9.5% to beets, 9.5% to fodder crops, and 11.8% to orchards. These areas also produce 82% of the total emissions of dust and gas pollutants in Poland, which amount to 5,464,000 tons per year. This figure includes 1,065,000 tons per year of fly ash which carries various heavy metals; these areas were responsible for 77% of the total fly ash emission in Poland. Emission of sulfur dioxide in these areas is also very high (1,987,000 tons per year) which is 81% of the SO_2 emission for the whole country. Obviously, these data indicate that the safety of food produced in these areas is much lower because of the heavy deposition of different pollutants.

Acid Deposition

Lack of effective means for limiting industrial emissions of sulfur and nitrogen oxides leads to progressive acidification of soils. The National

TABLE 1 Mg and Ca leaching in watersheds with different amounts of sulfur dioxide air pollution.

Watershed	Area (km)	Arable Land (percent)	Concentration of SO_2 in air (ug m^{-3})	Rate of leaching (kg ha^{-1} y^{-1}) S-SO_4	Mg	Ca
1	425	66	50	4.8	3.8	19.8
2	694	71	20	3.3	2.1	13.2

SOURCE: Unpublished information obtained from B. Karlik, Institute of Agrobiology and Forestry, Poznan.

Program for Natural Environmental Protection through 2010 estimates that by the year 2010, the SO_2 concentration in air will not be lowered below the officially accepted air quality standard value of 32 $\mu g \cdot m^{-3}$ (Ministry of Environment and Natural Resources Conservation, 1988). Because of acid rain, about 50% of Polish soils are expected to have a pH value of about 4 (Stigliani et al., 1988).

The gradually changing chemical composition of soil, including leaching of calcium, magnesium (Table 1), and other elements like phosphorus, potassium, and aluminium, will likely eliminate some plant communities and related microorganisms and soil animals (Cook, 1983). A decline of soil bacterial biomass is expected with considerable increase of fungi. The amount of biologically inactive or even dead fungal hyphae is also expected to increase with rising acidity. The biomass and variability of earthworms and enchytraeids is expected to decline which, together with changes among microorganisms, will lead to slower decomposition of plant residues and regeneration of humus. Many invertebrate species living in soil water may be threatened with extinction. On the other hand, a quantitative rise in some groups of insects should be expected, e.g., springtails (Coleman, 1981).

Although the role of animals and microorganisms in formation of humus is still not fully known, changes from acid deposition are likely to unfavorably affect the soil humus by decreasing the rate of decomposition of plant residues. This will be most dangerous in light, arable soils which are widespread in Poland and in which we have observed increased leaching of organic matter with more intensive farming practices (Zyczynska-Baloniak, 1980). Intense farming contributes to decreased amounts of humus because of increased mineralization of organic matter, intensified erosion, or the leaching of water soluble organic compounds.

Increased soil acidity will result in changing patterns of nutrient cycling. For example, soil acidity will gradually limit the activity of bacteria responsible for nitrification or even denitrification processes, which will considerably lower the rate of nitrogen cycling (Cook, 1983). Soil acidity will also decrease symbiotic nitrogen fixation by legumes (Aleksander, 1980) and will change conditions important to the formation and the functioning of mycorrhizae (Domanski et al., 1987).

Air Pollution

Agricultural production in Poland is limited by air pollution. Even though evaluation of losses in plant production have not been conducted for the whole country, air pollution limits crop yields in areas of ecological threat (Wartesiewicz, 1978). Also, the concentration of heavy metals in soils has increased, which leads to concerns about food contamination. The report of the Committee for Food Technology and Chemistry of the Polish Academy of Sciences indicated that the amount of mercury and cadmium in average diets is close to the health hazard limit (Barylko-Piekielna et al., 1985).

Groundwater Degradation

Increased applications of nitrogen fertilizers have increased crop yields, but also have led to increased leaching of nitrogen into underground and surface waters. According to data from the Main Statistical Office (1987), the seriousness of the situation is illustrated by the fact that as much as 63% of household wells, 45% of factory wells, and 52% of public wells were classified as unhealthy by the National Sanitary Inspection. Contamination of these rural wells was caused by nitrogen compounds leached from fields and contamination by farm sewage.

Water Shortages

Water deficiencies are a significant threat to all living organisms. Management of water resources in Poland is made difficult by both unfavorable climatic conditions and past management errors. Natural water regimes in Poland, particularly in the central areas, are the least favorable among all European countries. Annual precipitation ranges from about 500 mm in central Poland to over 1,000 mm in the mountains. Water shortages are common in the Polish lowlands, where the amount of transpiration exceeds total rainfall during the vegetative period and results in the lowering of the groundwater table (Niewiadomski, 1979; Maciak and Zawadzki, 1981).

Drainage projects have also lead to excessive drying of considerable areas of land and have lowered groundwater tables. Increasing needs for water by industry, agriculture, and municipalities have lead to further increases in water deficiencies. The area of excessively dried land in Poland amounts to about 4 million hectares, including a dramatic deficit area of 0.6 million hectares (Partyka et al., 1979).

Cropland has been augmented over time by conversion of grasslands, pastures, and even marshes which have been drained for agricultural use. However, reclamation projects based exclusively on technological criteria, and which neglect ecological considerations, have decreased water storage capacities in Poland. The channelization of rivers and construction of levees have lead to increased water flow rates and deepening river beds, resulting in a lower groundwater table in adjacent areas (Palys, 1985). The elimination of small ponds in fields and the drying of marshes also reduces water holding capacities and increases wasteful surface runoff.

Because of a poorly developed network of water reservoirs (only 6% of the mean annual outflow can be stored in Poland), considerable quantities of surface runoff waters are typically lost to the sea. Thus, water management errors often actually result in reductions of water from watersheds.

The role of non-systematic drainage—drainpipes open to small ponds which are unconnected with river system—is neglected, particularly in areas with steep slope. Water retention is greater on the areas which are drained non-systematically. The processes of self-purification of water are also more effective in these areas, especially when whole watersheds are considered (Kosturkiewicz, 1988). Increased drying of the central region of Poland (i.e., Konin, Wloclawek, Plock, Ciechanow, Lodz, Skierniewice, and Lublin districts) is also associated with high degrees of deforestation in those areas where forest cover is below 15%.

Water shortages also limit economic activity and the conservation of some organisms in Poland. This situation is aggravated due to contamination of surface waters by industry and municipal sewage discharges as well as by agricultural sources, e.g., fertilizers and pesticides leached from fields, farm sewage, and discharges from small food processing plants located in agricultural areas. The result is that pollution of surface waters is increasing. About 40% of all rivers in Poland carry water which is below established standards for use, even for industry. The inflow of pollutants into lakes leads to severe degradation. About 60% of Polish lakes are strongly polluted. Problems of solid waste disposal in rural areas is also leading to the filling-in of small ponds. Unfortunately, it is common practice to dump solid wastes into small ponds located in rural areas, turning them into scrap heaps over time. According to research carried out by the Institute of Agrobiology and Forestry, about 30% of the small ponds in the studied area (400 km^2) disappeared within the last 30 years (Karg and Ryszkowski, 1984).

Soil Erosion

Another environmental hazard associated with agriculture involves water and wind erosion of soil. More intensive agricultural technologies can accelerate erosion processes because they are often accompanied by elimination of shelterbelts and hedgerows as fields are made larger to accommodate increased mechanization. Changing crop rotations may also have significant impacts on erosion processes. For example, an increased proportion of row crops in crop rotation often increases erosion (Szumanski, 1977).

Three regions of Poland are particularly threatened by wind and water erosion (Ziemnicki, 1978). The northern region of Poland, the lakelands, is mildly affected by erosion. The southern uplands region has loess soils which are very strongly threatened by water erosion; as a result, many gullies have been formed there. In about 0.3% of the region, gully density is above 2 km per km^2 (Jozefaciuk and Jozefaciuk, 1988). The third region prone to erosion is the mountain area; here, slopes are steep, and annual rainfall often exceeds 1,000 mm per year.

Jozefaciuk and Jozefaciuk (1988) estimate that surface water erosion hazards exist on 39.3% of the country's area, and that wind erosion threatens an additional 10.8%. These estimates of erosion potential are based on the assumption that climatic conditions favoring erosion are uniformly distributed over the whole country. However, actual erosion hazards occur on about 10-15% of the country's area. In addition, increasing concentrations of carbon dioxide in the atmosphere (which could stimulate increased rainfall) as well as air pollution by sulfur and nitrogen oxides (which could destroy plant cover) could directly or indirectly enhance water erosion in the future. Potential for wind erosion is also increased when soils are dried out as a result of drainage projects in many regions of Poland.

NEED FOR ECOLOGICAL GUIDELINES FOR AGRICULTURAL DEVELOPMENT

Many of the sources of environmental degradation in rural areas mentioned above appear to be negative side effects of efforts to increase agricultural production. Such efforts include increased use of agricultural chemicals, greater intensity of mechnization, and introduction of other elements of "modern" technology for intensive plant and animal production. Negative effects of these efforts often result from an inadequate understanding of ecological processes in agroecosystems, suggesting that current economic development guidelines are too simplistic and do not address natural laws of ecosystem function.

Therefore, ecological guidelines for agriculture need to be developed on the basis of sound ecological knowledge and joint optimization of agricultural production and nature conservation. Such an approach or program must design waste-free technologies for plant and animal production which would incorporate recycling by-products (e.g., using slurry, food industry wastes, etc.) and would also take advantage of previously unused production factors. For example, a portion of applied fertilizers remains in the environment and may result in eutrophication of water bodies or pollution of drinking water. Modern ecology provides the means to control these negative environmental side effects and, at the same time, to develop sustainable agriculture.

ECOLOGICAL GUIDELINES FOR CONTROL OF ENVIRONMENTAL HAZARDS IN RURAL AREAS

Intensive cultivation prevents development of more complex plant communities. Animal communities in agroecosystems are more impoverished than in natural ecosystems, and bacteria start to predominate among soil microorganisms (Golebiowska and Ryszkowski, 1977; Kaszubiak and Kaczmarek, 1985; Ryszkowski, 1979, 1981). In addition, agricultural activity often leads to decreased amounts of humus in the soil. These phenomena result in the development of a less complex network of interrelationships among the components of an agroecosystem. As a consequence of this simplification, relationships among ecosystem components are altered so that there is less tie-up of local cycles of matter. This leads to increased leaching, blowing off, volatilization, and the eventual escape of various chemical compounds and materials from agroecosystems (SIDA/FAO, 1972; Frissel, 1977; Clark and Rosswell, 1981).

As an illustration of the above, research carried out in Poland by Borowiec et al. (1978) showed that landscapes with a higher proportion of arable fields had increased leaching of elements from watersheds (Table 2). On an average, there was 2.5 times more nitrogen, almost 3 times more phosphorus and potassium, and over 1.5 times more calcium and magnesium leached from a unit of agricultural watershed than from a unit of forest watershed. Fields were intensely fertilized in both types of watersheds. These results indicate lowered retention of various chemical compounds in arable fields. The same study also indicates that increases in watershed plant cover can also control migration of nutrients.

An important conclusion for both agriculture and environmental protection can be drawn from investigations on matter cycling in cultivated fields: short-term increases in crop yield can be achieved in simple agroecosystems, but such systems lead to decreased retention of chemical compounds in the long run. Such long-term losses can be prevented by better

TABLE 2 Influence of plant cover of watersheds on ion leaching.

Type of Watershed (WS)	No. of WS	Percent of area			Outflow of elements during two years ($g\ m^{-2}$)				
		arable	forests	grass-land	$N\text{-}NO_3$	$P\text{-}PO_4$	K^+	Ca^{2+}	Mg^{2+}
Larger contribution of cultivated fields	1	53	12	35	1.63	0.03	3.32	37.66	2.82
	2	62	8	28	1.43	0.04	2.99	35.42	3.30
	3	53	32	13	1.13	0.04	2.57	30.95	2.28
	4	51	21	27	0.71	0.02	2.91	32.40	2.70
	MEAN	55	18	26	1.22	0.03	2.36	34.13	2.78
Smaller contribution of cultivated fields	1	38	44	17	0.60	0.02	1.58	23.33	1.80
	2	29	45	26	0.48	0.01	1.00	25.63	1.44
	3	21	65	14	0.39	0.01	0.91	22.54	1.56
	4	32	47	21	0.44	0.01	0.91	15.26	1.20
	MEAN	30	50	19	0.48	0.01	1.10	21.69	1.50

SOURCE: Borowiec, Skrzyczynski, and Kucharska, 1978.

management of the agricultural landscape structure through use of shelterbelts, grasslands, and small water reservoirs to better control dispersion of various compounds from arable fields.

MANAGEMENT OF AGRICULTURAL LANDSCAPE

Agroecosystems are characterized by openness of matter cycling. This means that parts of the materials used in intensification of production are spread outside the boundary of the ecosystem. Thus, the management of material migration could be improved if agricultural regions include structures such as grasslands and small forest plots and ponds which could direct the cycles of matter and thus retard the accumulation or spread of harmful compounds or materials.

Interfield stretches of grassland and shelterbelts are biological barriers which can modulate the dispersal of various chemical compounds or materials from arable fields. Analyses of the concentration of various ions in groundwater flowing from arable fields through adjacent forests or shelterbelts show a considerable decrease in nitrates and a smaller decrease in concentrations of calcium, magnesium, and phosphorus (Table 3).

Decreased concentrations of nitrates in groundwater flowing under adjacent forested areas result not only from direct adsorption by root systems, but probably indirectly by denitrification as well (Peterjohn and Correll, 1984). Paliukevicius (1981) has also noted the importance of tree

TABLE 3 Mean element concentration (mg dm^{-3}) in groundwater under cultivated fields, forest, and shelterbelts adjancent to the fields, August 1982 - September 1986.

	AREA 1		AREA 2	
Elements	Cultivated Fields	Forest	Cultivated Fields	Shelterbelt
Nitrate nitrogen	22.2	1.0	37.6	1.1
Calcium	158.0	82.0	198.0	116.0
Magnesium	15.8	8.6	41.1	18.4
Phosphate phosphorous	0.21	0.11	0.8	0.06

SOURCE: Bartoszewicz and Ryszkowski (in press).

stands in controlling water-based migration of various chemical compounds from fields. Experimental defoliation and various methods of logging in watersheds have shown significant relationships between plant cover and thc chemistry of surface waters (Likens and Borman, 1972; Likens et al., 1977).

Nitrogen cycling in forests is relatively closed because of the high affinity of soil organisms and plants for this element. Thus, only a small amount of nitrogen is lost from forest ecosystems (Rosewall, 1976). Even so, the input of nitrogen often exceeds output, indicating that nitrogen is stored in forest ecosystems (Likens et al. 1977). When the system of interrelationships between soil, plants, and heterotrophs is broken due to human intervention, large quantities of organic nitrogen are mineralized and nitrates are leached from soil (Margowski and Bartoszewicz, 1976; Margowski, 1979; Ryszkowski, 1979).

Grasslands also provide barriers to prevent spreading of different compounds in agricultural landscapes. At a distance of 15 to 25 meters from the edge of a field, there are reductions of nitrates from 10.4 mg·dm^{-3} to 2.4 mg·dm^{-3} in groundwater flowing under grasslands (Bartoszewicz and Ryszkowski, in press).

Different kinds of cultivated plants also influence element leaching rates. In an analysis of water drained from fields having different cultivated plants, Borowiec (1986) showed high leaching rates of various elements under fields planted with row crops and maize, whereas lower leaching rates were associated with cereal cultivation (Table 4). Because Borowiec used as the base reference value the mean for each element calculated from all samples (including row crops and cereals), his analysis indicates only relative rates. Nevertheless, the conclusion is that the pattern of plant rotation influences the leaching of elements from soils of cultivated fields.

TABLE 4 Influence of winter cereals, row crops and maize cultivation on concentration of nutrients in drain water.

Nutrients	Mean concentration during 9 years (mg dm^{-3})	Change of concentration percent from mean due to cultivation: Row Crops and Maize	Winter Cereal
N - NO_2	8.7	+ 37	- 21
P - PO_4	0.14	+ 57	- 43
K	5.0	+ 22	- 30
Ca	157.0	+ 19	- 18
Mg	26.0	+ 3	- 6

+ = above mean - = below mean

SOURCE: Borowiec (1986).

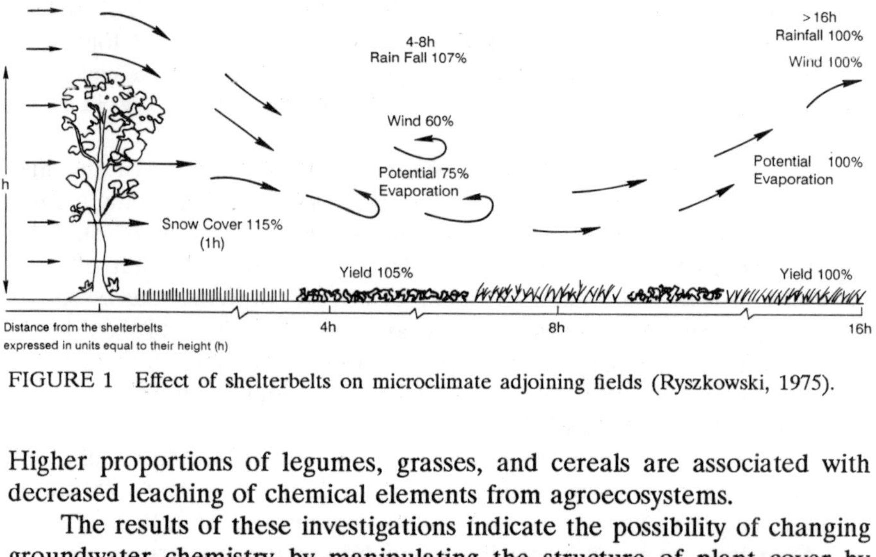

FIGURE 1 Effect of shelterbelts on microclimate adjoining fields (Ryszkowski, 1975).

Higher proportions of legumes, grasses, and cereals are associated with decreased leaching of chemical elements from agroecosystems.

The results of these investigations indicate the possibility of changing groundwater chemistry by manipulating the structure of plant cover by planting shelterbelts, grasslands, forests, and crops. Plant cover also decreases soil losses by water erosion. Preventing soil loss by wind erosion is another problem for contemporary agriculture. Shelterbelts can help control wind erosion. Long-term research carried out in the vicinity of Turew by the Institute of Agrobiology and Forestry of the Polish Academy of Sciences has provided results which show that shelterbelts can significantly modify wind velocity and consequently affect the spatial distribution of atmospheric deposition and decreases potential evaporation from adjacent fields (Ryszkowski, 1975; Ryszkowski and Karg, 1976). Figure 1 illustrates the general results from these studies.

TABLE 5 Parameters of heat and water balances for vegetation season agricultural landscape components.

Parameter	Units	Shelterbelt	Wheat	Grassland
Intercepted solar energy (Rn)	MJm^{-2}	1,730	1,536	1,494
Energy used for evapotranspiration (LE)	MJm^{-2}	1,522	1,090	1,250
Evapotranspiration	mm	609	436	500
LE : Rn		0.88	0.70	0.8

SOURCE: Ryszkowski and Kedziora, 1987.

Shelterbelts capture large amounts of incoming solar energy (because of low albedo values) and use nearly 90% of intercepted energy for evapotranspiration (Table 5). As a result of these thermodynamic characteristics, shelterbelts evaporate about 170 more liters of water per square meter during the growing season than fields with cultivated wheat (Ryszkowski and Kedziora, 1987).

There are at least two reasons for this difference, both of which are connected with the difference in structure of plant cover of these two systems. Trees have much better developed root systems than do wheat and other agricultural crops, allowing them to absorb water from deeper layers of soil. Thus, more water is within direct and indirect (capillary flow) reach of the tree roots. Shelterbelts also have a larger canopy roughness than wheat. Together with higher wind speeds and turbulence in shelterbelt canopies, this canopy roughness results in more intensive vapor exchange over shelterbelts (Ryszkowski and Kedziora, 1987). In this respect, shelterbelts function as powerful "natural water pumps" and, thus, influence groundwater chemistry when groundwater is within direct or indirect reach of the tree root system.

Because of higher albedo values, grasslands intercept smaller amounts of incoming solar radiation (Table 5). Managed grasslands are usually located in terrain depressions, often in or adjacent to areas of natural drainage. Here, the roots reach shallow groundwater and the plants use a high proportion of the intercepted energy for evapotranspiration. Thus, grasslands also affect the chemistry of underlying groundwaters.

The ion exchange capacities of soils under shelterbelts and under grasslands differ from those under cultivated fields. This also leads to different effects on the chemistry of groudwater. Taken together, these phenomena explain why shelterbelts and grasslands impact groundwater chemistry as discussed above.

Ponds and small reservoirs provide another kind of environmental barrier which can prevent the spread of chemical compounds. At present, the role of small ponds is almost totally neglected in development planning. These ponds can be effective in controlling matter cycling in agricultural landscapes. The sediment deposited on the bottom of ponds contains nutrients leached from the fields. Thus, sedimentation of nutrients in ponds decreases the amount lost from the landscape. Fertilizing fields with the sediment from small, shallow field ponds facilitates recovery of mineral plant nutrients otherwise lost from the agroecosystem. Thus, small water ponds or reservoirs located according to terrain relief to create barriers for surface runoffs can play an important role in shaping pathways for matter cycling in the landscape (Golley et al., 1978).

Gliessman (1978) reported a practical utilization of small catchment reservoirs in Yucatan, Mexico. Attempts to intensively cultivate crops on large fields led to a considerable intensification of soil loss by water erosion. Eventually, losses in soil fertility created conditions which led to abandonment of farms. However, a return to traditional farms—cultivating smaller fields (15 hectares or less in size) surrounded by shelterbelts and with catchments for collecting sediment carried by water during heavy monsoon rains—has lead to a return of a more prosperous agriculture. Soil nutrients collected in catchments are now returned to the fields after the monsoons are over.

Natural, compatible structures which assist in controlling matter cycling are of great importance for farming. Arable fields have an open type of matter cycling, with low utilization of mineral fertilizers. Strict agricultural use of applied nitrogen amounts to only 50 percent of input. The rest either escapes to the atmosphere or migrates to ground and surface waters causing various environmental problems. Providing different environmental barriers can reduce the loss or spreading of various chemicals and nutrients from farm fields. Shelterbelts and small patches of forest or grasslands distributed within the agricultural landscape can influence matter cycling over the entire area, facilitating enhanced environmental protection in rural areas.

OPTIMIZATION OF AGRICULTURAL PRODUCTION AND ENVIRONMENTAL CONSERVATION

Increased recognition of the natural laws of ecosystem function facilitates more objective evaluation of alternative technologies which seek to optimize agricultural production, nature conservation, and achievement societal needs at the same time. An agroecosystemic method of analysis is useful in developing a better understanding and better control of environmental threats in rural areas caused by industry and urbanization of the country, as well as for sustainable agricultural development.

A broad, systematic method of analysis of these phenomena helps to overcome the common view that maximum yields can be obtained by intensified farming on a large scale. Use of mixtures of shelterbelts, grasslands, open drainage canals with banks covered with trees or shrubs, and small ponds or catchment reservoirs are often viewed as obstacles to mechanized production. Thus, there is a tendency to avoid them on the part of farmers who want to take maximal advantage of mechanization, albeit at the expense of increased wind erosion and increased surface and subsurface runoffs.

This narrow and ecologically harmful point of view can be corrected by introducing a wider agroecosystem perspective. Of course, maintaining appropriate field mosaics—shelterbelts, grasslands, hedgerows, small water reservoirs, streams, canals, etc.—will require a reorientation in attitudes as well as the development of new and more appropriate agricultural and forest technologies. Agricultural and forest implements and machines should work effectively without destroying environmental barriers. Their designers should consider their maneuvering ability, utility in fulfilling multiple tasks, as well as limit their size, improve speed control, and allow flexibility in changing their area of contact with ground.

This ecosystem perspective would enable creation of regional or national programs of agricultural development with fewer negative environmental effects. Controlling humification and mineralization processes of soil organic matter is of fundamental importance. Knowledge about participation of various groups of organisms in energy flows in agroecosystems also can be of great use. Recognition of the laws of matter cycling would facilitate development of so-called "waste-free" technologies, and mineral fertilizers might be used more effectively, causing fewer unfavorable changes in water and soil quality. Ecological agriculture does not lead to a refutation of modern means of agricultural production, but to a more rational agroecosystem strategy which seeks to meet the objectives of food production and nature conservation.

In conclusion, increased emphasis on ecological farming stems from the societal need to reduce environmental hazards in rural areas, and not from a desire to return to a less productive, traditional economy. Ecological agriculture is an attempt to optimize output and achieve high yields while maintaining the long-term productive potential of soils. Thus, it can be viewed as a farming system which maximizes "social" profits by optimizing agricultural production and environmental conservation, using the laws of ecosystem function to avoid environmental hazards.

Acknowledgement

Partial financial support for this research was provided by Ministry of Education of Poland.

REFERENCES

Alexander, M. 1980. Effects of acidity on microorganisms and microbial processes in soil. Pp. 363-374 in Effects of Acid Precipitation on Terrestrial Ecosystems, T.C. Hutchinson and M. Hacas, eds. New York: Plenum Press.

Bartoszewicz, A., and L. Ryszkowski. (in press). Influence of shelterbelts and meadows on the chemistry of groundwater. In Dynamics of Agricultural Landscape, L. Ryszkowski, ed. New York: Springer Verlag.

Barylko-Piekielna, N., Cz. Kierbinski, and S. Tyszkiewicz. 1985. Estimate of food contamination levels as result of environment contamination. Mimeographed text. Polish Academy of Sciences, Warsaw. p. 45 (in Polish).

Borowiec, S. 1986. Influence of plant cover in agricultural landscape on nutrient concentrations in surface waters and nutrient leaching from soils. Proceedings of Conference on Problems of Environment and Landscape Protection in Light of Actual Needs of the Country. Towarzystwo Naukowe Organizacji i Kierowania w Poznaniu. Poznan. pp. 73-79 (in Polish).

Borowiec, S., T. Skrzynski, and T. Kucharski. 1978. Migration der mineralischen Bestandteile aus den Baden der Nizina Szczecinska. Polish Scientific Publishers, Warsaw. p. 68.

Clark, F.E., and T. Rosswall, eds. 1981. Terrestrial nitrogen cycles. Ecol. Bull. 33:1-714. Stockholm.

Coleman, D.C. 1981. The impact of acid deposition on soil biota and C cycling. Pp. 33-53 in Response of Agricultural Soils to Acid Deposition. Proceedings of the Meetings of the Electric Power Research Institute. Columbus, Ohio.

Cook, R.B. 1983. The impact of acid deposition on the cycles of C, N, P, and S. Pp. 345-364 in The Major Biogeochemical Cycles and Their Interactions, B. Bolin and R.B. Cook, eds. New York: Wiley.

Domanski, S., S. Kowalski, and T. Kowalski. 1987. Industrial emissions and pathogenic activity and biotrophic changes in fungi with special reference to Upper Silesia and Cracow industrial districts. Pp. 281-288 in Biological Reactions of Trees to Industrial Pollution, R. Siwecki, ed. Wydawnictwo Naukowe Uniwersystetu w Poznaniu, Poznan.

Frissel, M.J., ed. 1977. Cycling of mineral nutrients in agricultural ecosystems. Agroecosystems 4:1-354.

Gliessman, S.R. 1978. Ecological processes of traditional agroecosystems in the humid lowland tropics of southern Mexico. 135. Paper Prepared for the Second International Congress of Ecology. Abstracts. INTECOL, Jerusalem.

Golley, F., O.L. Louks, and L. Ryszkowski. 1978. Role of various elements of agricultural landscape in the functioning of watersheds. 137. Paper Prepared for the Second International Congress of Ecology. Abstracts. INTECOL, Jerusalem.

Golebiowska, J., and L. Ryszkowski. 1977. Energy and carbon fluxes in soil compartments of agroecosystems. Ecol. Bull. 25:274-283. Stockholm.

Jozefaciuk, C., and A. Jozefaciuk. 1988. Strategy of ecosystem and agricultural landscape protection against erosion. Manuscript of Report presented for Strategy of Living Resources Conservation in Poland. Institute of Agrobiology and Forestry, Poznan.

Karg, J., and L. Ryszkowski. 1984. Environmental threats in Koscian district. In Stan i Ksztaltowanie Wartosci Krajobrazowych Na Ziemi Koscianskiej, L. Ryszkowski and K. Zimniewicz, eds. Towarzystwo Milosnikow Ziemi Koscianskiej. Koscian 109-161 (in Polish).

Kaszubiak, H., and W. Kaczmarek. 1985. Differentiation of bacterial biomass in croplands and grasslands. Intecol Bulletin 12:29-37.

Kosturkiewicz, A. 1988. Program of water reclamations for protection of living resources. Manuscript of Report Presented for Strategy of Living Resources Conservation in Poland. Institute of Agrobiology and Forestry, Poznan.

Likens, G.E., and F.H. Borman. 1972. Nutrient cycling in ecosystems. Pp. 25-67 in Ecosystems Structure and Function, J.A. Wiens, ed. Corvallis, Oregon: Oregon State University Press.

Likens, G.E., F.H. Borman, R.S. Pierce, J.S. Eaton, and N.M. Johnson. 1977. Biogeochemistry of a forested ecosystem. New York: Springer Verlag. p. 146.

Maciak, F., and S. Zawadzki. 1981. Role and development trends of land reclamations in Poland. Zeszyty problemowe postepow nauk rolniczych. 248:13-29.

Main Statistical Office. 1984. Regions of ecological threat in Poland. Warsaw, p. 237 (in Polish).

Main Statistical Office. 1986. Yearbook 1986 (in Polish). Warsaw. p. 630.

Main Statistical Office. 1987. Environmental protection and water management. Department Rolnictwa i Gospodarki Xywnosciowej. Warsaw, p. 363 (in Polish).

Margowski, Z. 1979. Natural and agronomic factors limiting eutrophication of waters in the agricultural landscapes. Zeszyty problemowe postepow nauk rolniczych 228:65-76.

Margowski, Z., and A. Bartoszewicz. 1976. Infiltration of fertilizer components into groundwater. Komitet Ekologii PAN. Materialy konferencji: Nawoxenie a eutrofizacja wod. 75-97. Zielona Gora. Wyxsza Szkola Inxynierska.

Ministry of Environment and Natural Resources Conservation. 1988. The national program for natural environment protection until 2010. Wydawnictwa Geologiczne. Warsaw, p. 122 (in Polish).

Niewiadomski, W. 1979. Ecological consequences of the agriculture intensification. Zeszyty problemowe postepow nauk rolniczych 228:9-28.

Palys, E. 1985. Influence of river erosion on agriculture and water management. Wiadomosci melioracyjne i lakarskie 28:245-248 (in Polish).

Partyka, T., B. Szymanska, and E. Suwara. 1979. Changes of spatial structure of the agricultural and forest utilization of soil. Zeszyty problemowe postepow nauk rolniczych 217:45-62.

Pauliukevicius, G. 1981. Ecological role of the forest stand on the lake slopes. Vilnius: Pergale, p. 191.

Peterjohn, W.T., and D.L. Correll. 1984. Nutrient dynamics in agricultural watershed: Observations on the role of a riparian forest. Ecology 65:1466-1475.

Rosswall, T. 1976. The internal cycle between vegetation, microorganisms, and soils. In Nitrogen, Phosphorus, and Sulfur: Global Cycles, B.H. Svennson and R. Soderlund, eds. Ecol. Bull. 22:157-167. Stockholm.

Ryszkowski, L. 1975. Rewiew of studies carried out in Turew on influence of shelterbelts on environment of adjoining fields. Zeszyty problemowe postepow nauk rolniczych 166:71-82.

Ryszkowski, L. 1979. Croplands. Pp. 301-331 in Grassland Ecosystems of the World, R.T. Coupland, ed. Cambridge: Cambridge University Press.

Ryszkowski, L. 1981. Effects of agriculture intensification on fauna. Zeszyty problemowe postepow nauk rolniczych 233:7-38.

Ryszkowski, L., and J. Karg. 1976. Role of shelterbelts in agricultural landscape. Pp. 305-309 in Les bocages: Historie, ecologie, economie: J. Missonnier, ed. CNRS. Univ. dr Rennes. Rennes.

Ryszkowski, L., and A. Kedziora. 1987. Impact of agricultural landscape structure on energy flow and water cycling. Landscape Ecology. 1:85-94.

SIDA/FAO. 1972. Effects of intensive fertilizer use on the human environment. FAO Soils Bulletin. 16:360. Rome.

Stigliani, W., F.M. Brouwer, R.E. Munn, R.W. Shaw, and M. Antonovsky. 1988. Future environments for Europe: Some implications of alternative development paths. International Institute for Applied Systems Analysis. Laxenburg, Austria. p. 62.

Szumanski, A. 1977. Changes in the course of the lower San channel river in XIX and XX centuries and their influence on the morphogenesis of its floodplain. Studia Geomorphologica Carpatho-Balcanica 11:139-153.

Warteresiewicz, M. 1978. Sulfur dioxide effect on plants. Zeszyty problemowe postepow nauk rolniczych 206:29-42.

Ziemnicki, S. 1978. Protection of soils against erosion. Panstwowe wydawnictwo rolnicze i lesne. Warszawa, p. 183 (in Polish).

Zyczynska-Baloniak, I. 1980. Dissolved organic matter in water of lake and channel in the agricultural landscape. Polish Ecological Studies 6:155-166.

Ecological Problems Associated With Agricultural Development: Some Examples in the United States

WARREN E. JOHNSTON
University of California, Davis

The primordial symbiotic relationship between agriculture and environment is one that has been respected by centuries of traditional farmers. Modern agriculture, however, appears to be responsible for a number of significant environmental problems that defy easy explanation and/or mitigation.

Today's agriculture, engaging only three percent of the U.S. population, differs substantially from earlier structures which occupied the majority of the population in "hunter and gatherer" or in "husbander" roles. The agricultural sector now consists of a mosaic of specialized types of farms ranging in size and intensity from part-time operators of small farm units with only modest economic outputs to large-scale, industrialized farms with significant economic presence in the agricultural sector, and which are sometimes involved in nonagricultural production activities, as well. The sector's evolution to one dominated by large, economically efficient "industrialized" farms is revealed by the statistic that roughly one-eighth of America's farms produce two-thirds of the total U.S. agricultural output.

Technological innovations and national agricultural policies of this century have had significant impacts on agricultural productivity and on the structure of agriculture (Carter and Johnston, 1978). However, there are increasing sectorial and societal concerns about the longer-term ecological consequences of recent cultural, mechanical, and biological innovations and attendant scales of development in the U.S. agricultural sector (Johnston and Carter, 1983).

Agriculture is essentially a land-based enterprise, though it is dependent on the total endowment of natural and environmental resources to define both the set of production possibilities and effective natural constraints to production (e.g., soil fertility, growing season, drainage, climate, etc.). In the United States, all but 12% of nonfederal land was devoted to either agricultural or forestry uses in 1977 (USDA, 1981). Human actions have expanded the possibilities, reduced the influence of natural constraints, and increased productivity per unit of land and of other scarce or costly economic inputs. Unfortunately, human actions with respect to agricultural development have also resulted in certain unfavorable impacts, some of which were unanticipated in their severity or in the speed with which they have negatively impacted on the environment.

Ecological systems require diversity and balance in order to support the interactions of living organisms. Agricultural practices can alter ecological balances and diversity, as well as modify basic agroecosystem properties of productivity, stability, sustainability, and equitability (Altieri, 1986). Recent interests in organic farming and sustainable agricultural systems are efforts to examine alternative production systems (University of California, 1986; Carter, 1988).

Production systems which adversely affect sensitive ecological systems in severe and nonreversible ways burden both present and future generations. Ecological impacts can occur both on-site and off-site from agricultural land. On-site impacts generally result from land use patterns, including the decision to develop land for more intensive agricultural uses. These impacts are localized and generally result in reduced productivity or in the displacement of wildlife and native plant species due to altered site habitat. They may also cause long-term changes in agricultural productivity. Off-site impacts can occur both nearby and at great distances from the farming operation. For example, agriculture can change the ecology of waterways and groundwater basins via changes in rates of flows or in amounts of sediment and chemicals which may, in turn, contribute to turbidity, eutrophication, biochemical oxygen demand, or toxicity, and ultimately may affect many downstream users of water in the same drainage system.

AGRICULTURAL AND ENVIRONMENTAL POLICY ISSUES

There are a number of areas in which agricultural and environmental problems and policies seem to be in conflict. A recent Organization for Economic Cooperation and Development (OECD) draft report (1988) discusses major agricultural and environmental policy issues in developed economies under the following classifications:

- intensive crop production and the use of agricultural chemicals;
- intensive animal production and the management of animal manure;
- dryland farming, soil conservation and erosion; and
- changing landscapes, land-use patterns and the quality of rural landscapes.

Observers from the arid west of the United States might also add as an equally important issue:

- increased competition for scarce water supplies and water quality.

The traditional approach to many environmental problems associated with agriculture has been "effect" oriented (Young, 1988). As problems emerged, environmental policies were implemented to control and prevent the perceived negative impacts of agriculture on the environment. This often resulted in partial and piecemeal approaches. For example, prescriptions for testing and registration of agricultural chemicals and governmental regulations for application and handling of chemicals have been designed to protect the health of farm workers and consumers of food products. Public zoning policies were developed to restrict concentrations of animal production units and control the collection and disposal of animal wastes. Conservation of soil has been attempted by providing economic incentives and by mandated farming practices to reduce erosion. Land-use zoning by local and state governments often seeks simultaneously to maintain a viable agriculture sector and a pleasing rural landscape, but is often ineffective when confronted by decisions that either intensify agricultural uses or develop higher, economically valued nonagricultural uses. Resource policies to maintain air and water quality sometimes prohibit certain agricultural practices.

Throughout the recent experience in the United States, there has been an the uneasy awareness that linkages between agricultural price and income policies and resource and environmental policies have been inadequate. In fact, numerous examples can be cited in which independent agricultural and environmental policies have been in direct conflict (Phipps et al., 1986; National Research Council, 1974; National Research Council, 1982). For example, agricultural price policies and world market conditions in the 1970s increased incentives for farmers to both intensify production on cropped acreage and expand agricultural production onto marginal lands, leading to increases in chemical applications and soil erosion. At the same time, environmental policies sought increased water quality and soil conservation. A current assessment of environmental issues in agriculture in a period of apparent global agricultural surplus capacity is that this "effect-oriented" and uncoordinated approach to environmental issues has been inadequate. Unacceptable levels of agricultural pollution, soil erosion, and degradation in water quality continue, while landscape quality continues to decline.

Young (1988) notes four reasons for the failure of environmental regulations which were designed to protect the environment against impacts from agricultural development and production activities:

- failure to enforce existing regulations and to introduce regulations necessary to keep agricultural practices within sustainable limits;
- the reluctance of governments to use economic instruments to internalize the external costs of many agricultural practices;
- the existence of a wide range of price support, market intervention, and tariff policies which tend to stimulate and intensify agricultural production; and
- the dominance of agricultural policy instruments over environmental instruments and, in particular, the failure to develop integrated agricultural and environmental policies.

The remainder of this chapter discusses some of the more prominent issues in the recent U.S. experience, addresses the need for integration of environmental and agricultural policies, and presents a selected bibliography for the interested reader.

NEGATIVE IMPACTS OF AGRICULTURE ON THE ENVIRONMENT

The physical processes of erosion of agricultural lands and runoff of sediment and nutrient or pesticidal chemicals create impacts that are felt both on-site and off-site. Most individual farmers' decisions to invest in soil conservation measures are made primarily to avoid on-site productivity declines. Erosion generally affects soil productivity by depleting and/or degrading the inherent physical, biological, and chemical characteristics of the surface layer of the soil (USDA, 1986). Incentives to promote good soil management via conservation investments have, for much of the 1980s, been blunted by the depressed financial situation affecting much of U.S. agriculture. The problem is compounded by the off-site (and therefore externalized) water pollution and sedimentation problems which affect downstream users and ecological systems.

Agricultural Runoff

When the United States enacted the Federal Water Pollution Control Act in 1972, the nation's waterways were under tremendous assault from the industrial development that had occurred since World War II. The legislation set forth a mandate to improve water quality and reduce pollution. Sources of water pollution can be attributed to point sources (where there is an identifiable discharge) and to nonpoint sources that contribute to water-quality degradation through diffuse mechanisms. Regulations and standards set controls on the quantities and methods of disposing of waste products and discharges. The initial target for implementing the law addressed point source polluters, primarily industrial sources that were discharging wastes into water systems. Point source polluters, though still major sources of biological oxidation damage and dissolved heavy metals, are relatively minor contributors of other pollutants, suspended and dissolved solids, phosphorus, and nitrogen which stem mainly from nonpoint sources (USDA, 1981).

Erosion and runoff from agricultural lands are leading sources of nonpoint pollutants. They make up about 50% of the total sediment load carried by waterways, and contribute significant amounts of pesticides, fertilizers, salts, and metals that affect the ecology of both water and land environments (Clark, 1985). Pollution problems from nonpoint sources, nutrients, suspended solids, total dissolved solids (salts), pesticides, and bacteria affect nearly every region of the country.

Soil Erosion

Erosion generally occurs via one of four modes: sheet, rill, gully (i.e., water-induced modes), and wind erosion. In an attempt to estimate the amounts of soil erosion from various lands, the National Runoff and Soil Loss Data Center developed the Universal Soil Loss Equation (USLE). This formula estimates soil erosion loss as a multiplicative function of six variables: rainfall intensity and duration, soil erodibility, slope length, slope grade, vegetative cover, and tillage practices (USDA, 1981).

Additionally, in order to identify problem areas, the U.S. Department of Agriculture (USDA) has estimated the maximum annual soil losses that can be sustained on a land area without adversely affecting soil productivity. These are referred to as *tolerance values* (t-values) and usually range from one to five tons per acre per year depending on climate and soil characteristics. In 1977, the national average soil erosion loss from croplands was estimated to be about 4.8 tons per acre (TPA), equivalent to about 1.75 metric tons per hectare, which is approximately 1/30 of an inch (0.84 mm) of topsoil per year. Local erosion rates can range from insignificant amounts to well over 100 TPA (36.7 metric tons per hectare) in some areas, depending upon local conditions (Batie, 1983).

Soil Sedimentation

The USDA Soil Conservation Service's appraisal of soil and water resources in the United States, which was carried out in response to the Soil and Water Resource Conservation Act of 1977, identified agriculture as the primary cause of nonpoint water pollution for more than 68% of the nation's watershed areas (USDA, 1981). The most significant constituents with respect to volume were suspended solids, which contributed approximately 50% (i.e., about 760 million tons) of the total sediment load. Increased amounts of sediment decrease the viability of aquatic ecosystems and can produce both direct and indirect impacts on living systems. In very heavy concentrations, sediment can clog the gills, and thus, decrease the uptake of oxygen by fish. More likely, however, are indirect effects that arise from increased water turbidity. Increased turbidity reduces the penetration of sunlight and consequently the primary production of photosynthetic plants and algae, which in turn reduces the productivity of higher trophic levels. In addition to a general decrease in the biomass sustained in the system, shifts in the biological mix of species is also likely to occur in which sediment-tolerant and adaptive species replace less tolerant organisms.

METHODS FOR REDUCING SOIL EROSION AND RUNOFFS

Clearly, erosion from agricultural lands severely impacts the ecology of U.S. waterways. Negative off-site impacts affect recreation, flood control, municipal and industrial water use, navigation, and other water-based or water-dependent economic activities. However, these negative off-site effects are generally not considered by individual farmers or nonfarm owners of agricultural lands. Thus, they are not likely to utilize soil conservation techniques to as large an extent as is socially optimal. Such erosion is seen as an "externality" which adversely effects other members of society, but has no direct impact on the farmer who has a vested (ownership) interest in agricultural land. The "external" nature of this problem suggests that governments may need to develop incentives that will encourage erosion control and reduce agricultural runoffs. National agricultural policies in the 1970s encouraged production on marginal lands that were previously uncultivated. New land development activities, sometimes called "sodbusting" and "swampbusting," increased soil erosion and agricultural runoffs, and adversely impacted wildlife habitats (National Research Council, 1982).

Idling of Highly Erodible Marginal Lands

Some highly erodible lands should not be used for crop production and therefore should be returned to native grass or forest cover (Webb et al., 1986). Identification and retirement of marginal lands is a key component of current agricultural policy in the United States which also seeks to reduce excess productive capacity in U.S. agriculture. Supporting studies have indicated that soil conservation policies are effective in reducing off-site erosion problems and, simultaneously, in decreasing agricultural production and government expenditures for purchasing, storing, and disposing of crop surpluses.

Soil conservation policies include the Conservation Reserve Program, the Sodbuster Provision, and the Conservation Compliance Provision of the Food Security Act of 1985, which established current U.S. agricultural policy. While the law is in effect through 1990, it may be extended without major revision for another four- to five-year period.

The *Conservation Reserve Program* pays farmers annual rental payments and one-half the cost of establishing permanent cover to retire highly erodible cropland for 10 years. About 100 million acres are eligible for enrollment. The goal is to enroll 45 million acres, and approximately 28 million acres have been enrolled since the program began in 1986.

The *Sodbuster Provision* denies price support and deficiency payments, farm storage facility loans, crop insurance, disaster payments, and FmHA-insured loans to any person producing an agricultural commodity on highly erodible land converted since December 23, 1985, unless an approved conservation plan is adopted and implemented (USDA, 1988). This provision affects about 227 million acres with some potential for conversion.

The *Conservation Compliance Provision* of the Food Security Act requires that farmers with highly erodible cropland begin implementation of a conservation plan by 1990 and complete it by 1995 in order to retain eligibility for programs identified under the Sodbuster Provision. This policy could affect production possibilities and costs on up to 65 million acres, and as many as 10 million acres could drop out of production or out of government programs (USDA, 1988).

Erosion Control Practices

A number of physical techniques have been developed and utilized to control erosion and stabilize the movement of soil. These techniques include tillage, cropping, and structural measures that limit undesirable soil transport. Tilling of the soil is a primary factor that mobilizes soil particles and reduces or eliminates ground cover and cover crops that otherwise would stabilize the soil and decrease erosion, especially during rainy seasons. Conservation tillage practices ("no-till" and "minimum tillage" practices) have increased in terms of farmer acceptance and use over the past several years, for both economic and soil-conserving reasons. The effectiveness of conservation tillage techniques varies by region and by crop, and is also related to the amount of crop residues retained in fields. Studies have shown that sediment losses can be reduced by 15% to 90%. However, these conservation tillage practices often include increased use of herbicides to control weeds (Clark, 1985). Thus, reduced soil erosion practices may transfer the relative balance of agricultural runoff problems from sedimentation to higher concentrations of pesticides in runoffs.

Contour farming can also be used to substantially reduce erosion rates on lands that are sloping and are consequently more susceptible to erosion processes. The principle of contouring involves cultivating lands in a pattern to reduce the effects of the slope by farming along natural gradients and topography. Contouring can result in erosion reductions of 25% to 50%, with concomitant decreases in the transport of nutrients and pesticides (Clark et al., 1985). The practice of mechanical levelling of fields is also effective in reducing agricultural runoffs in irrigated areas and making possible the more efficient application of irrigation water. There are also benefits with respect to mechanization of cultural and harvesting operations.

Other structural possibilities include the construction of diversion channels and sediment basins that catch runoff and decrease sediment loads. Planting grass along waterways and irrigation canals also will reduce erosion losses from stream banks. Altered cropping rotations may also decrease soil erosion losses. Clark et al. (1985) found a decrease in soil loss from 19.7 tons per acre with continuous corn cultivation to 2.7 tons per acre with a corn, wheat, and clover rotation.

Soil conservation policies of the sort required for highly erodible lands under the Sodbuster and Conservation Compliance Provisions of the Food Security Act can include prescriptions that will mandate conservation tillage, structural modifications, and changes in cropping systems—singly or in combination—to reduce soil erosion and agricultural runoff. Batie (1985) estimates that soil erosion rates can be reduced by 60-95% by a combination of conservation tillage, contour planting, strip cropping, terracing, and other known conservation techniques. One USDA study concludes that modification of USDA commodity and conservation programs to achieve greater consistency in program objectives could affect one-third to one-half of the nation's cropland acres that are eroding at unacceptable rates (Reichelderfer, 1985).

CHEMICAL CONSEQUENCES OF AGRICULTURAL DEVELOPMENT AND PRODUCTION ACTIVITIES

Fertilizers and pesticides are important nonpoint pollutants in surface waters. More recently, however, concern about chemical pollution from agriculture has spread to include the nation's groundwaters. The seriousness of many of these problems is still not certain, though they have emerged rapidly as important environmental issues.

Accelerated erosion processes, in addition to carrying sediment into waterways, also bring other constituents of the soil including agricultural chemicals. The U.S. agricultural system is highly reliant on the use of chemical inputs to provide nutrients for growth and production as well as pesticides to control destructive plant and animal infestations. ("Pesticide" is the term used to describe general classes of chemical-controlled agents, such as herbicides, insecticides, fungicides, nematocides, and rodenticides). These chemicals can be carried into aquatic ecosystems either directly through runoff or by absorption to soil particles that are subsequently eroded.

The 1977 amendments to the Federal Water Pollution Control Act of 1972 expanded the regulation of pollutants in groundwater, surface water, and coastal waters by providing stronger incentives to protect water quality from agricultural sources of pollution than had been provided by any previous national legislation (Crowder et al., 1988). The Act extends emphasis beyond point sources of pollution to nonpoint agricultural sources. In areas where nonpoint source pollutants endanger water quality, farmers could be subject to state or local restrictions on land use and agricultural chemical use (USDA, 1988).

Fertilizers

Farming accounts for roughly 97% of all fertilizer use. Fertilizers containing nitrogen, phosphorus, and potassium are often applied to soils to increase crop yields. Total application of nitrogen and phosphate in 1983 was in excess of nine and four million tons, respectively (USDA, 1986). In the United States, fertilizer use in kilograms of plant nutrient per hectare of arable and permanent cropland increased by 50% in a 20-year period, from 63 kilograms in 1964-66 to 94 kilograms in 1981-83 (World Resources Institute, 1986). Corresponding values for Poland were 84 kilograms per hectare in 1961-63, rising by nearly threefold to 237 kilograms in 1974-76, and declining somewhat to 220 kilograms in 1981-83.

The application of large amounts of fertilizers in intensive agricultural production activities creates a situation where water transport of soluble chemicals can easily occur. Once carried into an aquatic ecosystem, these nutrients are available for aquatic plant and algae growth. Phosphorus is the most common limiting nutrient in aquatic systems. Increased amounts of phosphorus and nitrogen increase biological production and the aging process (eutrophication) in lakes and reservoirs. Eutrophication is a natural process in many lakes and streams, but the influx of nutrients from agricultural sources greatly accelerates the process. Eutrophication of streams, lakes, and reservoirs usually results in excessive growth of aquatic weeds and algae, which in turn can create toxins and remove available oxygen, thereby killing fish and greatly reducing the recreational value of lakes and reservoirs (Clark et al., 1985). The USDA estimates that between 15% and 54% of all nutrients applied to agricultural lands reach surface water systems. This tremendous inflow of agricultural nutrients creates a significant impact on the nation's waterways, its surface-water supplies, and the ecological composition and maturation of water bodies.

Pesticides

Agriculture accounts for about 75% of all pesticide use in the United States, equivalent to about 725 million pounds of active ingredients (USDA, 1986). Prior to World War II, pest control was largely accomplished through a variety of cultural, mechanical, and tillage practices. However, the advent of the chemical pest-control era brought abrupt changes in agricultural systems which resulted in increased production, increased quality, and greater efficiency in the production of food.

Almost all crops are subject to attack by diseases, insects, and weeds. This susceptibility often increases when mixed croping is replaced by continuous monocultures (Altieri, 1986). Efforts to control pests have included the application of various toxic chemicals. More than 1,800 biologically active compounds have been developed to protect agricultural crops. In 1977, herbicides were applied to more than 200 million acres, insecticides to over 75 million acres, and fungicides to 8 million acres in the United States. Eichers (1981) estimated that 5% of applied herbicides and insecticides eventually reach surface waters, but the USDA has estimated that, under normal rainfall conditions, pesticide losses in runoff tend to average no more more than 0.5% of the quantities applied (USDA, 1986). However, the latter source notes differences in losses depending on type of substance, with wettable powders having runoff rates of up to 5.0%.

Pesticides may enter other environments from spray drift, soil erosion, precipitation and irrigation runoff, soil moisture seepage, groundwater flow, direct contact with animals and humans, or by other means (USDA, 1986). Monitored concentrations of pesticides in waterways has generally been low except when heavy rains follow applications (USDA, 1981). However, this observation can be deceptive due to the relatively high toxicity of these chemicals and the unknown or unmonitored potential of a number of these compounds to accumulate at very high concentrations as they rise through the food chain. While pesticide use has become prevalent in modern agricultural systems, it is not uniformly applied to all crops. For example, Eichers (1981) estimates that cotton represented the largest share of insecticide use in 1976, accounting for nearly 40% of the total, while corn accounted for another 20%. He also estimated that corn accounted for 52% of total herbicide use while soybeans accounted for 20%.

Environmental effects of pesticides can vary substantially. Regulations often favor less persistent and more selective compounds in order to be less environmentally disruptive. In general, insecticide use has trended away from the persistent, biomagnifiable organochlorine compounds that are suspected of causing chronic diseases to nonpersistent substances such as carbonate and organophosphate products (USDA, 1986). The Federal Insecticide, Fungicide, and Rodenticide Act (FIFRA) empowers the U.S.

Environmental Protection Agency (EPA) to curb the use of a pesticide if, among other affects, it poses undue risk to human health or is an imminent hazard in the environment because of its persistence or toxic effects (Crowder et al., 1988; Chapter 6, this volume).

Salinity and Heavy Metals

In addition to sediment and chemical intrusion into surface water systems, suspended solids such as salts and heavy metals are also carried into waterways by erosion. Salinity problems are common in the arid western United States and can significantly affect the aquatic environment. For example, it is estimated that between 24% and 41% of the total salt load in the Colorado River results from percolation of agricultural drainage water through salt-laden soils (USDA, 1981). Alterations in diversity and patterns of species dominance can also occur with the replacement of less salt-tolerant species by more salt-tolerant species. These changes can also negatively affect the food chain and the higher trophic levels that depend on the aquatic systems for food.

While the erosion of heavy metals from agricultural lands does not appear to be a widespread problem, regional soil differences and irrigation patterns can create potential heavy-metal problems in water systems. For example, elevated soil selenium levels in the Westlands area of the San Joaquin Valley in California—combined with high soil salinity, saline irrigation water, distinct irrigation patterns, and a high water table—have created a situation where high concentrations of selenium have accumulated in drainage water and have consequently severely affected the ecosystem at Kesterson reservoir. Impacts have included deformities in birds and significant declines in fish reproduction (University of California, 1987). Interim policies have led to the cessation of off-farm drainage flows, to increased on-farm investments in drainage ponds, and to more precise management of applied irrigation water and drainage flows.

Groundwater Contamination

Although groundwater has many sources of contamination, evidence suggests that agricultural pesticides and fertilizers are significant sources (Nielsen and Lee, 1987). The United States relies heavily on its groundwater. Over 97% of rural drinking water comes from groundwater sources and 40% of the population served by public water supplies (i.e., nearly 74 million people) use groundwater sources.

The distribution of potentially affected groundwater areas is influenced by the magnitude, extent, and duration of contamination, and conditioned by land use, agricultural practices, climate, hydrogeology, soil characteristics, net aquifer recharge rate, depth to the water table, and characteristics of the unsaturated zone and the aquifer (Nielsen and Lee, 1987). Characteristics of the potential pollutant (i.e., water solubility, adsorption, and persistence) strongly affect its ultimate fate. About 800 of the 3,000 counties in the United States have potential for contamination by pesticides; another 300 have potential for nitrate contamination; and 300 more have potential for both types of contamination. More than 50 million people rely on groundwater for drinking in these 1,400 potentially affected counties (USDA, 1988).

As mentioned previously, the 1977 amendments to the Federal Water Pollution Control Act applies to groundwater and specifies that agricultural practices may be subject to state or local restrictions on land and agricultural chemical uses. Under FIFRA, more attention has generally been given to pesticides that are known to leach into groundwater than to chemicals which primarily run off cropland. The Safe Drinking Water Act also deals with the possibility that nonpoint sources could contaminate groundwater sources of public wells (Crowder et al., 1988). The rising concern about contamination of groundwater—especially because contamination can persist for many years and cleanup costs can be prohibitively expensive—has heightened awareness of rural and urban inhabitants.

REDUCING CHEMICAL DEPENDENCE

Since intrusion by agricultural chemicals is a one of the serious off-site impacts affecting surface water and groundwater quality, a reduction in the use of environmentally damaging chemicals should help enhance water quality. A number of techniques have been developed that can, in some cases, reduce use and need for chemicals, as well as encourage the use and safe application of less environmentally damaging chemicals. These techniques include biological control, host resistance, cultural control, and physical and mechanical control, in addition to use of chemical pesticides. Additionally, application modes can also have a significant effect on the efficiency of chemical use and reducing off-site migration.

Integrated pest management (IPM) techniques have increased in use as farmers realize the economic and environmental advantages of combining biological pest controls and management practices with reduced chemical use. IPM seeks to control pests in an economically efficient and environmentally sound manner by maximizing the use of natural control agents (i.e., predators, parasites, weather, crop varieties, tillage, etc.) before resorting to the use of chemical controls (Smith and Pimentel, 1978; Pimentel, 1981).

In many instances, IPM is a viable alternative to chemical pest management. This is true especially because of rising chemical costs, increased genetic resistance of pests to pesticides, and the presence of significant external impacts. IPM can often achieve equal yields at reduced costs, thereby increasing the profitability of the operation (Archibald, 1984; Bottrell, 1979). IPM uses both biological and economic criteria to gauge the ability of a crop to tolerate pests. The economic threshold sought in the control of pests is the density of the pest population below which the cost of applying control measures exceeds the losses caused by the pest. These economic threshold values are derived by assessing the potential value of the pest damage and the ecological, social, and economic costs of controls.

Successes in the use and implementation of IPM have occurred with many crops including cotton, citrus, walnuts, almonds, fruits, soybeans, and alfalfa. In Texas, it has been demonstrated that cotton (the major target of insecticides in the United States) can be produced with 50-75% less insecticides, thus increasing farmers' profits from $62 to $170 per acre (Bottrell, 1979). Archibald (1984) has also demonstrated the efficacy of IPM as an optimal strategy for pest management in California cotton, especially in light of the dynamic nature of insect resistance to chemical controls. The results of IPM research indicate that it is both possible and economically feasible to pursue environmentally sound agricultural practices with lesser applications of agricultural chemicals.

Social concern about adverse consequences of agricultural technologies, including but not restricted to heightened levels of agricultural chemical applications, has lead to recent emphasis on "agricultural sustainability." Carter (1988) discusses the various perceptions of "sustainable" agriculture and notes that other terms for agricultural sustainability include alternative, regenerative, low-input, ecological, environmentally sound, and even organic agriculture. He notes:

> These terms are used by people interested primarily in alternative systems of farming that will feed expanding populations while minimizing potential negative effects whatever they might be. Defining negative effects essentially separates or categorizes the various proponents of sustainable agricultural systems.

Other definitions of sustainability place emphasis on resource stewardship, rural community sustainability, food self-sufficiency, and energy conservation. Use of these terms illustrate the social, ecological, economic, and emotional connotations of concern about current agricultural practices. Harwood (1987) identifies the following dimensions of the agricultural sustainability concept:

- time;
- social sustainability;
- economic sustainability;

- maintenance of soil and genetic resource bases;
- minimization of environmental pollution; and
- lowered use of industrialized inputs.

The USDA has recently developed the Low-Input Sustainable Agriculture (LISA) program which supports research and education programs in alternative farming systems that reduce the farmer's dependence on certain kinds of purchased inputs in ways that increase profits, reduce environmental hazards, and ensure a more sustainable agriculture for generations to come. The goals of this program include the following:

- develop economically viable crop and livestock systems to reduce reliance on off-farm purchased inputs (especially synthetic chemical pesticides and fertilizers that may pose environmental or human health hazards);
- maintain and enhance soil productivity;
- reduce soil erosion and loss of water and nutrients;
- conserve energy and natural resources; and
- minimize environmental contamination.

PROSPECTS

We currently find ourselves in the situation of having globally over-responded in our efforts to meet the need for additional food and fiber production which arose in the 1970s. All developed nations and many less developed nations, as well, now find agricultural production capacities and commodity markets to have shifted from "shortfalls" to "surpluses." This achievement was possible in part due to food and agricultural policies which both intensified production on existing lands with the aid of purchased inputs, and expanded or developed additional cropland acreages, often with less than due regard for environmental and ecological consequences. However, the current costs of governmental policies to support agriculture via supply control, export assistance, and other price and income support programs are very large and are of concern not only in the United States, but also in the nations of the European Common Market, and to diverse members of the Cairns Group of developed and developing nations which seeks fundamental change in agricultural and trade policies within the current General Agreement on Tariffs and Trade (GATT) negotiations.

The cumulative effects of past agricultural development and production activities have brought into sharper focus some of the adverse environmental impacts which now seem to arise more quickly and more severely than had been anticipated. It would appear that under such conditions of surplus capacity and environmental threat, there is opportunity for greater integration of agricultural and environmental policies. In fact, a recent report by the Organization for Economic Cooperation and Development

(OECD) (1987) gives considerable attention to prospects for better integration of agricultural and environmental policies among OECD nations. Young (1988) summarizes the OECD recommendations by noting that in the development of new agricultural, environmental and related regional development policies ". . . consideration needs to be given to a trilogy of three factors:

- the need to enhance the positive contribution which agriculture can make to the environment;
- the need to reduce agricultural pollution; and
- the importance of adapting all agricultural policies so that they take full account of the environment."

The latter factor would involve targeting agricultural policies to be more effective by simultaneously reducing surpluses and agricultural pollution while enhancing environmental quality. Future examinations of ecological problems associated with agricultural development will be less critical if such holistic policy strategies are met with popular global support as well as farmer and taxpayer acceptance.

REFERENCES

Altieri, M.A. 1986. Ecological diversity and the sustainability of California agroecosystems. Proceedings of the Symposium on Sustainability of California Agriculture. University of California, Davis.

Archibald, S.O. 1984. A dynamic analysis of production externalities: Pesticide resistance in California cotton. Ph.D. Dissertation, Agricultural Economics, University of California, Davis.

Batie, S.S. 1985. Environmental limits: The new constraints. Issues in Science and Technology II(1):134-143.

Batie, S.S. 1983. Soil erosion: Crisis in America's croplands. The Conservation Foundation, Washington, D.C.

Bottrell, D.R. 1979. Integrated pest management. Council on Environmental Quality. Washington, D.C.: U.S. Government Printing Office.

Carter, H.O. 1988. The agricultural sustainability issue: An overview and research perspective. Invited Paper Prepared for the Seminar on the Changing Dynamics of Global Agriculture: Research Policy Implications for National Agricultural Research Systems, Feldafing, Federal Republic of Germany, September 1988.

Carter, H.O., and W.E. Johnston. 1978. Agricultural productivity and technological change: Some concepts, measures and implications. University California Priced Publication 4085, Berkeley.

Clark, E.H. II, J.A. Haverkamp, and W. Chapman. 1985. Eroding soils: The off-farm impacts. The Conservation Foundation, Washington, D.C.

Crowder, B.M., M.O. Bibaudo, and C.E. Young. 1988. Agriculture and water quality. Agricultural Information Bulletin Number 548, U.S. Department of Agriculture, Economic Research Service, Washington, D.C.

Eichers, T.R. 1981. The use of pesticides by farmers. In the CRC Handbook of Pest Management in Agriculture Vol. II, D. Pimentel, ed. Boca Raton: CRC Press.

Harwood, R.R. 1987. Low-input technologies for sustainable agricultural systems. In Policy for Agricultural Research, V.W. Ruttan and C.E. Pray, eds. Boulder, Colorado: Westview Press.

Johnston, W.E., and H.O. Carter. 1983. Policy issues. In a Guidebook to California Agriculture, A. Scheuring, ed. Berkeley: University of California Press.
National Research Council. 1974. Productive Agriculture and A Quality Environment: Food Production, Living, Recreation and the Rural Urban Interface. Committee on Agriculture and the Environment, Division of Biology and Agriculture, National Academy Press, Washington, D.C.
National Research Council. 1982. Impacts of Emerging Agricultural Trends on Fish and Wildlife Habitat. Committee on Impacts of Emerging Agricultural Trends on Fish and Wildlife Habitat, National Academy Press, Washington, D.C.
Nielson, E.G., and L.K. Lee. 1987. The magnitude and costs of groundwater contamination from agricultural chemicals: A national perspective. U.S. Department of Agriculture, Natural Resource Economics Division, Economic Research Service, Washington, D.C.
Organization for Economic Cooperation and Development (OECD). 1987. Opportunities for the integration of environmental and agricultural policies. Draft Final Report, ENV/AGR/87.16. Ad Hoc Group on Agriculture and Environment, Environment Committee, Paris.
OECD. 1988. Draft Report on selected policy issues relating to agriculture and environment. ENV(88)11. Environment Committee, Paris.
Phipps, T.T., P.R. Crossman, and K.A. Price, eds. 1986. Agriculture and the environment. National Center for Food and Agricultural Policy, Resources for the Future, Washington, D.C.
Pimentel, D., ed. 1981. CRC Handbook of Pest Management in Agriculture. Vol. III. Boca Raton: CRC Press.
Reichelderfer, K.H. 1985. Do USDA farm program participants contribute to soil erosion? Economic Research Service, U.S. Department of Agriculture, Washington, D.C.
Smith, E.H., and D. Pimentel. 1978. Pest Control Strategies. New York: Academic Press.
University of California. 1986. Sustainability of California Agriculture: A Symposium. Proceedings, Sustainability of California Agriculture Research and Education Program, Davis, California.
University of California. 1987. Resources at risk: Agricultural drainage in the San Joaquin Valley Vol.1. Agricultural Issues Center, Davis, California.
U.S. Department of Agriculture (USDA). 1981. 1980 Appraisal: Soil, water, and related resources in the United States: Status, condition, and trends. Washington, D.C.: U.S. Government Printing Office.
USDA. 1986. U.S. Department of Agriculture Acreage Adjustment Programs. Draft Environmental Impact Statement. Agricultural Stabilization and Conservation Service, Washington, D.C.
USDA. 1988. Agricultural Chemicals and the Environment. Special Reprint of May 1988, Agricultural Outlook. Economic Research Service, Washington, D.C.
U.S. Library of Congress. 1979. Agricultural and Environmental Relationships: Issues and Priorities. Congressional Research Service, Environmental and Natural Resources Policy Division, Washington, D.C.
Webb, S., C.W. Ogg, and W.Y. Huang. 1986. Idling erodible cropland: Impacts on production, prices, and government costs. Natural Resource Economics Division, Economic Research Service, U.S. Department of Agriculture. Washington, D.C.: U.S. Government Printing Office.
World Resources Institute. 1986. World Resources 1986: An assessment of the resource base that supports the global economy. International Institute for Environment and Development. New York: Basic Books Inc.
Young, M.D. 1988. The integration of agricultural and environmental policies. 18th European Seminar of Agricultural Economists on Economic Aspects of Environmental Regulations, Copenhagen.

Impacts on Aquatic Ecosystems

Assessment of the Trophic Impact on the Lake Environment in Poland: A Proposal and Case Study

ANNA HILLBRICHT-ILKOWSKA
Institute of Ecology
Polish Academy of Sciences

AIMS AND CONSTRAINTS OF IMPACT ASSESSMENT

The proper assessment of anthropogenic impacts on the environment is the initial and fundamental step in any effective program of environmental protection and management. In general, an assessment system is based on research data on the responses of the population, community, ecosystem, and landscape components to differing intensities of impact factors (James and Evison, 1978; Best and Haeck, 1982; Gower, 1980; Myers and Shelton, 1980). These assessments are based on field observation, controlled experimentation, and laboratory simulation, and are designed to be used in management practices. Compromises must be made between the ecological complexity of the given phenomena and the need to simplify the information as is usually done in any classification system.

The specific aims of environmental assessment are:

- to predict impacts over short and long time periods, with special emphasis on situations in which disturbances are irreversible under real technical and economical conditions; and
- to assess accurately the stimulus/response relationship and to indicate the kind of protective measures and/or treatment to be undertaken.

The proposed assessment system discussed below concerns trophic impacts on a lake ecosystem and the process of eutrophication as the response of its biota to nutrient loading.

Eutrophication—the uncontrolled increase in the fertility and productivity of aquatic ecosystems—still remains the most common and widespread process of anthropogenic impact on lake environments. In fact, it is the only process of significant disturbance in the belt of Baltic lakelands in northern and Eastern Europe which are managed chiefly for touristic purposes. This post-glacial landscape consists of numerous small- and medium-sized lakes which are surrounded by arable and/or forested land. According to Wuhrman (1984), the further eutrophication of most European lakes is inevitable, even after the elimination of point sources of nutrient input (e.g., sewage discharge and industrial effluents), since inputs from surface runoff and atmospheric depositions are sufficient to stimulate the "algal bowl."

There are many systems which can be used to quantify lake eutrophication and the water quality resulting from this process (Kudelska et al., 1981, 1983). Most of these systems are based on a few easily-measured physicochemical properties of water (e.g., oxygen, BOD, nutrients, and chlorophyll concentrations), or they utilize one of the saprogenic systems (Sladecek in James and Evison, 1978).

The proposed assessment system is based on a more holistic approach to the lake ecosystem as a component of the land/water/atmosphere complex. Specifically, it assesses:

- trophic impact in terms of phosphorous nutrient loading from external sources;
- impact of the watershed on the sources and transport of nutrients;
- natural resistance of lakes to trophic and pollution impacts; and
- responses of biological communities and basic ecological processes in the lake ecosystem across increasing intensities of trophic impacts.

Each of these four assessment components is based on a range of values of selected parameters converted into a three- to four-point scale to reflect the intensity of the process involved or the change in the response. The final result gives the position of the lake in each of the four scales. This combination provides a basis for the choice of proper management and protection treatments for a particular lake and its watershed. It also indicates the highest risk of further degradation.

RATE OF TROPHIC IMPACT AND THE ROLE OF THE WATERSHED IN SUPPLYING AND TRANSPORTING NUTRIENTS

Vollenveider's (1976) concept of permissible and dangerous phosphorous (P) loads as related to the mean depth and residence time of water in a lake and determining the amount of total phosphorus (*TP*) in the spring appears to be useful for management of most P-limited lakes. It is still used in many assessment systems and models.

There is a relationship between the annual loading rate of *TP* from external sources (L_{TP} mg year^{-1}) and the average concentration of *TP* in the water column (mg m^{-3}) during the spring overturn. Furthermore, *TP* is related to the transparency (*SD* m) and chlorophyll concentration (*Chl* mg m^{-3}) in the summer season. For a group of 35 Polish lakes located mainly in the Masurian Lake District, the relevant equations were determined by Uchmanski and Szeligiewicz (1989) as follows:

$$(1)\quad TP_{\text{spring}} = \frac{(L_{TP})}{(V \cdot \sigma)} \cdot \left[1 - \frac{1}{1+0.687 \cdot \sqrt{\sigma}}\right]$$

$$(2)\quad \log SD = -\ 1.09 \log \text{TP} + 2.015$$

$$(3)\quad \log Chl = 0.792 \log \text{TP} - 0.172$$

where V = lake volume, in cubic meters, and
σ = annual exchange rate

These equations are valid for P-limited lakes—those with positive retention of P—with an annual exchange rate higher than zero, and without significant internal loadings. These include lakes that are moderately eutrophic in which *TP* in spring is below 250 $\mu g\ l^{-1}$ and 0.100 $\mu g\ l^{-1}$ in equations (2) and (3), respectively.

The annual total loading to a lake includes at least the sum of the phosphorus load from point sources (e.g., sewage, polluted effluents), nonpoint sources (e.g., surface runoff from direct watershed), and bulk precipitation. The behavior of these phosphorous inputs were studied in special experiments (Hillbricht-Ilkowska et al., 1981; Hillbricht-Ilkowska and Kawacz, 1983, 1985).

For the initial assessment, calculations were made relying on such average regional- or site-specific parameters as export rates from different land-use areas, concentrations of *TP* in precipitation or in sewage with known treatment, and total bulk loadings. However, these calculated values do not indicate the seasonality of the inputs, and they ignore short-term events like storm effects in runoff, variations in sewage input, and the bioavailability of phosphorous. This last factor often requires bioassay studies but, in general, the contribution of available phosphorous (dissolved and/or phosphate phosphorous) in different inputs can be estimated.

The three categories of endangerment (*CatEND*) in terms of phosphorous loadings are proposed as a scale of trophic impact (Hillbricht-Ilkowska, 1984, 1985). When the actual annual *TP* load (i.e., total input from sewage, runoff, precipitation, and tributaries) is below permissable levels according to Vollenvieder's (1976) criteria, the lake is in *CatEND 1*. *CatEND 2* means that the actual *TP* load is equal to or higher than the permissible level, and *CatEMD 3* means that the *TP* load is equal to or exceeds dangerous levels (Table 1). The percentage of *TP* due to sewage inputs is considered important supplemental information, since this part of *TP* is ultimately controllable. The distribution of *CatEND 1-3* in the representative sample of Polish lakes is presented by Cooper (Chapter 18, this volume).

TABLE 1 Four-point classification of watershed impacts on lakes (ClaWI).

PARAMETER	NUMBER OF POINTS 0	1	2	3
Ohle's index	<10	10-40	40-150	>150
Water balance	--	with outflow	without outflow	flowthrough
Drainage density (km - km^{-2})	<0.5	0.5-1.0	1.0-1.5	>1.5
Average slope (%)	<5	5-10	10-20	>20
Share of depression (%)	>60	45-60	20-45	<20
Geological substratum				
loam sand-loam	loam-sand	sand		
Land use	forest-swamp	forest-arable	arable	arable with urban areas

ClaWI	*Mean Point Value*	*Watershed Impact*
0	≤ 1	very weak
1	1.1 - 1.4	weak
2	1.5 - 1.9	moderate
3	>2.0	strong

SOURCE: Bajkiewicz-Grabowska, 1987.

The physiographic properties of the watershed influence its transport function and have been assessed by a four-part watershed impact classification system (*ClaWI*) as shown in Table 1. The system is based on the range of values for selected properties that impact the processes of source and transport of nutrients (Gower, 1980; Myers and Shelton, 1980). The drainage density gradient (i.e., the ratio of the length of watercourses in the watershed to its surface), average sloping, and the ratio of watershed to lake area (Ohle's index) are factors that are positively correlated with the rate of surface transport and accumulation of matter. In contrast, the contributions of marshy areas (without surface outflow) and forested areas (as opposed to arable and urban land) are factors that negatively influence input to the lake. In addition, the dominance of clay as opposed to sand or gravel in the geological substratum is negatively correlated with the rate of underground transport to the lake. Circumstances of strong impact (i.e., *ClaWI 4*) would be watersheds with very high values for at least three or four parameters, such as density gradient, ratio of watershed to lake surface, mean slope, and an area covered by sandy substratum without clay intrusions and at least 10% arable and/or urban areas in the lake watershed.

TABLE 2 Four categories of natural lake resistance (CatLR).

PARAMETER	NUMBER OF POINTS 0	1	2	3
Mean depth (m)	>10	5-10	3-5	<3
Ratio: lake volume (10^3 m^3) to shoreline length (m)	>5	3-5	1-3	<1
Fraction of unmixed layer in lake volume (%)	>35	20-35	10-20	<10
Ratio: bottom area in epilimnion (m^2) to its volume (m^3)	<0.10	0.10-0.15	0.15-0.30	>0.30
Annual water exchange rate	>10	5-10	1-5	<1
Schindler's index	<10	10-30	30-100	>100

CatLR	*Mean Point Value*	*Lake Resistance*
0	≤0.8	high
1	0.9-1.6	moderate
2	1.7-2.4	weak
3	>2.4	very weak

SOURCE: Kudelska et al., 1983; Bajkiewicz-Grabowska, 1987.

ASSESSMENT OF NATURAL RESISTANCE OF LAKES TO TROPHIC IMPACT

Nutrient loading of comparable intensity will produce different effects in lakes with different morphometry and turnover rates. To assess these conditions, four categories of lake resistance (*CatLR*) were proposed by Kudelska et al. (1981,1983) and modified by Bajkiewicz-Grabowska (1987) (Table 2).

The morphometry and flow-through properties of a lake are the factors involved in the accumulation of the external *TP* load, the release of internal loading, and the recirculation of nutrients. Important factors are mean lake depth, the ratio of lake volume to length of shoreline, the percentage of the unmixed layer in lake volume, the ratio of sediment area covered by epilimnion to the epilimnion volume, and the ratio of the sum of watershed and lake areas to lake volume (known as Schindler's ratio). Decreasing or increasing values for these indices indicate four categories of lake resistance. *CatLR 1* means that the lake should be relatively insensitive to the impact of non-point sources (e.g., watershed and precipitation). It usually includes deep, large lakes and lakes with high flushing rates. *CatLR 4* means lower resistance, and usually includes shallow, well-mixed lakes with small outflow and well-developed shoreline.

TABLE 3 Trophic classification of lowland temperate lakes.

	Concentration in summer, surface layer			
	TP $\mu g \cdot l^{-1}$	SD (m)	Chl a $\mu g \cdot l^{-1}$	Phytopl. biomass $mg \cdot l^{-1}$
Dimictic lakes:				
1. mesotrophic	≤50	≥3	≤10	<5
2. moderately eutrophic	≤100	<3	≤30	≤20
3. strongly eutrophic	>100	≤2	>30	>20
Polymictic lakes:				
1. mesotrophic and/or moderately eutrophic	≤100	≤2	≤30	≤30
2. strongly eutrophic	≤300	≤2	≤100	≤100
3. hypertrophic	>300	≤1	>100	>100

SOURCE: Kajak, 1983a, b; Hillbricht-Ilkowska, 1984.

THE ECOLOGICAL RESPONSE OF LAKES TO TROPHIC IMPACT

Research was carried out on about 50 Masurian lakes (including the Great Masurian Lakes) to evaluate trends and magnitudes of changes in lake ecosystems along the trophic continuum and under known conditions of P- loading (Kajak, 1983; Zdanowski et al., 1984). The relationship between the principal discriminant of trophic state (i.e., summer concentration of *TP* in surface layer) and numerous qualitative and quantitative indices of ecosystem processes and communities were determined (Kajak, 1983) (Table 3). This system was proposed for the quantitative assessment of ecological responses of lowland, temperate lakes experiencing differing levels of eutrophication (Hillbricht-Ilkowska, 1984, 1985).

Three trophic classes of lakes (*ClaLT*) are defined according to *TP* concentration in summer, with different ranges for dimictic lakes (D-lakes), which are of sufficient depth to be permanently stratified in summer, and for polymictic lakes (P-lakes), which are shallow and permanently or frequently mixed to the bottom. The generally very high correlation between internal and external *TP* loading in polymictic lakes is the main reason for dividing the whole system into two subsystems, with different ranges of values for the same set of parameters. In polymictic lakes, the relation "L_{TP} - *TP* in spring" does not exist because of unpredictable inputs of *TP* from sediment during the entire vegetation season. The boundary values for successive classes (1-3) were set as follows: $\leq$ 50, $\leq$ 150, > 150 $\mu g\ l^{-1}$ for D-1, D-2, and D-3 lakes, respectively; and $\leq$ 100, $\leq$ 300, > 300 $\mu g\ l^{-1}$ for P-1, P-2, and P-3 lakes, respectively.

The two other parameters considered as secondary discriminants of trophic state are water transparency (*SD*) and chlorophyll concentration

(*Chla*), and are closely related to *TP* in surface waters during the summer. The respective correlations (*r*) between *TP* vs. *SD* and *TP* vs. *Chla* are 0.768 (0.05) and 0.743 (0.01) for dimictic lakes, and 0.484 (0.05) and 0.730 (0.01) for polymictic lakes (Kajak, 1983b).

The indices and parameters related to nutrient availability, abundance and composition of producers and consumers, and transfer efficiencies in the plankton food chain were found to be strongly correlated with *ClaLT* and, therefore, are useful indices for both biotic and functional responses to trophic impacts (Hillbricht-Ilkowska, 1985). These indices of nutrient availability are:

- *TN*:*TP* mass ratio in summer-surface layers as an indicator of the nutrient deficiency to algae demands;
- the fresh weight of phytoplankton biomass in summer;
- the blue-green algae as a percentage of the total biomass and as the ratio to chlorophyll; and
- the percentage of small algae (nanoplankton-size, less than 30 microns) as a rough indicator of the amount of edible items vs. nonedible producers.

The biomass of rotifers, daphnids, and all invertebrate plankton predators, as well as biomass ratios of zooplankton to phytoplankton, *Cyclopida* to *Cladocera*, daphnids to blue-green algae, and predatory to nonpredatory species were found to demonstrate both decreases in the efficiency of the grazing food chain with eutrophication and increases in the efficiency of consumer links. Some of these values could also indicate the influence of fish predation on the zooplankton community. The clogging effect of excess biomass of blue-green algae on daphnid biomass is also apparent. The extreme values of these parameters and indices usually occur in D-3 lakes and/or in P-2 and P-3 lakes which are the most eutrophic and hypertrophic deep and shallow lakes.

The above three trophic classes of lakes are compatible with the four classes of lake water quality (*ClaWQ*) developed by Judelska et al. (1981, 1983) and Cyzdik and Soska (1988), and are based on spring and summer values of 18 parameters which include *TP*, *TN*, *SD*, *Chla*, and oxygen (Table 4). Some events, such as massive fish kills or high numbers of *E. coli* titre, will determine the classification irrespective of other parameters. This classification system is widely used in Poland by state environmental agencies to control the quality of lake water. Relevant statistics are given in Chapters 18 and 19 (this volume).

The combination of these five evaluation systems provides the basis for the proper choice of correction measures in lake management practices. For instance, D-1 lakes with very good water quality (*ClaWQ 1* or *2*) are rather insensitive to watershed inputs. Lakes in *CatLR 1* or *2* are

TABLE 4 Lake quality evaluation system in Poland using water purity indices for three classes of lake water (n = stratified; ns = non-stratified lakes).

Nº	PARAMETER	PERIOD LAYER	CLASS OF WATER PURITY 1	2	3
(s)	Hypolimnetic oxygen saturation (mean %)	summer	≥40	≥20	≥5
(ns)	Oxygen content (mg O_2 1^{-1})	summer, over bottom	≥4.0	≥2.0	≥1.0
(s,ns)	C.O.D. dichromate method (mg O_2 1^{-1})	summer, surface	≤20	≤30	≤50
(s,ns)	B.O.D.$_5$ (mg O_2 1^{-1})	summer, surface	≤2	≤4	≤8
(s,ns)	B.O.D. (mg O_2 1^{-1})	summer, over bottom	≤2	≤5	≤10
(s,ns)	P-PO (mg 1^{-1})	spring, surface	≤0.02	≤0.04	≤0.08
(s)	P-PO (mg 1^{-1})	summer, over bottom	≤0.02	≤0.04	≤0.08
(s)	Total P (mg 1^{-1})	summer, over bottom	≤0.06	≤0.15	≤0.60
(s,ns)	Total P (mg 1^{-1})	spring and summer (mean value), surface	≤0.050	≤0.100	≤0.200
(s,ns)	Inorganic N (N_{NH4} + N_{NO3})(mg 1^{-1})	spring, surface	≤0.20	≤0.40	≤0.80
(s)	N-NH_4 (mg 1^{-1})	summer, over bottom	≤0.20	≤1.00	≤5.00
(s,ns)	Total N (mg1^{-1}) (mean values), surface	spring and summer	≤1.0	≤1.5	≤2.0
(s,ns)	Specific conductance (us cm^{-1})	spring, surface	≤250	≤300	≤350
(s,ns)	Chlorophyll (mg m^{-1})	spring and summer (mean value) surface	≤8	≤15	≤25
(s,ns)	Dry mass of (mg 1^{-1})	spring and summer (mean value), surface	≤4	≤8	≤12
(s,ns)	Secchi Disc (m)	spring and summer (mean value)	≥4	≥8	≥12
(s,ns)	Fecal coli titre	spring and summer surface and above bottom (the worst result)	≥1.0	≥0.1	≥0.01
(s,ns)	Biological field observations	all year, whole lake	occurrence of fish kills or mass mortality of other aquatic organisms (both in littoral and pelagial); puts the lake "out of the class" irrespective of the other parameters		

SOURCE: Kudelska et al., 1981; Cydzik et al., 1986; cited in Hillbricht-Ilkowska, 1984.

usually large or deep and are situated in forested areas. These are the most valuable bodies of water and require special protection in order to maintain their quality. Protective measures are needed immediately if the lake appears to be *CatEND 2* or *3*, since the actual external *TP* load is already dangerously close to threshold level and its present state could deteriorate rapidly. Some change in land use in the watershed might also be needed if *ClaWI* is close to *3* or *4*, e.g., as a result of deforestation.

Another extreme example is that of a rather shallow P-2 lake with very bad water quality (*ClaWQ 4*, or "out of classification"), a very dangerous *TP* loading (*CatEND 3*), a naturally moderate resistance to degradation (*CatLR 2*) associated with a high flow-through regime, and a mostly forested watershed (*ClaWI 1 or 2*). This situation is the clear result of long-term discharge of sewage, and not the effect of improper watershed management. Probably the sewage diversion, together with the restoration of the lake by the inactivation of internal loading, will be sufficient to restore water quality.

ASSESSMENT OF TROPHIC IMPACT ON LAKES IN THE MASURIAN LANDSCAPE PROTECTED AREA AND SOME PROPOSALS FOR THEIR PROTECTION

The assessment systems cited above are intended for use in scientific management and were applied to lakes situated in a protected area of about 700 km^2 in the Lake District in Poland known as the Masurian Landscape Protected Area (Figure 1). This area was established in 1977 to protect the land forms, forest complexes, marsh and bog habitats, rivers, and lakes of the postglacial Masurian landscape. The largest lakes in Poland are located in this area: Lake Sniardwy, a biosphere reserve of 110 km^2, and Lake Luknajno, with a population of mute swans. Other small lakes and marshes in the area form reserves for the protection of rare plant communities and water fowl colonies. There are 24 lakes larger than 50 hectares, ranging in depth from 0.5 to 30 m, and most are mesoeutrophic in character.

Each year as many as 80,000 tourists visit the area. It is protected from industry but is open to limited farming and forestry activities in addition to tourism. Because of the good natural connection between the lakes and the river Krutynia, the area is one of the most popular in Poland for canoes and sailboats.

The area was divided into principal watersheds associated with the larger lakes, and data was collected in order to evaluate watershed impact, natural lake resistance, water quality, and trophic state (Figures 1 and 2) (Hillbricht-Ilkowska, 1988). For most of the lakes, including the deepest and largest ones, conditions of *Cat END 3* were assessed, i.e., the actual *TP*

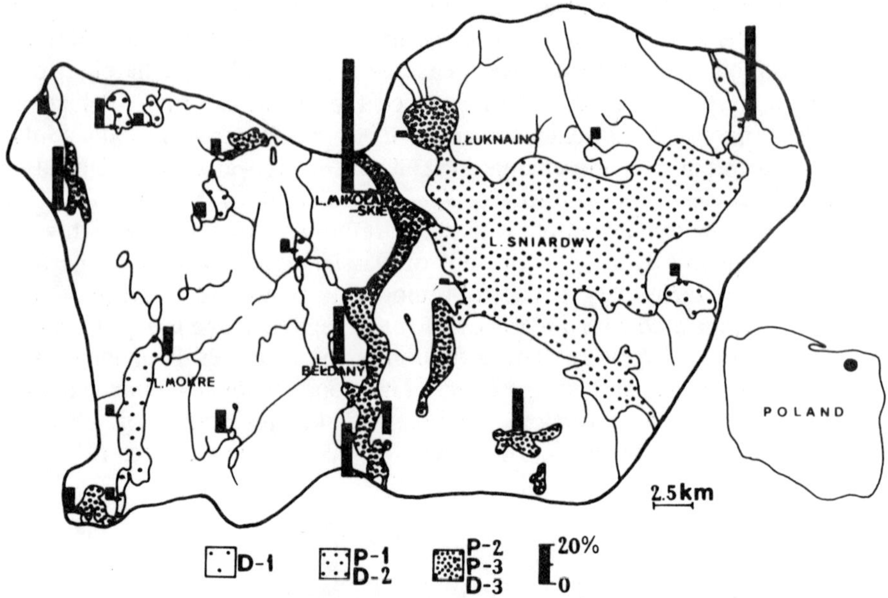

FIGURE 1a Masurian Landscape Protected Area and its lakes. Lake Sniardwy (A=110 km^2; z=5.8 m; z_{max}=23.4 m); Lake Mikokajskie (A=5 km^2; z=11.2 m; z_{max}=25.9); Lake Beldany (A=9.4 km^2; z=10.0 m; z_{max}=46.0 m), Lake Mokre (A=8.5 km^2, z=12.7 m; z_{max}=51.0 m); Lake Luknajno (A=6.8 km^2; z=0.6 m; z_{max}=3.0 m). D-1, P-1 and D-2, and P-3 and D-3=trophic groups of lakes according to *TP* content in summer (surface layer), i.e., < 50, < 150 and > 150 $\mu g\ 1^{-1}$ for D-lakes, and < 300 and > 300 $\mu g\ 1^{-1}$ for P-lakes in above three groups. The percent (%) contribution of *TP* in sewage is indicated next to each lake (vertical black bars).

load exceeded danger levels many times over. Only one deep lake (Lake Mokre) receives an annual *TP* load within permissible limits (*CatEND 2*).

The *TP* input from tributaries, as in the case of Lake Sniardwy, or from rivers, like Lake Beldany, predominates in annual *TP* loadings. In only a few lakes does sewage contribute from 30-70% of total inputs; Lake Mikolajskie, Lake Beldany, and some smaller lakes experience these inputs. This means that for the rest of the lakes, dangerous levels of inputs are attained only from nonpoint sources of *TP* and from tributaries which are uncontrollable sources. In this situation, high water quality in these lakes should be a temporary condition. Eutrophication rates appear to decrease due to relatively high natural resistance to watershed impact (mostly *CatLR 1* and *2*) resulting from the depth, shape, and flow-through regime, as well as by moderate watershed impacts (*ClaWI 2* and *3*) resulting from intermediate or lower value density gradient, prevalent forest landscape, and the clay substratum (Figure 2). The most threatening conditions are

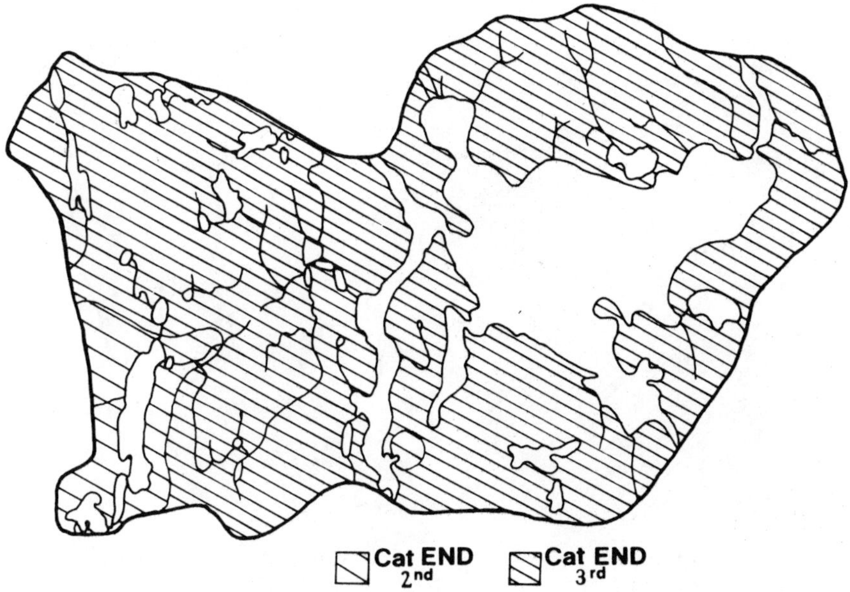

FIGURE 1b *CatEND*—category of endangerment when annual *TP* loading is at permissible levels (2) or above dangerous levels (3) according to Vollenweider's criteria (1976).

the narrow and deep lakes (Mikolajskie and Beldany) and several smaller lakes which have a higher position in all five ranking systems, including the category of watershed impact (Figures 1 and 2).

Based on these findings, three principal methods of protection and management were proposed to maintain the actual status of some of the lakes and to control the further eutrophication and hypertrophication of others (Figure 3).

For the central part of the area around Lake Mikolajskie and Lake Beldany, a program of lake restoration was proposed utilizing deep water aeration, inactivation of *TP* internal loading, and diversion of principal point sources of pollutants (Figure 3). The intensive protection of the watershed and shore zone includes the recultivation of near-lake slopes and shore zones around whole lakes, including anti-erosion installations and reforestation. An immediate ban was placed on the further development of year-round and seasonal tourist centers, and touristic paths were redistributed in order to disperse them over the area. Full protection was provided for small, marshy areas and forest fragments, which included reforestation of sections of plowed land in contact with the lake shore, and control of fertilization and cattle breeding in nearby arable land.

For the western part of the Masurian Landscape Protected Area—in

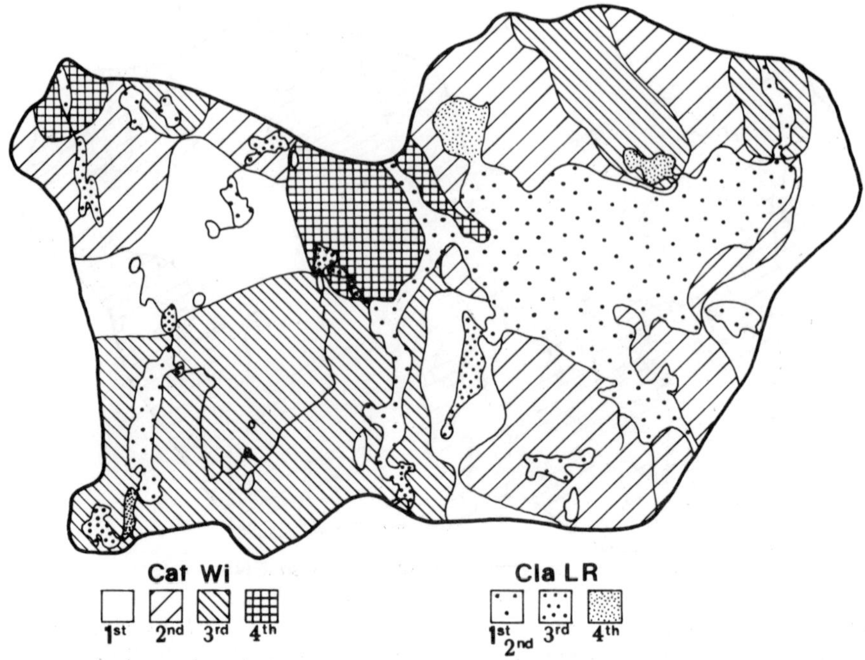

FIGURE 2 Categories of watershed impact on lakes (*CatWI*) from 1 to 4 (lowest to highest) and classes of natural lake resistance to eutrophication (*ClaLR*) from 1 and 2 (high and moderate resistance) to 4 (very low resistance) for lakes and watersheds in Masurian Landscape Protected Area (Bajkiewicz-Grabowska, 1987).

which the deep mesotrophic Lake Mokre and other mesotrophic shallow lakes are located—special protection of watersheds and lakes was established to preserve their actual state, including control of fishing operations and a ban on the introduction of planktivorous fishes. Full protection of riparian forests included a ban on wood harvesting. Deconcentration of tourist activity and a ban on its further development was required, as was a ban on the liming and fertilizing of small lakes for fishery purposes.

For the eastern part of the area, including Lake Siniardwy (Figure 3), some of the same suggestions were made concerning fishery, forestry, and agricultural operations, and somewhat more relaxed suggestions were made concerning tourism. In two cases, proposed watershed and lake protection measures were to be fully subordinated to the needs of waterfowl protection, including a ban on all tourist activity in the vicinity of the lake.

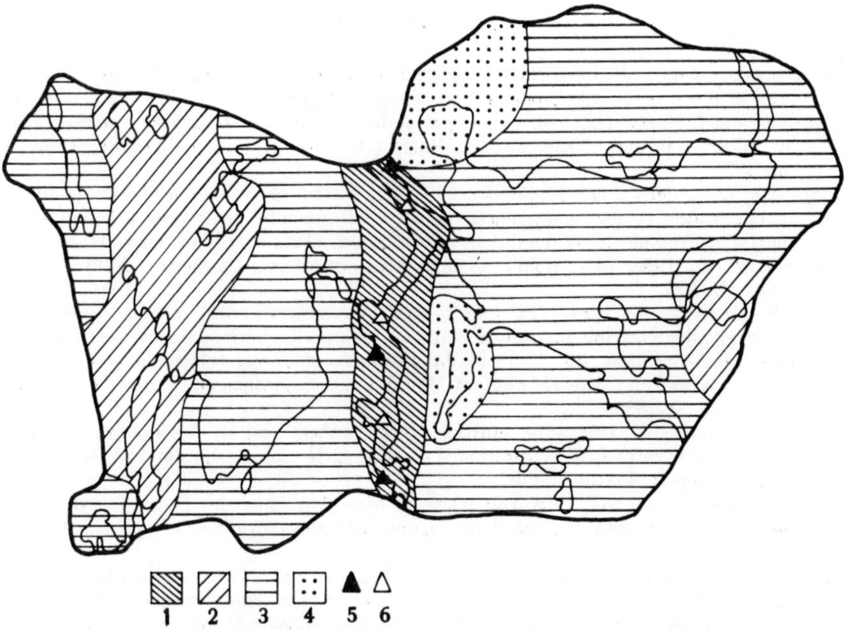

FIGURE 3 Recommended trends in watershed and lake protection for Masurian Landscape Protected Area. 1=intensive protection with recultivation of land around the lake and lake restoration; 2=special protection in order to maintain the status quo of mesotropic lakes and their watersheds; 3=partial protection of watersheds in order to control further eutrophication of lakes; 4=protection of lake and shore area with regard to requirements of waterfowl protection; 5=diversion of point source of pollution; 6=localization of deep water aeration installation in lakes.

CONCLUSION

Eutrophication remains the primary cause of anthropogenic disturbances of natural lakes and their environments, even after the elimination of point sources of nutrient inputs. It is also the primary cause of changes in the numerous small- and medium-sized natural lakes situated in arable regions (e.g., the North European postglacial lakelands) which function as important tourism centers.

An assessment system of trophic impacts based on the role of the watersheds in the supply and transport of nutrients, the natural resistance of lakes to eutrophication, the response of the lake biota and its internal processes along the trophic continuum was proposed as the operational system for management purposes. This system, which is composed of relevant classes and categories, was applied to the case of the 24 large lakes of the Masurian Landscape Protected Area, and methods for different forms of protection of watersheds and lakes were recommended.

REFERENCES

Bajkiewicz-Grabowska, E. 1987. Natural degradation ability of lakes and the role of drainage area in the process. Ekol. pol. 33:280-289 (Polish/English Summary).

Best, E.P., and J. Haeck, eds. 1982. Ecological indicators for the assessment of the quality of air, water, soil, and ecosystems. Dordrecht (Boston): D. Riedel Publishing Company, pp. 407.

Cydzik, D., D. Kudelska, and H. Soszka. 1986. The system of evaluation of lake quality and its application by the field agencies of the environmental protection in Monitoring of the Lake Ecosystems, A. Hillbricht-Ilkowska, ed. Ossolineum, Wroclaw, 11-20.

Gower, A.M., ed. 1980. Water quality in catchment ecosystems. Chichester: John Wiley and Sons, pp. 335.

Hillbricht-Ilkowska, A. 1984. The indices and parameters useful in the evaluation of water quality and the ecological state of temperate, lowland lakes connected with their eutrophication. Pp. 55-69 in the Proceedings of the International Conference, "Conservation and Management of World Lake Environment." Otsu Shiga, Japan, August 24-30, 1984.

Hillbricht-Ilkowska, A. 1985. The eutrophication process and its control in Poland and the perspectives of lake restoration. Pp. 362-372 in Lake Pollution and Recovery, R. Vismare and R. Marforio, eds.

Hillbricht-Ilkowska, A., ed. 1988 The lakes of Masurian Lakeland Park: Eutrophication, protection, and management. Ossolineum, Warszawa-Workaw. (in press) (in Polish with English summary).

Hillbricht-Ilkowska, A., W. Grozczynska, and M. Planter. 1981. Nonpoint sources of nutrients to the lake watershed of river Jorka, Masurian Lakeland, Poland. Pp. 152-183 in Impact of Non-Point Sources on Water Quality in Watersheds and Lakes, J.H.A.M. Steenvoordenand and W. Rast, eds. Amsterdam, The Netherlands, May 11-14, 1981.

Hillbricht-Ilkowska, A., and W. Kawacz. 1983. Biotic structure and processes in lake system of river Jorka watershed, Masurian Lakeland, Poland. I. Land impact loading and dynamics of nutrients. Ekol. Pol. 31:539-585.

Hillbricht-Ilkowska, A., and W. Kawacz, eds. Factors affecting nutrient budget in lakes of the river Jorka watershed, Masurian Lakeland, Poland. Ekol. Pol.33(2):171-318.

James, A., and L. Evison, eds. 1978. Biological indicators of water quality.

Kajak, Z., ed. 1983a. Ecological characteristics of lakes in northeastern Poland versus their trophic gradient. Ekol. Pol. 31:239-530.

Kajak, Z. 1983b. Dependence of chosen indices of structure and functioning of ecosystems on trophic status and mictic type of 42 lakes. Ekol. Pol. 31:495-530.

Kudelska, D., D. Cydzik, and H. Soszka. 1981. A proposal of lake water quality classification. Wiad. Ekol. 27:149-173 (in Polish with English summary).

Kudelska, D., D. Cydzki, and H. Soska. 1983. Lake water quality classification system in Polish. Publ. Inst. Environment Management, Warsaw, pp. 44.

Myers, W., and R.L. Shelton, eds. 1980. Survey methods of ecosystem management. New York: John Wiley and Sons, pp. 403.

Uchmanski, J., and W. Szligiewicz. 1989. Models for predicting the water quality in lakes. Ekol. Pol. (in press).

Vollenweider, R.A. 1976. Advance in defining critical level for phosphorous in lake eutrophication. Mem. Inst. Ital. Idrobiol. 33:53-83.

Wuhrman, K. 1984. Lake eutrophication and its control. Pp. 26-37 in the Proceedings of the International Conference, "Conservation and Management of World Lake Environment," Otsu Shiga, Japan, August 24-30, 1984.

Zdanowski, B., A. Korycka, and J. Zachwieja. 1984. Thermal and oxygen conditions and the chemical composition of the water in the Great Masurian Lakes. Ekol. Pol. 32:651-678.

Aquatic Research and Water Quality Trends in the United States and Poland

WILLIAM E. COOPER
Michigan State University

This chapter addresses trends in water quality in the United States and Poland and the ecological research associated with environmental assessments of the impacts of these trends. These impact analyses are generally based on risk assessments that involve transport, fate, exposure, and toxicology of various materials. The research activities discussed below are those that contribute directly to these processes.

Trends in water quality parameters include both surface and groundwater resources. The nutrient concentrations are limited to phosphorus and nitrogen. The toxic chemicals include the most commonly detected classes of chlorinated organic compounds. The U.S. data were obtained mostly from state and federal monitoring programs, and the discussion here uses the Laurentian Great Lakes as a prototype watershed. The Polish data were obtained primarily from the Board of Environmental Protection and Water Management in Poland.

ECOLOGICAL RESEARCH

Traditionally, environmental scientists have subdivided integrated ecosystems into a media-specific taxonomy. Research, teaching, and regulatory activities are organized into categories such as soil, groundwater, surface water (i.e., freshwater lakes and streams, marine-inshore, and bluewater), and atmosphere. The basic concepts of ecology are thought to be generic to all media. The differences are associated with the physical constraints that limit the expression of ecological processes to a subset of the total array possible. These physical constraints are usually media-specific.

Impact assessments of environmental and ecological resources are currently required by many state and federal agencies. Many of these regulations require the assessments to be anticipatory. The analyses are, therefore, based on predictive models, laboratory determinations of critical processes and rates, microcosm experiments with simplistic ecosystems, or field validations with small-scale test systems. These activities involve the use of scientific assumptions involving reality, precision, and scale. Substantial research activity has been directed toward understanding and measurement of the fate, effect, transport, and exposure pathway of toxic chemicals in the environment. Even so, there is a great deal more that we need to learn before impact assessment models will work generically under field conditions.

If biological communities and transport pathways of chemicals were unique to each media, the existing organizational structure would be ideal. Unfortunately, there is considerable documentation that organisms with complex life histories occupy different media during their life span, and most chemicals cycle through ecosystems involving many media. For example, lake sediments are both sources and sinks for nutrients, and toxicants cycle in the planktonic community. Erosion is a major transport mechanism between terrestrial and aquatic communities. Most shallow groundwaters emerge as surface water through seeps, springs, and emergent water courses. Rain and snow scavenge materials from the atmosphere, resulting in wet deposition to both terrestrial and aquatic systems. Gravity does the same for larger particles as dry deposition.

The pathways that couple media-specific subsystems do exist. The major issues involve the intensity of the transport processes relative to the processing capability of the media, given the concentrations of the materials and the retention times within the media receiving the input. This section will focus on those biological and chemical processes that affect the fate, transport, effect, and exposure of materials that are utilized in performing ecological risk assessments in freshwater environments.

Predictions of ecological events can be based on stimulus/response behaviors of interacting species populations. The flows of the requisite energy and material resources required to maintain viable populations involve a wide variety of biological processes. Multispecies interactions mediated through such processes as competition, predation, and succession determine which constellation of organisms are predominant under a designated set of abiotic conditions. There is a dynamic structure to ecological systems which is both opportunistic and adaptive. This hierarchical structure is both temporally and spatially distributed, and its behaviors are often nonlinear, with an array of process-specific time lags. The control systems are entirely feedback and are decentralized, as they are often associated with the processes themselves.

These characteristics make it very unlikely that useful predictions of specific behavior resulting from the aggregation of multistage processes will result from simple numerical solutions. Most ecological predictions result from dynamical simulation models that contain a number of components; each component is represented by a set of state and response equations; and each equation involves a series of specific parameters. The values of these parameters are represented as distributive functions with coefficients of variation that usually exceed 30%. The variations in parameter values include elements of genetic, physiological, and behavioral variability as well as sampling errors. There currently is no way to calculate confidence intervals around the final predictions from such complex ecological interactions.

An illustration of the multifactor interactions that relate to the fate of toxic materials in the environment can be obtained from the anaerobic degradation of halogenated organic compounds. Bouwer et al. (1981) presented results of microbial degradation of several halogenated compounds under anaerobic conditions. Cultures seeded with methanogenic bacteria showed significant reductions in the parent compounds within a 16-week period. Only the original compounds were analyzed, so there is no information on the extent of mineralization. Thus, the methanogenic pathway of carbon flow does result in the dehalogenation of several organic toxicants.

Given that the mechanism does exist, the issue becomes the importance of this pathway of carbon flow in natural ecosystems. There are two major pathways for carbon flow in the anaerobic sediments of warm-water lakes. The methanogenesis pathway involves an array of microbes that utilize electron acceptors (i.e., HCO_3) generated or regenerated internally within the zone of anaerobic metabolism. The sulfate reduction pathway involves a different array of bacteria whose metabolic rates are limited by electron acceptors (i.e., SO_4) generated or regenerated externally to the anaerobic sediment zone. Lovley and Klug (1986) presented a more detailed model of the two competing pathways. Both pathways utilize acetate, which arises from the leaching and hydrolysis of particulate organic matter followed by anaerobic fermentation. The factors that control the production of particulate organic matter involve nutrient availability, competition, and predation in the epilimnetic portions of a lake.

The profundal sediments in oligotrophic and mesotrophic lakes have respiration indices (RI = $^{14}CO_2/^{14}CH_4$) of approximately 1.0 in the upper 4 cm. The eutrophic lakes—with higher organic inputs to the sediments and limited SO_4 recharge from the water column—have RI indices between 0.2 and 0.4. The dominance of either pathway is the result of a competitive interaction of two very different bacterial communities mediated by community events in the adjacent water column (Lovley et al., 1982; Lovley and Klug, 1982; Lovley and Klug, 1983).

Many of the diagnostic parameters of chemicals are measured under simple two-phase conditions. Volatilization rates are measured as transfers through an air/water interface. Solubilities are measured as solute and solvent ratios in pure water. Bioaccumulation rates are related to octanol/water partition coefficients that are measured with two phases in pure state. However, the behaviors of chemicals in natural systems cannot usually be predicted from these idealized parameters. Natural systems are "multiphasic" and the processes of adsorption, absorption, volatilization, hydrolysis, and photodegradation are not easily characterized by simple rate coefficients.

A good example of the importance of chemical processes is the current dilemma with the Story/Ott/Cordova site in Muskegon, Michigan. The groundwater is a contained aquifer; the only outlet is surface flows through Bear Creek. The aquifer is contaminated with many thousands of kilograms of organic compounds, including benzene, dichloroethanes, toluene, and vinyl chloride. The parent company, Corn Products Corporation, has recommended an out-of-court settlement that involves monitoring, financial compensation for impairment of the resource, and air stripping through natural volatilization as the groundwater flows to the surface. However, environmental groups have demanded a multimillion dollar purge and incineration alternative. The analysis of trade-offs involves risks, benefits, and uncertainties which depend upon physical parameters that control the fate and transport of these volatile organic compounds.

The behavior of contaminants in the environment is determined by two fundamental groupings of processes. Transport processes serve to move chemicals through the environment. This may simply involve flow within a medium—such as the movement of the constituents in a wastewater discharge—downstream with the flow of water. It can also involve the movement of chemical species between media, e.g., when the wastewater discharge contains a volatile compound which moves into the atmosphere as it travels downstream. Such intermedia transport processes result in a distribution of the contaminant, necessitating consideration of flow processes in two media. Consideration must also be given to transformation processes—reactions that serve to alter the nature of environmental contaminants. Transformation processes include such environmentally significant reactions as hydrolysis, protolysis and oxidation/reduction.

The fates of organic and inorganic chemicals in the environment are highly influenced by biological, particularly microbial, processes (Alexander, 1981). These processes can involve mineralization, cometabolism, activation, or incomplete biodegradation. Recently, there has been an increasing research effort to document the mechanisms, environmental constraints, and requisite microbial flora that produce these alternative scenarios. Priority areas for research fall into several general categories. Synthetic organic compounds often do not have natural structural analogs. Evolution is not preadaptive, so there is no guarantee that an appropriate biochemical pathway is present in any given microbial community. The susceptibility of synthetic compounds to mineralization in various media—such as anaerobic groundwater, aerobic surface water, and saturated soil—is a critical area requiring further investigation.

Often, portions of a metabolic sequence that would lead to mineralization are present. Limitations such as the lack of an adequate electron acceptor, insufficient concentrations of the parent compound to support a healthy microbial community, or ecological interactions like predation or competition from other organisms, can restrict the expression of some idealized process of mineralization. Since natural microbial communities evolved to exploit a vast array of natural organic compounds under many environmental conditions, there is an enormous number of combinations of "compound × microbial community × ecosystem" that are important to understand. The daughter products that arise from incomplete degradation present the ecologist with an impact assessment task of increasing complexity and uncertainty.

Many toxic compounds exist in nature at trace levels of concentration. Even if the potential for microbial processing exists, the concentrations of organics are too low to supply sufficient energy to maintain adequate populations of the microbe (Boethling and Alexander, 1979). "Piggybacking" energy resources with degradation processing appears to occur through cometabolism. In addition, bacteria produce polysaccharide substrates that bind dissolved organic matter and increase their concentrations in microsites. We need to understand these natural reactions and investigate ways of enhancing these activities in both freshwater and marine aquatic ecosystems.

Several types of natural organic substrates are quite resistant to microbial breakdown (Alexander, 1973). Leaves with high levels of condensed tannins, the humic components of many soils, and wood materials containing cellulose, lignins, and hemicelluloses all have been demonstrated to be recalcitrant. These substrates are also prime binding sites for trace organic toxicants. The accessibility of compounds for microbial degradation is often reduced when they are attached to the surfaces of large substrates.

All of these factors are critical issues when one starts with ambient concentrations of some toxicant and attempts to assess the ecological impact in some specified environment. Once the fate, distribution, and concentration of the parent compound and daughter products are characterized, the remaining factor is the exposure rate for each biological species of concern. Exposure rates under field conditions are the most difficult measurements to obtain. This involves a blending of analytical chemistry, life history aspects of the organism, and the physiological processes of assimilation, depuration, and passive uptake. There are very few organisms of social and/or economic interests for which these rates are well documented.

Over the past decade, an increased emphasis on study of the fate of chemicals in groundwater systems has developed. More than 40% of the U.S. population uses groundwater for drinking, often with no treatment other than disinfection in municipal systems. Rural sources of groundwater are often used for potable water without treatment.

Therefore, the attention of both researchers and regulators has been focused on the problems of widespread use of substantial quantities of chlorinated organic solvents (more than 2.0 million tons/year in the United States) and the discovery of literally thousands of sites across the country where these materials, along with petroleum derivatives, have been detected as contaminants in freshwater aquifers. Contamination of groundwater is a serious problem because subsurface aquifers do not have the same natural degradation mechanisms that are present in surface water systems. Although groundwater systems are substantially less active both biologically and chemically than are surface water ecosystems, there are still a large number of processes that affect the transport and fate of trace contaminants in subsurface aquifers. Questions of interest in understanding the environmental fate of trace organics include:

- What are the mechanisms of removal or transformation?
- What are the intermediate and end products of the transformations?
- What are the transport kinetics of the chemical contaminants in relation to the general flow of the aquifer?
- What manipulations can be performed on contaminated areas most effectively to remove or reduce contaminant levels?

Many of the same physical characteristics of the contaminant molecules that influence transport in surface waters are also important in groundwater movement. Water solubility and the adsorption potential of a chemical are among the most significant. Chemicals or chemical mixtures with low water solubilities and densities less than 1.0 generally exist as free-phase layers on top of the surficial aquifer. Depending upon the rate of infiltration and vapor pressure of the material, there may be significant losses to the atmosphere as well. The most common example of this behavior is the case of losses of gasoline or fuel oil from catastrophic spills or leaking underground storage tanks. In such cases, recovery systems can often be installed to withdraw the supernatant layer off of the aquifer and remove the bulk of the polluting material. Trace concentrations of organic contaminants often remain in the system at the limits of solubility and in the unsaturated overlying soils. The dynamics of the environmental transport and fate of these contaminants are therefore of great interest.

TABLE 1 Trends in water quality, 1975-1981. Trends represent the number of streams showing significant differences at the 90% confidence level.

	Total Phosphorus	Inorganic Nitrogen
Increasing Trend	35	72
Decreasing Trend	29	24
No Change	245	152

SOURCE: USGS Stream Quality Accounting Network, 1984

TRENDS IN WATER QUALITY IN THE UNITED STATES

Nutrients

Nutrient loadings to surface waters were a major focus of the 1972 Federal Water Pollution Control Act. Phosphorus and nitrogen were the two elements associated with cultural eutrophication. The geochemical cycling of phosphorus is highly influenced by physical parameters, while the nitrogen cycle is strongly regulated by microbial activity.

The U.S. Geological Survey (USGS) has maintained a network of stream monitoring stations since 1975 (USGS, 1984). Table 1 presents the direction of trends that are statistically significant between 1975-1981. The pattern is obviously mixed, with the southern coastal areas indicating increasing concentrations and the majority of the interior section indicating a reduction in phosphorus concentration.

In the Great Lakes Region, the trend reflects a systematic reduction in phosphorus concentration. The 1985 Report on Great Lakes Water Quality (Figure 1) presents total phosphorus concentrations in the surface waters of Lake Ontario. Between 1970 and 1983 there was almost a 50% reduction in concentration. The major changes in loadings resulted from a ban on the sale of high-phosphate detergents and the precipitation of phosphorus in the secondary treatment phase of municipal wastewater treatment plants (Figure 2).

Nitrogen, on the other hand, has demonstrated a trend of increasing concentrations in surface stream waters during the period from 1975-1983 (Table 1). This trend is also apparent in the Great Lakes surface waters (Figure 3). The sources of nitrogen are varied, and the transport processes more complex. On a regional basis, the most likely source of increasing inputs are human and animal wastes and mineral fertilizer from agricultural and domestic activities.

The trend in nutrient concentrations in groundwaters is basically associated with nitrogen. Table 2 presents the nitrate-nitrogen concentrations of about 124,000 wells categorized by depth. The only significant increase in

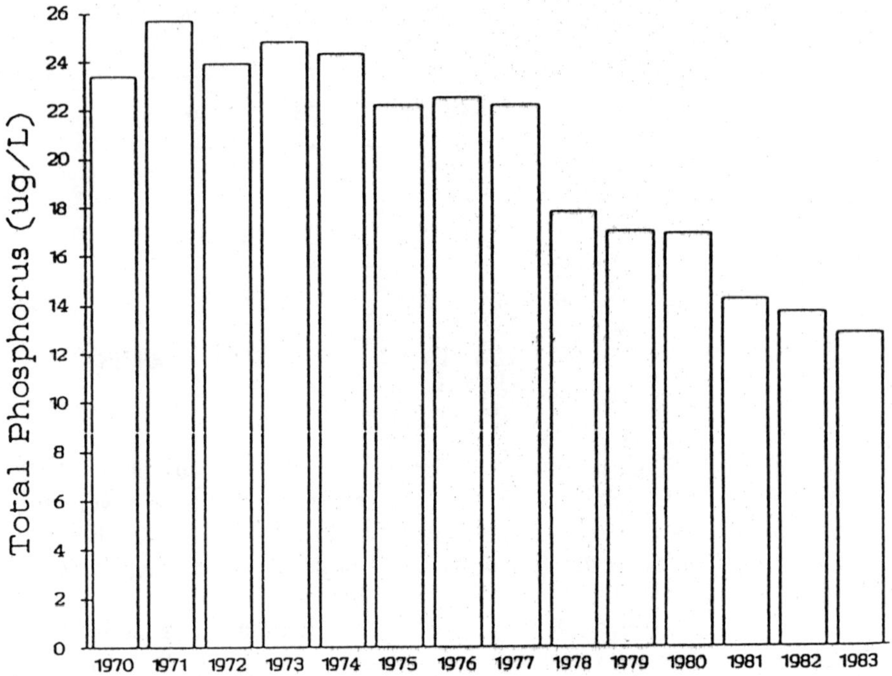

FIGURE 1 Area-weighted mean whole lake spring total phosphorus concentrations in the surface waters (1 m) of Lake Ontario, 1970-1983. Data from Environment Canada (12), 1985 Report on Great Lakes Water Quality, I.J.C.

nitrogen concentration was observed in shallow aquifers less than 100 feet (30 m) in depth. The most likely sources of increasing inputs are animal production facilities, human septic fields, and mineral fertilizers. Hallbert (1987) reported that nitrate concentrations in groundwater for two well systems increased from less than 10 to over 200 mg/l during the period from 1934-1984. This correlates very well with the pattern of nitrogen fertilizer used in Iowa during the same period, which increased from small amounts to over 1.1 million tons per year. The regulation of nitrogen inputs into both surface and groundwaters is still a major problem that demands a high research priority.

Toxic Chemicals

Recent developments in analytical chemistry have expanded the array of toxicants that can be extracted for analysis from organisms, water, and sediments. Technical developments have also drastically lowered the level of detectability. For example, dioxin analyses can be conducted

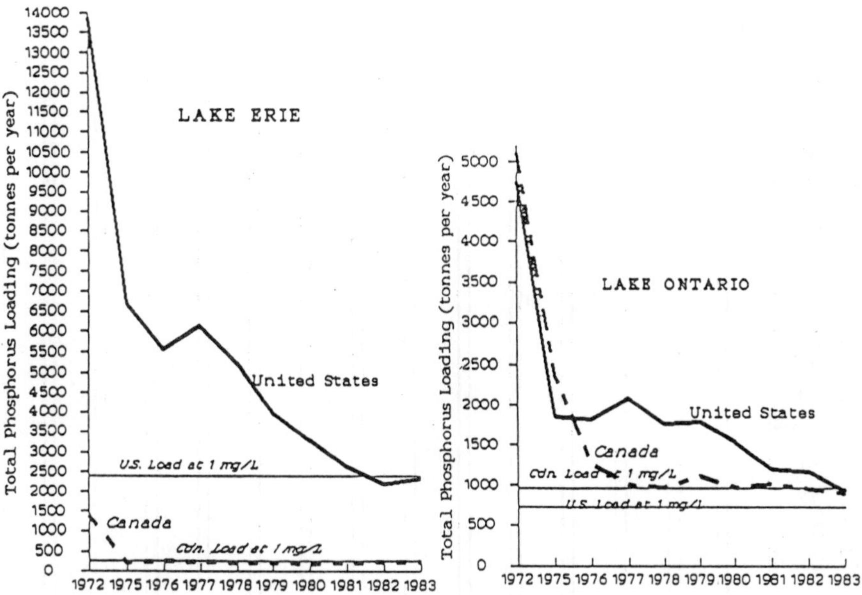

FIGURE 2 Municipal phosphorus loadings to the lower Great Lakes, 1972-1983. Data from the 1985 Report on Great Lakes Water Quality, I.J.C.

with quality control at levels of 1×10^{-12}, with water samples utilizing high-volume filtration. As a result of these developments, there is an ever-increasing list of toxicants being detected in various environmental media. In 1983, the Great Lakes Water Quality Board reported finding over 1,000 anthropogenic chemical substances in surface waters. A general scan of organic compounds in the flesh of lake trout from Lake Michigan indicate some 160 identifiable organic compounds. There are, however, only a relatively few compounds that exist at concentrations high enough to cause concern for human health and the natural resource base.

The compounds of most concern include PCBs, DDT, dioxins, and pesticides like dieldrin. Most of these compounds were identified as environmental contaminants back in the early 1970s. Measurements were also made of survivorship of young fish and eggs from both natural populations and hatchery stocks. Two facts are apparent: the body loads of PCBs and DDT are consistently decreasing in adult lake trout; and fry and egg survival was not reduced by ambient concentrations of these two compounds. In general, the levels of toxicants in surface waters restrict the economic use of these living resources, but they do not threaten the survival of the natural aquatic populations.

Most of the organohalogenated compounds of major concern have

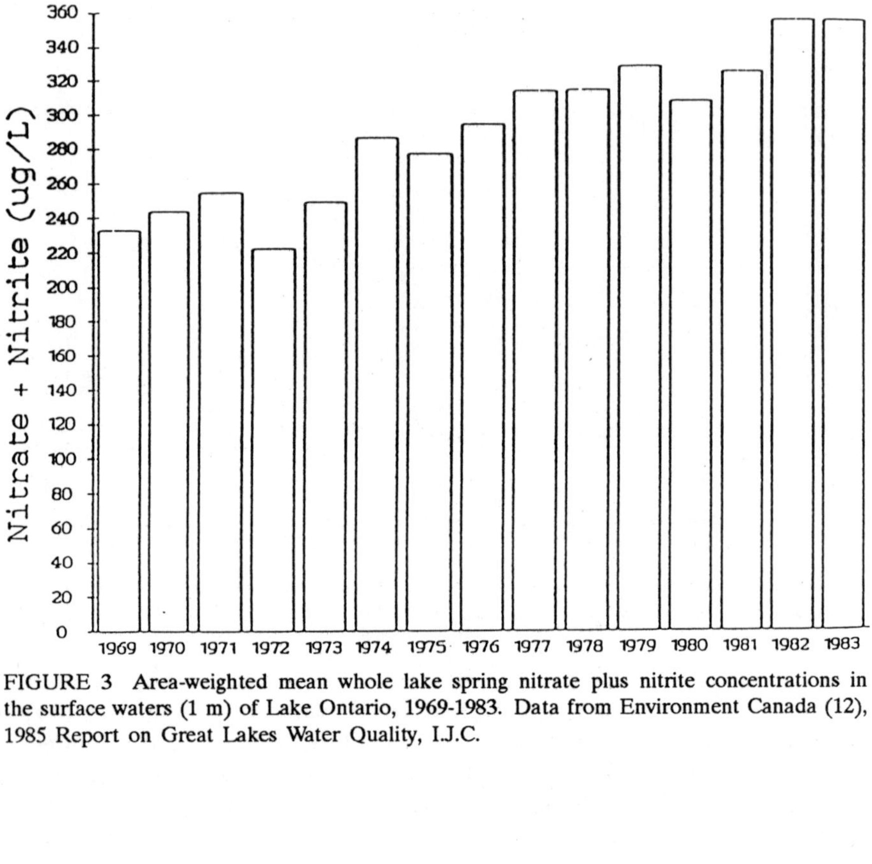

FIGURE 3 Area-weighted mean whole lake spring nitrate plus nitrite concentrations in the surface waters (1 m) of Lake Ontario, 1969-1983. Data from Environment Canada (12), 1985 Report on Great Lakes Water Quality, I.J.C.

TABLE 2 Groundwater contamination (as percentages).

		Nitrate-Nitrogen Concentration		
Depth of Well (feet)	Fraction of Wells Sampled	$<3mg.1^{-1}$	$3mg.1^{-1}$ $-10mg^{-1}$	$>10mg.1^{-1}$
<100	38	32	52	68
101 - 200	22	21	22	18
200 - 300	12	14	11	6
>300	28	33	15	8

SOURCE: USGS, 1984

partitioning coefficients between 10^4 and 10^6. Therefore, the compounds are tightly bound to particulates and do not occur in the aqueous phase. When collecting data to determine temporal trends in ambient concentrations, one should analyze sediments, not water, and deterimine sediment transport, not water flow. This has not routinely been part of the U.S. monitoring program, so available trend data do not exist before the early 1970s.

There has been a considerable amount of effort spent on determining trends in concentrations of these organochlorine compounds in humans and natural fish and wildlife populations. Although there has been a rise in U.S. production of pesticides in the last 20 years, the concentrations of DDT dropped significantly, and the levels of dieldrin remained low in human diets during this same time interval (Conservation Foundation, 1984). Figure 4 shows the trend for DDT and dieldrin for fish and bird populations. Again, there is a consistent reduction, except for dieldrin in waterfowl. The 1985 Report on Great Lakes Water Quality presented data for bloater chubs, lake trout, and herring gull eggs. The reductions in body concentrations are correlated with sharp reductions in U.S. production and use of these compounds. The majority of the inputs of PCBs into the surface waters of the Great Lakes are airborne materials emanating from municipal and industrial incinerators which are currently recycling through the ecosytem.

In general, once a toxic compound is found in concentrations that raise immediate concerns for public health or the integrity of the ecological resource, actions have been taken that have reduced the ambient concentrations of the toxicant. Often, this means a ban or a very restricted use of the product. For those products that remain in use, the industrial expenditures on pollution control are considerable.

A form of environmental pollution in the United States of particular social concern is groundwater contamination. With both state and federal ("Superfund") resources, all states have instituted major programs to characterize the quality of their groundwater resources. The USGS monitors the distribution of rural groundwater contamination in the United States. Each year the number of aquifers identified as contaminated has systematically increased. However, this may not reflect an increase in the frequency of new pollution events; rather, this trend is the result of the increasing effort being spent on groundwater surveys. Most presently contaminated aquifers have contained pollutants for decades.

Many of the organic contaminants found in surface waters are not major constituents in groundwater. The partitioning coefficients for these compounds are high, and the compounds are bound to the particulates in the soil matrix and are generally not found in the aqueous phase. This limits their mobility through soils and protects aquifers from surface

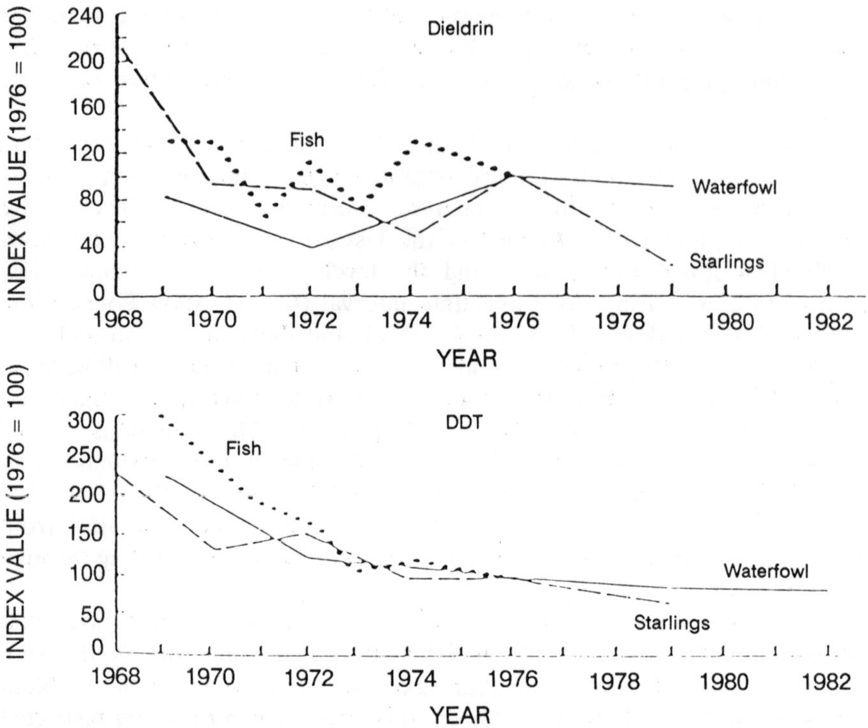

FIGURE 4 Trends in levels of dieldrin and DDT in wildlife, 1968-1979. Data from U.S. Fish and Wildlife Service. State of the Environment, 1981; a report from the Conservation Foundation.

sources. The most common organic contaminants in groundwater are organic solvents. PCBs and dioxins move through the soil matrix only when they are comingled with organic solvents like TCE or DCE.

Klepper et al. (1987) summarized the major sources of contaminants to groundwater (Table 3). Underground gasoline storage tanks and poorly designed landfills are dominant sources of organic contaminants. Uncovered storage of salt for ice removal in the winter months is a major inorganic source. In some areas, the production and transport of petroleum products are also major sources. Landfills are obviously the source of the greatest array of chemical contaminants that enter groundwaters.

TABLE 3 Rural groundwater contamination. Major contaminants associated with known groundwater contamination incidents, excluding urban-heavy manufacturing incidents.

Contaminant	LF	DMP	GUST	AGR	CUST	AGT	OIL	PET	SALT
Benzene	X		X			X	X	X	
Xylenes	X		X			X	X	X	
Toluenes	X		X		X	X	X	X	
Ethylbenzenes			X			X	X	X	
Cadmium	X								
Chromium	X	X							
Copper	X								
Lead	X		X						
Nickel	X								
Zinc	X								
Cyanide	X	X							
Arsenic	X	X							
Phenols	X	X				X			
Dichloroethanes	X	X			X	X			
Trichloroethanes	X	X			X	X			
Trichloroethylene	X	X			X	X			
Tetrachloroethylene	X	X							
Naphthalenes	X								
Chloroform	X								
Hexachlorobenzene	X	X							
PCBs	X								
Phthalates	X								
Paint Residues	X	X							
Nitrate				X					
Pesticides				X					
Salt/Brine									X

LF = Landfill; DMP = Dump; GUST = Gasoline Underground Storage; AGR = Agriculture; CUST = Chemical Underground Storage; AGT = Above Ground Tanks; OIL = Oil Transport; PET = Oil Field Operations; SALT = Salt Storage.

Data from Klepper et al., 1987.

TRENDS IN WATER QUALITY IN POLAND

In Chapter 19 (this volume), Gromiec presents the current river classification for Poland (see Figure 6). A majority of the major rivers in Poland are currently experiencing serious degradation of water quality. Untreated municipal waste streams, salt-water discharges from coal mines, direct industrial discharges, and nonpoint inputs from agriculture are all identified as major sources (see Chapter 19, Figure 1). As approximately one-half of the areas of the major rivers are classified as either Class III (industrial use

TABLE 4 Distribution of 230 natural lakes in Poland studied in 1974-1983, in categories of lake susceptibility to degradation.

Category	Number of lakes	Percent (%)
1st	24	10
2nd	103	45
3rd	66	29
out of 3rd category	37	16
TOTAL	230	100

SOURCE: Cydzik et al., 1982; Cydzik and Soszka, 1988.

only) or nonclassified (pretreatment required before any direct use), pollution of the riverine resources is currently a major environmental problem in Poland.

In Chapter 17 (this volume), Hillbricht-Ilkowska presents several classification systems that are presently being used to assess trends in water quality for fresh water lakes in Poland. Hillbricht-Ilkowska also provides data for trend analyses for several of the larger lakes in Poland.

Table 4 presents the results of classifying 230 natural lakes in terms of their susceptibility to anthropogenic degradation. The parameters that determine the ranking include lake morphometry, persistence of stratification, residency time of water, presence of point-source pollution, and the intensity of land use in the riparian watershed. Category 1 represents lakes that are expected to be most resistant to eutrophication, and "out of Category 3" represents lakes that are seriously threatened. About 45% of Polish lakes are moderately susceptible to eutrophication, and an additional 45% are rather highly susceptible. Many of these lakes are in the Masurian Lake District and support an important tourist industry.

Another classification system was utilized to examine 221 Polish lakes in terms of existing conditions of water quality (Table 5). This system relies on spring and summer measurements of 18 parameters like nutrient and dissolved oxygen concentrations, transparency (SD), chlorophyll, and other values (see Chapter 17, Table 4).

Approximately one-quarter (26%) of the volume of the examined lakes are currently mesotrophic, and these include most of the largest and deepest lakes in Poland (these data refer to lakes examined in the years 1973-1979). Fifty-six percent of the lakes by number, and 41% of the surface waters by volume, are currently experiencing a serious level of water quality degradation. This results in frequent fish kills, permanent

TABLE 5 Distribution of 221 Polish lakes, 1974-1983; classified by state of water quality.

Class	Number of Lakes		Lake Volume	
	number	percent (%)	volume ($10^3 m^3$)	percent (%)
1. 1st and intermediate between 1st and 2nd	20	9	2 126 701	26
2. 2nd and intermediate between 2nd and 3rd	77	35	2 606 131	33
3. 3rd	70	32	1 808 469	24
4. Out of classification	54	24	1 308 469	17
TOTAL	221	100	7 898 907	100

SOURCE: Cydzik et al., 1982; Cydzik and Soszka, 1988

TABLE 6 Percentage distribution of the number of 50 Polish lakes according to the yearly load of total P (L_{TP}, g m^{-2} lake area yr^{-1}), share (%) of the point sources in L_{TP}, and relation to the permissible and dangerous load based on Vollenweider's criteria.

L_{TP} (g m^{-2} yr^{-1})	% of lakes	% of point sources in L_{TP}	% of lakes	L_{TP} is:	% of lakes
≤ 0.1	20	0-10	53	< permissible	26
0.2-0.9	54	20-50	17	≥ permissible < dangerous	20
≥ 1.0	26	≥ 60	30	≥ dangerous	54

SOURCE: Hillbricht-Ilkowska, 1984.

anaerobic conditions in the hypolimnion, and dense blooms of green and blue-green algae in surface waters.

Although 45% of the lakes are considered highly susceptible to eutrophication (Table 4), 56% of the lakes have already experienced serious degradation (Table 5). A primary cause of this increased rate of lake eutrophication is presented in Table 6. The loading of total phosphorus (L_{TP}) is considered to be the most important stimulus of lake degradation. Fifty-four percent of the lakes are currently experiencing dangerous levels of inputs and 60% of the L_{TP} is from point sources in 30% of the lakes. Those are potentially treatable sources that should be given immediate attention.

Trend data exist for three lakes in Poland (Table 7). Lake Hancza is the

TABLE 7 Trends in transparency (SD) in three Polish lakes*

Lake Hancza		Lake Mikolajskie		Lake Mamry	
Date	SD (m)	Date	SD (m)	Date	SD (m)
925	7.5	1950	2.8-3.5	1950	4.6
1931	6.5	1957	3.0	1956	4.3-6.5
1935	8.2	1958	2.7	1957	5.1
1955	8.3	1967	1.5-2.2	1958	3.3
1956	7.9	1971	1.0-2.0	1968	5.5
1957	8.0	1972	1.0-2.0	1972	5.5
1958	5.5	1976	1.3-1.7	1976	5.1-5.8
1977	-0-	1977	1.1	1977	5.5
1984	8.5	1984	1.5	1978	3.8
		1986	1.3	1986	4.3
		1987	1.2		

*Data selected from Hillbricht-Ilkowska, 1988; Zdanowski et al., 1984; Hillbricht-Ilkowska, in press.

deepest lake in Poland (108.5 m) and has a surface area of 311.4 ha. The annual total phosphorus loading is 0.066 g $m^{-2}y^{-1}$ (Hillbricht-Ilkowska, in press), which is below the permissible level. The lake has maintained a mesotrophic state for the last 60 years. Consistent high transparency measurement and high summer hypolimnetic oxygen concentrations of 8.0-10.0 ppm support this nondegredation trend.

Lake Mikolajskie, on the other hand, is 28 m deep, with a surface area of 460 ha. This lake currently receives an annual phosphorus loading of 0.77 g $m^{-2}y^{-1}$, which is four times the permissible load based on Vollenweider's criteria. The predicted load for 1990 increases to 1.63 g $m^{-2}y^{-1}$ due to estimated increases in tourism and agricultural uses of fertilizers (Giercuszkiewicz-Bajtlik et al., 1983). The transparency data (Table 7) indicate an eutrophic state existed in 1950 and has degraded even further in the last four decades. Chorophyll concentrations have also increased from 14.2 $\mu g\ l^{-1}$ in 1973 to 51.8 $\mu g\ l^{-1}$ in 1986 (Hillbricht-Ilkowska, 1988).

Lake Mamry is one of the largest lakes in Poland, with an area of 2504 ha and a maximum depth of 43.8 m. The total phosphorus loading is 0.06 g $m^{-2}yr^{-1}$, which is well below the estimated permissible load (Giercuszkiewicz-Bajtlik et al., 1983). The transparency data show a consistent trend of mesotrophy over the 36-year period (Table 7); however, according to Gliwicz and Kowalczewski (1981), there is a visible increase of oxygen deficit in the hypolimnion.

The trend data support the notion that external phosphorus loading from anthropogenic sources is the primary stimulus of lake eutrophication in Poland. Unfortunately, there are no data on heavy metals and synthetic organics to evaluate the toxic chemical loadings to these lakes. Thus, there is real need to expand the lake monitoring program to include toxic chemicals in addition to the cultural eutrophication.

REFERENCES

Alexander, M. 1973. Nonbiodegradable and other recalcitrant molecules. Biotechnology and Bioengineering (XV):611-647.

Alexander, M. 1981. Biodegradation of chemicals of environmental concern. Science 211(9):132-138.

Boethling, R.S., and M. Alexander. 1979. Microbial degradation of organic compounds at trace levels. Environ. Sci. Technol. 13:989-991.

Bouwer, E.J, B.E. Rittman, and P.L. McCarty. 1981. Anaerobic degradation of halogenated 1- and 2-carbon organic compounds. Environ. Sci. Technol. 15:596-599.

Conservation Foundation. 1984. State of the Environment. Washington, D.C.

Cydzik, D., D. Kudelska, and H. Soszka. 1986. The system of evaluation of lake quality and its application by the field agencies of the environmental protection. Pp. 11-20 in Monitoring of the Lake Ecosystems, A. Hillbricht-Ilkowska, ed. Ossolineum, Wroclaw.

Cydzik, D., and H. Soszka. 1988. Atlas of water quality of lakes studied in 1980-1984. Wydawnictwo geologivine, Warsaw (in Polish).

Environment Canada, Water Quality Branch, Burlington, Ontario.

Giercuszkiewicz-Bajtlik, M., R. Lessow, and C. Mientki. 1983. The means for the lake protection and recultivation. Pp. 55-74 in The Protection of Lakes, B. Cie onska-Sikorska, ed. Epoka (in Polish).

Gliwicz, Z.M., and A. Koalezewski. 1981. Epilimnetic and hypolimnetic symptoms of eutrophication in Great Masurian Lakes, Poland. Freshwater Biology 11:425-433.

Great Lakes Water Quality Board. 1985. Report on Great Lakes Water Quality. Report to International Joint Commission (I.J.C.).

Hallbert, G.R. 1987. Nitrates in Iowa groundwater. Pp. 23-69 in Rural Groundwater Contamination, D'Itri and Wolfson, eds. Chelsea, Michigan: Lewis Publishers, Inc.

Hillbricht-Ilkowska, A. 1984. The indices and parameters useful in the evaluation of water quality and the ecological state of temperate, low-land lakes connected with their eutrophication. Pp. 55-69 in the Proceedings of the International Conference, "Conservation and Management of World Lake Environment." Otsu Shiga, Japan, August 24-30, 1984.

Hillbricht-Ilkowska, A., ed. 1988. The lakes of Masurian Landscape Protected Area—Eutrophication, protection, management. Ossolineum, Warszawa, (in Polish, with English summary). (in press).

Hillbricht-Ilkowska, A. (in press). Some properties of the functioning of low-land lakes preserving their mesotrophic character. Internationale Revue der Gesamten Hydrobiologie.

Klepper, G., G. Carpenter, and D. Gruben. 1987. Groundwater contamination from landfills, underground storage tanks, and septic systems. Pp. 147-160 in Rural Groundwater Contamination, D'Itri and Wolfson, eds. Chelsea, Michigan: Lewis Publishers, Inc.

Kudelska, K.D., D. Cydzki, and H. Soszka. 1981. A proposal of lake water quality classification. Wiad. Ekol. 27:149-173 (in Polish with English summary).

Lovley, D.R., and M.J. Klug. 1982. Intermediary metabolism of organic matter in sediments of a eutrophic lake. Appl. Environ. Microbiology (March 1982):552-560.

Lovley, D.R., and M.J. Klug. 1983. Methanogenesis from methanol and methylamines and acetogenesis from hydrogen and carbon dioxide in the sediments of a eutrophic lake. Appl. Environ. Microbiol. (April 1983):1310-1315.

Lovley, D.R., and M.J. Klug. 1986. Model for the distribution of sulfate reduction and methanogenesis in freshwater sediments. Geochimica et. Cosmochimica Acta (Vol. 50):11-18.

Lovley, D.R., D.F. Dwyer, and M.J. Klug. 1982. Kinetic analysis of competition between sulfate reducers and methanogens for hydrogen in sediments. Appl. Environ. Microbiol. (June 1982):1373-1379.

Vollenweider, R.A. 1976. Advance in defining critical level for phosphorous in lake eutrophication. Mem. Inst. Ital. Idrobiol. 33:53-83.

United States Geological Survey. 1984. Hydrologic events: Selected water-quality issues and groundwater resources. Water Supply Paper 2275.

Zdanowski, B., A. Korycka, and J. Zachwieja. 1984. Thermal and oxygen conditions and the chemical composition of the water in the Great Masurian Lakes. Ekol. Pol. 32:651-678.

River Water Quality Assessment and Management in Poland

MAREK J. GROMIEC
Institute of Meteorology and Water Management
Warsaw

BACKGROUND: WATER RESOURCES AND WATER DEMANDS

Poland is 312,520 km^2 in area, and its total water surface covers 5,000 km^2 or 1.6% of the total area of the country. Poland has about 9,300 lakes, covering an area of 3,200 km^2. For example, the Masurian Lake District has 1,063 lakes, the largest of which have areas of about 11,000 hectares. Lakes and artificial reservoirs have a capacity of 33 km^3, and a large number of ponds hold an additional 1 km^3. The two most important rivers in the country are the Oder River, which has a basin area of 110,000 km^2, and the Vistula River, with a basin area of 194,000 km^2.

The water balance during a normal annual cycle in Poland is presented below. The average annual amount of rainfall is 597 mm, equivalent to 186.6 km^3 of water per year over the whole country. Since tributaries from outside Poland yield an additional 5.2 km^3 of water annually, the total input of water is 191.8 km^3. Underground water resources have been estimated at 33 km^3 per year for an area of 272,520 km^2, since the remaining 13% of the total area is waterless. The annual dynamic underground water resources have been evaluated at 9.2 km^3. However, rivers and streams discharge only about 58.6 km^3 of water into the Baltic Sea during a mean low-flow year, and about 34 km^3 in a mean dry-weather year. Obviously, only a portion of this volume is available. About 10 km^3 is necessary as a minimum flow to maintain biological life and for sanitary reasons. Therefore, the available flow is only 24 km^3 of water.

Poland belongs to the group of European countries most deficient in water resources, ranking 22nd overall. Average annual water resources in Poland—estimated on the basis of atmospheric inputs and the number of population—amount to 1,800 km^3 per inhabitant, compared with 2,800 km^3

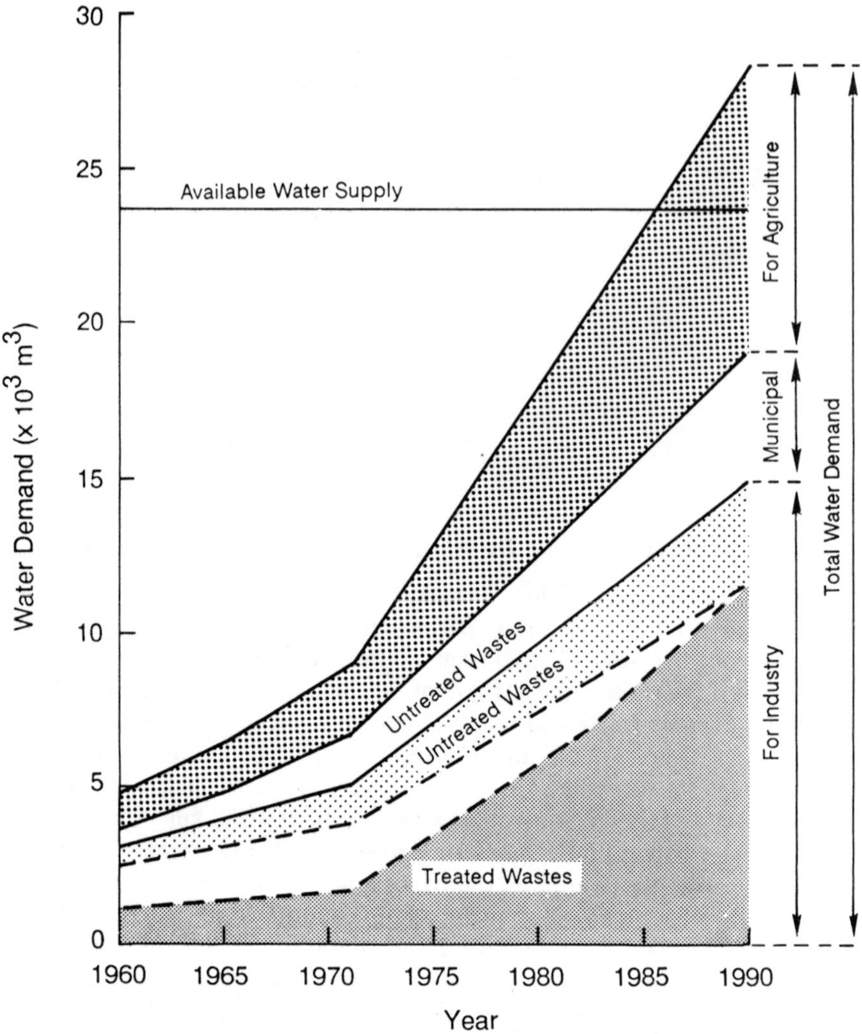

FIGURE 1 Water demand and wastewater discharge in Poland (Oleszkiewicz and Oleszkiewicz, 1978).

per inhabitant for Europe. Annual water demand for Poland in 1990 is anticipated to be about 28 km^3 (compared to 13 km^3 in 1976), with about 5 km^3 for municipal supply (2 km^3 in 1976) and 9 km^3 for agriculture (4 km^3 in 1976). Most water for agriculture is taken during the summer months. Water demands and wastewater discharges are shown in Figure 1. Available water volume compares unfavorably with the water demand anticipated in the year 1990.

LEGISLATION AND ADMINISTRATIVE ASPECTS

Poland has a sixty-year history of legislation for water pollution control. The Water Quality Act was issued in 1922 and revised in 1962. Presently, the basis for legal action in the field of water protection against pollution is a new version of the Water Law Act issued by the Polish Parliament in 1974. In 1975, on the basis of the Water Law, the Council of Ministers announced regulations concerning classification of waters and determination of of effluent standards, as well as financial penalties for the effluent discharges that do not meet the requirements specified in the regulations. These regulations are set up for biological oxidation demand (BOD), chemical oxidation demand (COD), ether extracts, polychlorinated biphenyls (PCBs), various metals, and pesticides. The following classes of surface water quality were established:

- *Class I* waters are those used for municipal and food processing supply purposes, and for salmon fish growth.
- *Class II* waters are intended for use as recreational waters, including swimming, and for growth of fish other than salmonidae.
- *Class III* waters (the lowest class) are only used as industrial water supplies and for irrigation purposes.

Water-quality standards are tailored to meet appropriate use of surface waters. For example, permissible concentrations of selected constituents for the different classes are shown in Table 1. In addition, the following provisions are laid down by the Water Law:

- Industrial plants and other operations which discharge wastewaters to water or to land are obliged to construct, maintain, and utilize wastewater treatment facilities.
- Without simultaneous operation of wastewater treatment systems, no industrial plant or any other plant from which wastewater is discharged can be started up.
- A permit is required to to maintain wastewater discharge.

In order to promote administrative measures for overall conservation of the environment, a Ministry of Environmental Protection and Natural Resources was established with environmental offices on a *voivodship* (district) level.

WATER-QUALITY SURVEILLANCE SYSTEM

The water-quality surveillance system is composed of conventional and automatic monitoring stations. The conventional monitoring with manual sampling at the district level was established in 1957. At present, the country is covered by a network of about 3,000 conventional stations and

TABLE 1 Examples of permissible concentrations of selected pollutants in surface freshwater according to the Polish Water Law.

PARAMETER	UNIT	WATER CLASS I	II	III
DO	mg O/dm^3	6	5	4
BOD_5	mg O_2/dm^3	4	8	12
COD	mg O_2/dm^3	40	60	100
Saprobic index		oligo to betamezo	betamezo to alfamezo	alfamezo
Chlorides	mg Cl/dm^3	250	300	400
Sulphates	mg SO_4/dm^3	150	250	250
Hardness	$mval/dm^3$	7	11	14
Dissolved solids	mg/dm^3	500	1000	1200
Suspended solids	mg/dm^3	20	30	50
Temperature	^oC	22	26	26
$N\text{-}NH_4$	mg NNH_4/dm^3	1.0	3.0	6.0
$N\text{-}NO_3$	mg NNO_3/dm^3	1.5	7.0	15
N-organic	mg $Norg/dm^3$	1.0	2.0	10
Total iron	mg Fe/dm^3	1.0	1.5	2.0
Manganese	mg Mn/dm3	0.1	0.3	0.8
Phosphates	mg PO_4/dm^3	0.2	0.5	1.0
Cyanides	mg CN/dm^3	0.01	0.02	0.05
Phenols	mg/dm^3	0.005	0.02	0.05
Lead	mg Pb/dm^3	0.1	0.1	0.1
Mercury	mg Hg/dm^3	0.001	0.005	0.01
Copper	mg Cu/dm^3	0.01	0.1	0.2
Zinc	mg Zn/dm^3	0.01	0.1	0.2
Cadmium	mg Cd/dm^3	0.005	0.03	0.1
Chromium	mg Cr/dm^3	0.05	0.1	0.1
TOTAL HEAVY METALS	mg/dm^3	1.0	1.0	1.0

established cross-sections located along 30,000 km of streams. The sampling frequency depends on the purpose for which data are recorded, ranging from a minimum of bimonthly sampling up to daily sampling at some points. The sampling of water is performed simultaneously with the rate of flow measurements.

For the continuous monitoring of the Oder and Vistula, the automatic water-quality monitoring stations (AWQMS) have been installed at important cross-sections, connected mostly with water intakes. The AWQMS are located either in laboratory buildings or on barges. The automatic network, one of the first in Europe, has been operating since 1968. It was organized under the auspices of the World Health Organization (WHO), and initially was based on imported monitoring equipment. At present, domestic automatic monitors (Aquamers) are being used. These monitors are capable of measurement and teletransmission of such parameters as water temperature, pH, conductivity, dissolved oxygen (DO), chlorides, turbidity, water level, and meteorological data. Additional parameters will be added after suitable sensors are developed.

However, it should be stressed that the scarcity of automatically measurable parameters limits the efficient application of AWQMS. The development of automatic measuring devices for other important parameters such as heavy metals, organo-chlorine compounds, oxidized nitrogen, soluble organic content, and others is greatly needed. At present, the automatic stations are also used for manual bioassay tests and fish tests. The future surveillance development program calls for the Basic Water-Quality Monitoring Network (BWQMN) based on a limited number of stations (with extensive measurements) as supplementary to the conventional water-quality surveillance system (Zielinski et al., 1988).

COMPUTERIZED DATA ANALYSIS

The river monitoring network provides a large number of observed data. These data are analyzed by a statistical method based on the assumption that, at a given cross-section, some correlation exists between the pollutant concentration and the rate of flow (Manczak, 1973). Three basic types of curves representing this correlation are applied (Figure 2). The shape of the curve depends on many factors, such as the degree of water pollution, the type of pollutant, hydrological characteristics of the river, its self-purification capacity, the distance between monitoring stations, and others. Figure 3 provides an example of the correlations between BOD, permanganate value (PV), dissolved oxygen (DO), and phenols for the Oder River at one of the monitoring stations. These relationships have been derived from about 300 observations within a year, with temperatures ranging from 0.1°C to 27°C. Inorganic compounds, such as chlorides and

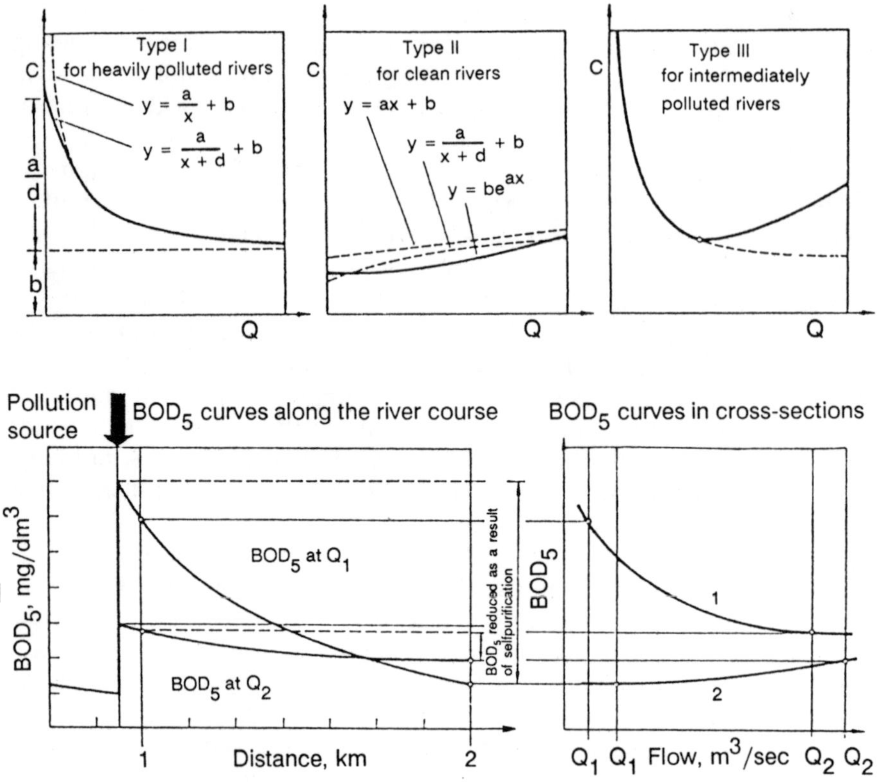

FIGURE 2 Relationship between concentration of pollutants and rate of flow (Manczak and Florczyk, 1971).

sulfates, are usually described by correlations defined by curves I and II in Figure 2; the effect of temperature is not included (Figure 4).

From these relationships between stream flow and concentrations of water-quality constituents, so-called *indicative concentrations* (IC) for a design flow are established. The mean low streamflow (MLQ) has been selected as the design streamflow at each site, based on the assumption that higher streamflows will result in higher DO concentrations and better water quality. In other countries, a similar approach has been taken. For example, in the United States the design flow is the minimum average 7-day consecutive flow expected once every 10 years. However, this is an extremely low streamflow which is exceeded more than 99% of the time.

The IC values are plotted along the river for various water-quality constituents. Final interpretation is based on these hydrochemical profiles (Figure 5), and the overall river classification is performed after all

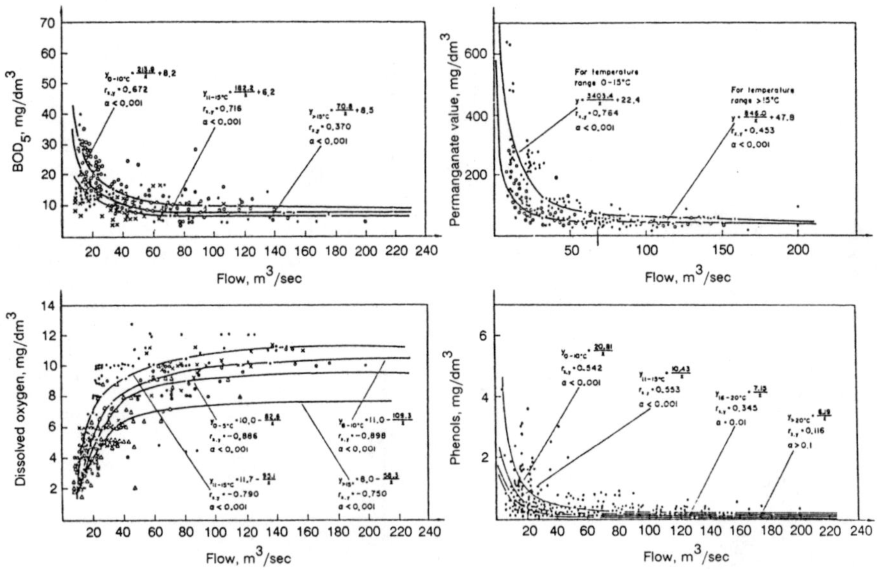

FIGURE 3 Relationships between BOD, PV, DO, phenols, and flow for the Oder River a given location, including water temperature effect (Manczak and Florczyk, 1971).

measured water-quality constituents are compared with standards. A compendium of hydrochemical profiles for major rivers and streams is prepared each year by the Institute of Meteorology and Water Management which serves as an overall river classification system (Figure 6).

WATER-QUALITY INDEX

The ability to assess the quality of river water in quantitative terms becomes imperative as concern for pollution control grows. The need for a water quality index (WQI) was recognized in 1981 by the Institute of Meteorology and Water Management. In the development of WQI, the following points have been considered:

- the choice of parameters to be included in the index;
- the proper weights of different parameters;
- the fitting of the mathematical formulae for calculation of the standardized values of the parameters; and
- the method of averaging standardized values of all parameters.

The need for calculation of standardized values of water-quality parameters was met by the imposition of mandatory water-quality standards in Poland. It can be assumed that, on the standardized scale of water quality (ranging from 0 to 100), the admissible value of any parameter for

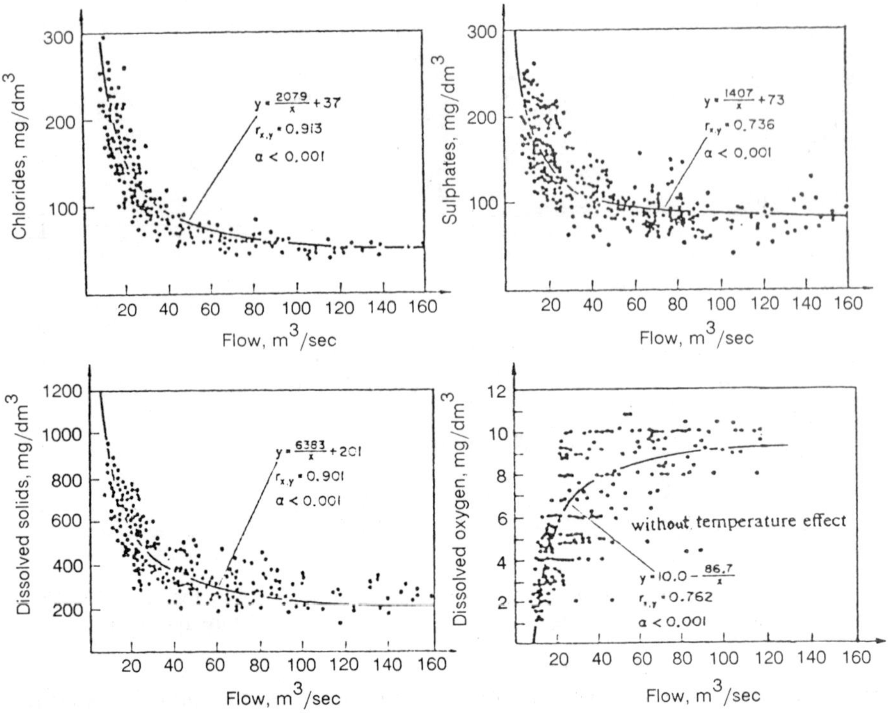

FIGURE 4 Relationships between chlorides, sulfates, dissolved solids, DO, and flow for the Oder River at a given location, without water temperature effect (Manczak and Florczyk, 1971).

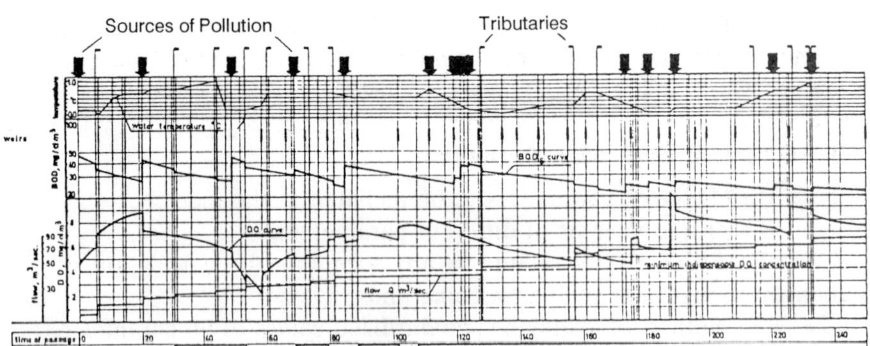

FIGURE 5 Example of a hydrochemical profile for the Oder River: winter (Manczak, 1969).

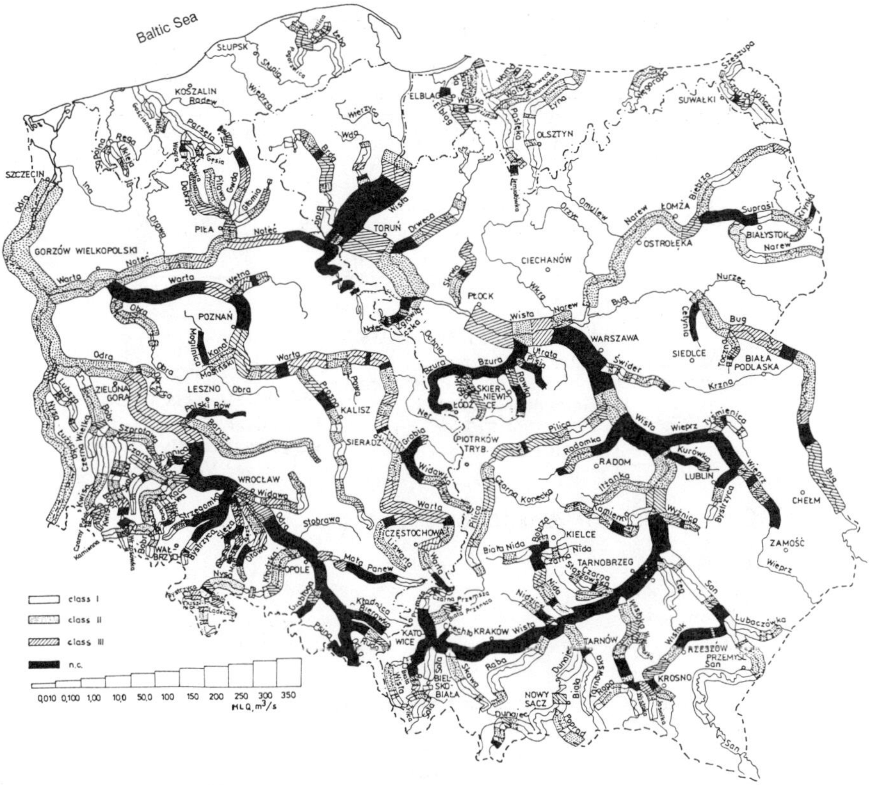

FIGURE 6 Overall river classification system in Poland (CSO, 1988.)

the first class always corresponds to 75 points, with the respective values for second and third class always corresponding to 50 and 25 points. Additionally, the water-quality standards established by the Council for Mutual Economic Assistance (COMECON) were used; therefore, five-point curves were obtained and corresponding mathematical functions were proposed for 31 water-quality parameters.

A harmonic mean index was selected with the form:

$$WQI = \left(\frac{n}{\sum_{i=1}^{n} \frac{1}{x_i^2}} \right)^{0.5}$$

in which

WQI = water-quality index, a number between 0 and 100;

n = number of the set of data;

x_i = the standardized value of the ith water-quality parameter.

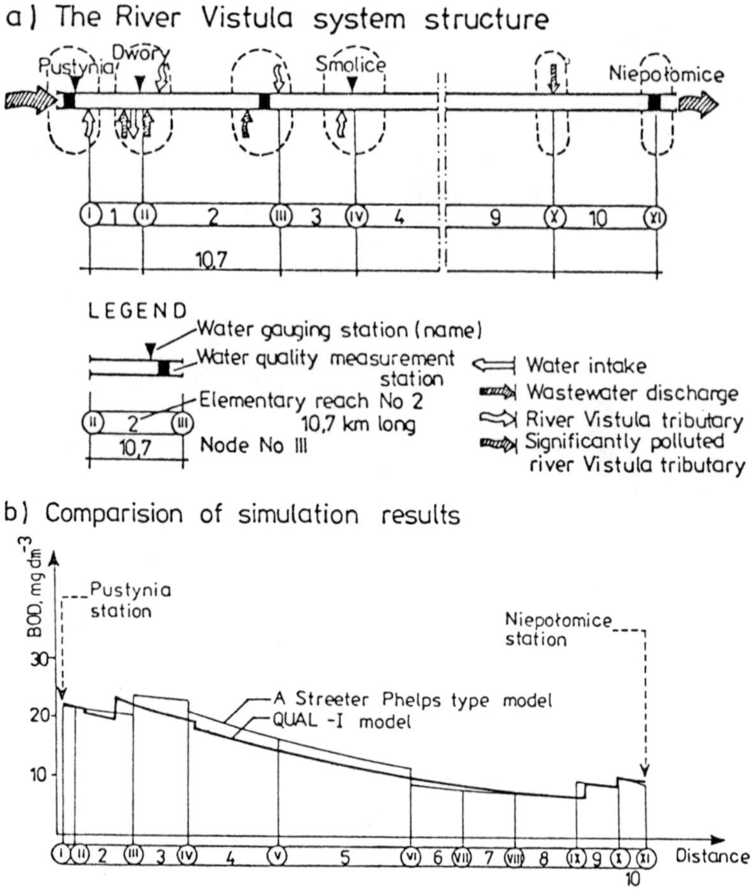

FIGURE 7 Comparison between BOD simulation results obtained by application of simple Streeter-Phelps type model and QUAL-I to a section of the Vistula River (Adamczyk et al., 1978).

This index is being used for the Vistula River basin.

MATHEMATICAL SIMULATION AND MANAGEMENT MODELS

Mathematical modeling of river systems in Poland has become an integral part of water resources planning and water-quality management. These models can be used to aid water-quality surveillance and to predict future water-quantity/quality conditions (Gromiec et al., 1983). Various computerized models have been applied for water-quality simulations in the Oder and Vistula rivers. As an example, a Streeter-Phelps type model

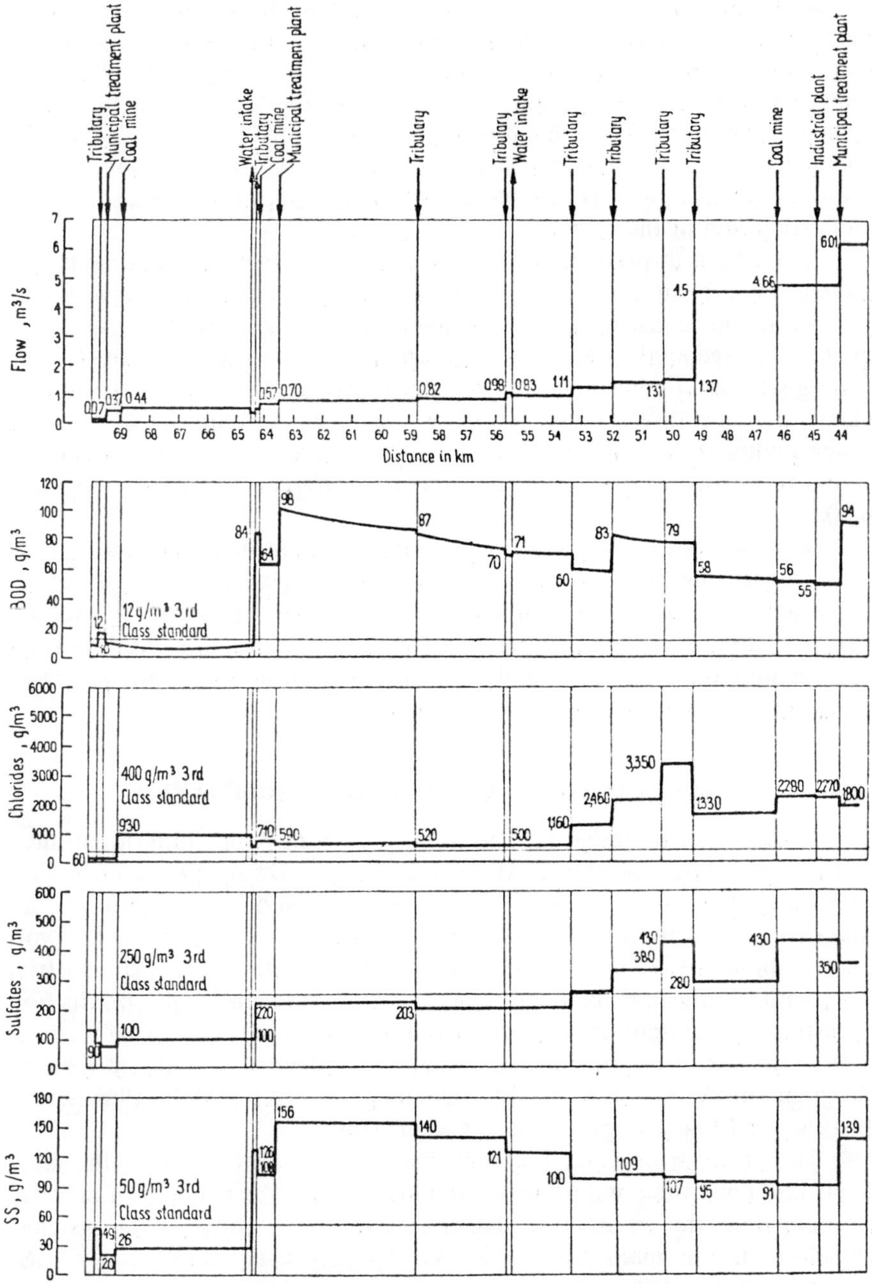

FIGURE 8 Hydrochemical profile of Klodnica River in 1980 (EPAC, 1979).

and QUAL-I model were used to evaluate concentrations of BOD in the Vistula River reaches (Figure 7). The first model is designed to simulate the spatial and temporal variations in BOD under various conditions of flow and temperature. The second model is capable of routing BOD, DO, and temperature through a one-dimensional, completely mixed branching river system. These BOD-DO models are representative of non-conservation coupled models. It should be stressed that the predictions obtained from these models are only as reliable as the input data, proper measurement, and estimation of the various model parameters.

In addition to prediction of water quality by simulation, mathematical models have been serving as the basis for determination of investment policies in the construction of treatment plants. For example, a water-quality management model has been applied to the Klodnica River. The main goal was to determine, for the given system of wastewater treatment plants, the level of efficiency necessary to achieve the required standard of water quality at the least cost. The quality of the Klodnica River catchment at the respective stages of the management program is shown in Figures 8-10.

It should be stressed, however, that these management models are only tools in assisting management decision-making processes. Final decisions are usually not made only on the basis of their predictions. Additional data, including socio-political factors, are taken into account. Still, in spite of their limitations, these models are the only reasonable means presently available for the prediction of water quality.

WATER-QUALITY MANAGEMENT SUMMARY

Processes of intensive urbanization, growth of population, intensification of agriculture, and the growth of industry have resulted in deterioration of surface water resources, despite the introduction of legal, technical, and financial measures for water pollution control (see Chapter 22, this volume). About 80% of wastewater comes from 600 cities and 2,800 industrial sites, with chemical, machinery, mining, power, food, and paper production industries as the main polluters. Currently, only about 60% of all wastewater is treated. Recognizing that water is a valuable resource, a Central Program on Water Resource Management was instituted in 1985 at the Institute of Meteorology and Water Management.

The problem of water pollution control has become one of the most important environmental problems in Poland, since the majority of industry is situated in the south near the origins of the country's river systems. In addition, the main rivers—the Vistula and Oder—are heavily used for municipal and industrial water supply, agricultural irrigation, cooling purposes for power plants, and navigation; however, these rivers also

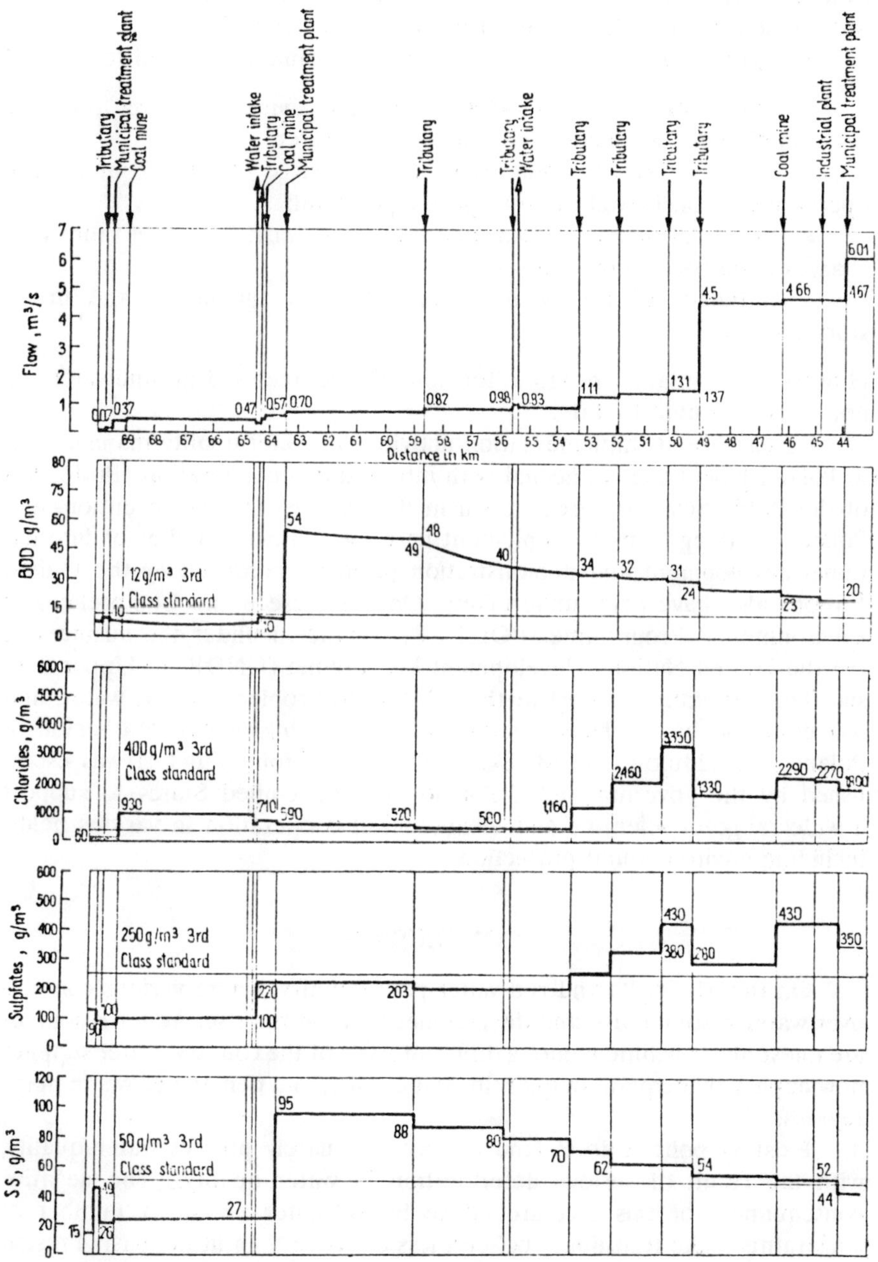

FIGURE 9 Hydrochemical profile of Klodnica River in 1985 (EPAC, 1979).

receive discharges of wastewater with varying degrees of treatment and runoffs. These multiple uses impose competing demands on waters, and water resource management must protect many desirable uses.

The principal water-quality problems in Poland are related to:

- effects of *dams and other water management structures* resulting from phenomena associated with impounded water;
- effects of *power plants*, since the discharge of heat from cooling operations is considered to be a specific pollutant;
- effects of *municipal and industrial wastewater*, including saline discharges from coal mines; and
- influence of *non-point sources*, such as agriculture and urban stormwaters.

However, a regulatory program for nonpoint sources and groundwater has not been instituted to date.

Research and implementation activities in water-quality management in Poland have been connected with international cooperation. Protection of the Baltic Sea was agreed upon in the 1974 Helsinki Convention, and Poland is taking steps to implement its provisions. A number of international development and demonstration projects sponsored by the United Nations also have been undertaken. One of these is the Comprehensive Environmental Programme (POL/CEP) conducted under the auspices of the the United Nations Development Programme (UNDP) in Upper Silesia. This program is aimed at the solution to problems of air, water, and soil pollution in this heavily industrialized and highly polluted region of Poland. In addition, the M. Sklodowska-Curie Joint Fund II was established by the governments of Poland and the United States to support a wide range of scientific and technological cooperation in various fields, including environmental protection.

CONCLUSION

On the whole, Poland is a water-poor country, where various conflicts over water resource use and development are on the rise. Water resources are unevenly distributed among different parts of the country, water supplies now appear inadequate in quantity and quality, and demand grows in many regions.

Existing policies in Poland do not adequately protect water quality, with the result of serious deterioration in water quality. The negative consequences of this fact are borne by all water users. A number of constraints make traditional policies less effective than in the past. Present policies are based on the concept of regional management. In addition, many traditional strategies for solving water-quality problems, such as

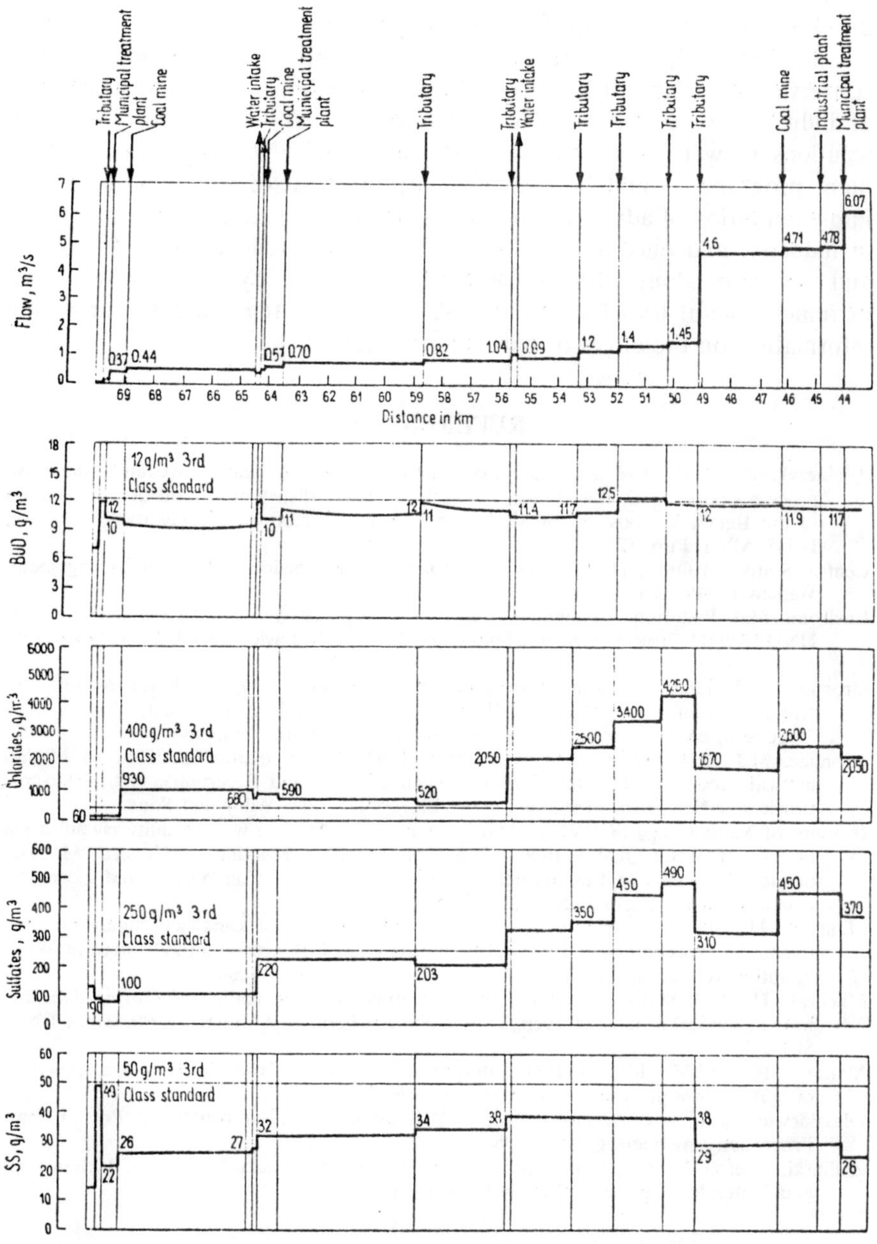

FIGURE 10 Hydrochemical profile of Klodnica River in 1990 (EPAC, 1979).

downstream pollution control, are extremely costly. The National Program for Protection of the Natural Environment to the Year 2010 calls for water pollution control investments of between 4,000 billion zlotys (variant A) to 2,590 billion zlotys (variant B) in 1986 constant prices.

In view of this commitment, river basin oriented approaches, water conservation and pollution prevention approaches, and the restructuring of industry are clearly needed to assure use of the most cost-effective solutions to water resource problems in Poland. Water-quality improvement programs should be aimed at improvement of treatment methods, implementation of advanced treatment processes, recovery and water reuse in industry, and encouraging the production of biodegradable detergents and pesticides along with use of dry technologies. Systems analysis is an extremely useful tool for the analysis of river systems and for providing information on the effect of particular policies.

REFERENCES

Adamczyk, Z., et al. 1978. A simple mathematical model of quantitative and qualitative process occurring in the stream channel for water distribution control. Proceedings of the Baden Symposium on Modelling of Water Quality of the Hydrological Cycle, IAHS-AISH, Pub. 125.

Central Statistical Office (CSO). 1988. Environmental Protection and Water Management. Warsaw (in Polish).

Environmental Pollution Abatement Centre (EPAC). 1979. Environmental Protection. UNDP/WHO Project POL/RCE-001-Final Report. Katowice: Slask Publishing Company.

Gromiec, M.J. 1983. Biochemical oxygen demand-dissolved oxygen: River models. In Application of Ecological Modelling in Environmental Management, Part A, S.E. Jorgensen, ed. Amsterdam: Elsevier Scientific Publishing Company.

Gromiec, M.J., P.D. Loucks, and G.T. Orlob. 1983. Stream quality modeling in Mathematical Modeling of Water Quality: Streams, Lakes, and Reservoirs, G.T. Orlob, ed. Chichester-New York-Brisbane-Toronto-Singapore: John Wiley and Sons.

Institute of Meteorology and Water Management. 1987. Surface water-quality classification by use of Water-Quality Index. Paper presented at Meeting of General Methodological Problems in Environmental Statistics. Statistical Commission and Economic Commission for Europe. June 2-5, Warsaw, Poland.

Manczak, H. 1969. The course of self-purification process of canalized highly polluted rivers. In Proceedings of the Fourth International Conference on Advances in Water Pollution Research, S.H. Jenkins, ed. Oxford: Pergamon Press.

Manczak, H. 1973. Statistical method for estimation of water-quality monitoring data and hydro-chemical profiles of rivers. United Nations Econ. Soc. Council, CES/SEM.6/ENV./SEM. 1.

Manczak, H., and H. Florczyk. 1971. Interpretation of results from the studies of pollution of surface flowing waters. Water Research 5:575-584.

Oleszkiewicz, J.A., and A. Oleszkiewicz. 1978. Water pollution control in Poland. Env. Protection Engineering 4(1):27-48.

Zielinski, J., et al. 1988. State monitoring of surface flowing waters. Institute of Meteorology and Water Management, Warsaw (in Polish).

Environmental Management Case Studies

Environmental Policy in Eastern Europe

PRZEMYSLAW TROJAN
Polish Academy of Sciences

Editors' Note: This chapter is a compilation of ideas deriving from a two-year comparative analysis of concepts, principles, and recommendations for the continuing evolution of environmental policy in several of the socialist countries of Eastern Europe. Please note the great similarity between the ideals presented in this chapter and those which guide similar developments in many democratic societies.

During 1987 and 1988, a Committee for the Study of World Socialist Systems of the Polish Academy of Sciences (PAN) was asked to prepare an analysis of environmental policy in several of the socialist countries of Europe, including the USSR, the German Democratic Republic, Czechoslovakia, Hungary, and Poland. This study was designed to provide a comparative analysis of the concepts and methods used in the development of environmental policy in these countries. The results of these studies, comprising eleven parts, are now in print (Dobrowolski et al., 1989). Such analyses have a long tradition in Poland and are published in reports by the Polish Academy of Sciences (Mcihajlow, 1976; Polityka, 1989). This chapter is a brief summary of the principal findings from this two-year study.

The first task of analysis was to try to explain why conflicts are so often seen between environmental protection and industrial development. Such conflicts are not a requirement of theory (Zagladin and Frolov, 1986). But they are all-too-frequent consequences of a lack of holistic thinking in the development of sound policies for protection of ecological values under conditions of continuing economic development.

DEVELOPMENT OF ENVIRONMENTAL POLICY

Social pressure for the proper resolution of environmental problems is growing in many East European countries. Official recognition of this idea was expressed very well in the conclusions of an international conference on "Socialism in the Context of Contemporary Global Problems" held in Prague in 1985 (Anonymous, 1985). The conclusions read as follows:

- Ecological crises have expanded to the whole domain of relationships between humans and nature.
- These crises exert strong influence on all aspects of social life and are not restricted to problems of industrial production.
- Reconstruction of all socioeconomic systems is necessary. The isolated improvement of single elements of real socioeconomic systems is meaningless.

Ecological conflicts are always present in human existence. These conflicts affect many aspects of our daily lives and lead to intense discussions about the place of humans in nature. An ecological viewpoint is developing in which humans are seen as a product of continuing biological evolution. That is, human life, like the life of all other living things, is ultimately constrained within the limits of the sustainable productive capacity of the land, available energy, and the other natural resources necessary to maintain human existence. Resolving our present ecological crisis is now the task of a global strategy for society.

The general purposes of ecological policy in Eastern Europe are easy to define. They are similar to those of many other countries; only the methods and mechanisms governing the course of environmental events are different.

Three kinds of activity are required in the development of wise environmental policy: prevention, control, and restoration. The outcome of these activities is determined by the following elements of ecological policy:

- *political activities* of governments which are influenced by social demands;
- the *system of law* which defines the rights and obligations of both citizens and the government in relation to natural and human resources;
- the activities of *environmental agencies* within the organizational structure of government;
- *economic instruments*, especially those related to economic and spatial planning;
- *foreign policy;*
- *educational systems and scientific investigations*; and
- *social movements.*

Lack of data in one or more of these elements restricts our ability to understand their role in contemporary East European societies. Nevertheless, some preliminary conclusions can be drawn and are presented here.

During the past 40 years, economic development in Eastern Europe has brought to these societies both progress in industrial production and many unintended effects. One of the most important of these effects is the extension of ecological crises over more and more territory in the region. The primary reason for these unfortunate results is overall inefficiency and/or lack of appropriate mechanisms for the protection and restoration of the environment and natural resources. Thus, the importance of environmental problems is growing with time. These problems exert very strong influences on many aspects of political, economic, and social activity.

Politics

The political parties in most East European countries have included problems of ecological policy in their programs. For example, in the Party Congresses of 1984 and 1986, issues relating to environmental policy were much more evident than they had been in previous Party Congresses. A comparison of ecological policies outlined in the documents which resulted from these meetings leads to the following conclusions:

- None of the documents includes analysis of the environmental situation in their respective countries. Both the diagnosis and the extent of change in environmental quality should constitute the basis for formulation of objectives and tasks between successive Congresses.
- Programs of ecological policy are best formulated in those countries where the exploitation of natural resources is most economical, e.g., the German Democratic Republic and Czechoslovakia.
- These documents indicate that the most important factors influencing environmental protection include enforcement of laws, pollution control, deliberate management of environmental quality, economic development policies, ecological education, and progress in science and technology.
- The documents from all Party Congresses disregard the prevention of environmental problems. This is especially evident in the lack of coordination between economical and spatial planning before decisions are made. All too frequently, the principles governing the processes of planning and decision making are too vague. In many cases, environmental interests are not taken into account.

Legislation

There are no great differences in the judicial solutions suggested by various East European countries for the resolution of environmental problems. The system of environmental law is generally well developed in all of these countries (Radecki, 1985), but judicial regulations are not administered effectively. There are also obvious contradictions between laws pertaining to the environment and laws pertaining to other aspects of society.

Three general problems need require solutions in order to improve judicial codes for environment protection:

- There is little enforcement of existing environmental law. This is due to so-called "economic necessity" and the widespread practice of following regulations that are contrary to existing environmental law.
- There are very few connections between judicial codes, e.g., between the laws for environmental protection, agricultural lands, mining, and nature protection. The lack of necessary connections between these related laws causes chaos in decision making and essentially precludes the realization of sound ecological policy.
- Many regulations in environmental protection codes are not economically sound. Some regulations provide the possibility to shut down a factory, but in other cases, fines or other penalties are so low that it is more economical to pay them than to make the larger investment in protecting the environment.

Environmental Management

In many East European countries, several different government organizations are concerned with environmental protection. Generally speaking, however, protection of the environment is an object of state policy determined by parliamentary resolutions and executed by the government and its various agencies. The following describes the environmental policies of East European countries and the types of government agencies that take responsibility for the fulfillment of environmental goals:

- Environmental protection policy should consist of the rational use of natural resources, ecosystem and species protection, protection of society and nature against harmful effects of industrial and agricultural activities, and environmental quality control.
- Many different linkages are needed among governmental and nongovernmental organizations. The responsibility for wise use of the environment is distributed among government, industry, and other groups in society that utilize natural resources.

• Effective management of environmental quality control must include central, regional, and local centers for State Inspection and Environmental Protection. In some cases, however, local authorities do not correctly use their power against environmentally degrading activities and against pollution sources that exceed official standards; this is particularly true for large coal-fired power plants and for large iron works. Thus, an efficient system is needed to facilitate decisions about the location of industrial plants in terms of local and regional ecological situations and risks.

• A system of protected areas is growing in all East European countries, consisting of national parks, nature reserves, regions of protected landscape, and landscape parks (Chapter 22, this volume). Within these protected areas, economic activity should be adjusted to fit existing environmental conditions and programs of nature protection.

• In all East European countries, statistics pertaining to environmental quality are very unsatisfactory. Complete information is lacking on pollution emissions from factories, soil contamination and degradation, and quality of groundwaters. Also lacking are unified models of organization for the efficient solution of contemporary problems. Many suggested solutions are strongly criticized.

Economics

A system of economic and planning instruments has been developed in most East European countries. Economic solutions are topics of strong debate (Anonymous, 1983, 1986). General restructuring of these systems is necessary. The following conclusions were drawn from an analysis of these economic and planning instruments:

• In all East European countries, the system of national economies is changing. Administrative methods of management executed particularly by governmental authorities are being replaced by market mechanisms. This is also true for environmental protection.

• A prerequisite for environmental management by market methods is to establish priorities for the goals and services entering the national economy under changing economic and environmental circumstances. Without this, the low level of efficiency in environmental protection will continue.

• The prices paid for natural resources at the time of use should reflect the costs of environmental protection.

• Nature protection costs should be unified throughout Eastern Europe. This is especially important in developing regulations for industrial and agricultural enterprises that have significant impacts on environmental quality. This unified system of economic/environmental calculations should serve as a basis for compensation in cases of transboundary pollution.

- The system of economic instruments for environmental protection in Eastern Europe should include consideration of methods of funding, tax policy, credit policy, and investment policy. Contradictions between economic and environmental goals often make choices difficult.
- East European countries generally lack coherent plans for maintaining the productivity of renewable natural resources. These plans should include regulations for maintaining a rational balance between economic exploitation of natural resources and their conservation for future generations. Such regulations should also provide guidance for maintaining the essential environmental and social functions of renewable natural resources, regardless of whether these resources have significant economic value.

Foreign Policy

At present, international cooperation among East European countries in solving environmental protection problems is not effective. Large amounts of pollutants are transported from country to country in contaminated air masses and rivers. Efforts must be made in all countries to decrease the international exchange of pollutants in water and air. This can be achieved by:

- international agreements governing the location and amounts of emissions from different types of industrial activity. By the year 2000, major decreases in the transboundary flow of pollutants must be achieved.
- import/export analysis of the exchange of pollutants between these countries. These analyses should provide the basis for estimates of economic compensation payments for pollutants received.

Education and Scientific Research

The educational system in Eastern Europe includes ecological education, but efforts in this area must be increased substantially to be more effective. While ecological education is given high priority in all East European countries, no country has elaborated an adequately comprehensive system of ecological education. In addition, large differences exist between the formal systems of ecological education in various East European countries.

Therefore, modern and comprehensive ecological education programs must be developed for all students at all educational levels, including nursery school, elementary school, and high school. Universities, and particularly polytechnic institutes, should develop programs for the ecological education of their students, as their professional activities will exert great influence on environmental events and hazards. Care must be taken to organize these programs so that desirable outcomes are achieved. At present, for example, schools of engineering contain no obligatory ecological curricula and the present optional programs are often not selected. As a result, lack of environmental awareness among graduate engineers often compounds the environmental problems in society. Finally, some East European countries have developed elaborate programs of continuing education for adults. These programs offer great possibilities for the ecological education of society as a whole.

With regard to research, scientific investigations in the field of environmental protection are often conducted in connection with the systems for economic cooperation among socialist countries. The following observations are pertinent to these programs:

- The current system of cooperation provides good possibilities for comparison of results from scientific investigations in participating countries. However, the effectiveness of information transfer between countries is much greater within a given discipline than it is between disciplines, even within the same country. It is therefore necessary to increase the degree of integration across different disciplines of environmental investigations.
- The most urgent task is to increase the translation of scientific results into economic practices within industrial and agricultural enterprises.
- In international cooperative research programs, the division of effort should be arranged so as to capitalize on the special scientific strengths of individuals and institutions within the cooperating countries.

Social Movements

Social movements in the field of environmental protection are growing in numbers throughout Eastern Europe. They are exerting strong, constructive influence on central and local authorities in developing solutions for local and regional environmental problems. Analysis of the activities of these movements leads to the following conclusions:

- Volunteers participating in ecological movements have often provided effective surveillance of the status of nature, the exploitation of natural resources, and environmental decision making. In many cases, these surveillance activities have led to the discovery and subsequent solution of important environmental problems. These movements often provide ecological educational programs for children and adults and sometimes undertake concrete projects for environmental protection.
- Many successes have been achieved when local authorities and ecological movements work together in the analysis and implementation of solutions to environmental problems. All too often, however, the activities of ecological movements have formed the basis for political arguments directed against the administration and political system.

- Cooperation between ecological movements and governments or industries has been particularly successful at the lower levels of administration, i.e., in small towns, or in identifiable communities within larger towns.
- Social movements in the field of environmental protection have been very effective in improving the aesthetic quality of many areas such as public gardens and "green areas" within towns, extracting useful raw materials from accumulated wastes, and planting of trees and other vegetation on waste beds or drastically disturbed lands.

CONCLUSION

A great potential exists for improvement of environmental policy in many East European countries. These improvements can be encouraged through:

- better organization of central and local administrative authorities for environmental decision making and management of natural resources;
- better division of responsibility among East European countries, such as in the production of environmental protection devices and technologies;
- increasing the effectiveness of planning systems, economic development authorities, and legal instruments to provide both economic incentives and regulations that channel human activity in directions which are consistent with aims and programs of sound ecological policy;
- improving systems of ecological education at all levels, from nursery school through university, and including programs for adult education;
- translating scientific understanding of environmental problems into economically viable systems for improvement of industry, agriculture, and management of waste disposal systems;
- connecting programs of economic development and social movements so that their outcomes are consistent with sound ecological policy.

Although the tasks are formidable, success in implementing these recommendations will enable the countries of Eastern Europe to do their part in solving global ecological problems.

Acknowledgement

This chapter was written in cooperation with K. Dobrowolski, A. Jankowska-Klapkowska, B. Prandecka, W. Radecki, and J. Sommer of the Committee for the Study of World Socialist Systems of the Polish Academy of Sciences. Editorial assistance from E. Cowling is also acknowledged with appreciation.

REFERENCES

Anonymous. 1983. Ekonomiczne problemy ochrony srodowiska (Economic problems of environment protection). Liga Ochrony Przyrody:63 (in Polish).

Anonymous. 1986. Ekonomiczne i spoleczne problemy ochrony srodowiska, (Economic and social problems of environmental protection). Wydawnictwo Uczelniane SGPIS:207 (in Polish).

Anonymous. 1986. Socialismus a globalni problemy soucastnosti: Materialy z mezinarodni konference. (Socialism in the context of contemporary global problems: Lectures of the international conference). Ustav pro filozofii a sociologii CSAV, Part 1:238; Part 2:225 (in Bohemian).

Dobrowolski, K.A., A. Jankowska-Klapkowska, B. Prandecka, W. Radecki, J.Sommer and P.Trojan. 1989. Polityka ekologiczna wybranych krajow socjalistycznych (Ecological policy of selected socialist countries) in Ecological Policy, Materialy Strdialne 3:1-95 (in Polish).

Mcihajlow, W. 1976. Srodowisko i polityka (Environment and policy), p. 176. Warsaw: Ossolineum (in Polish).

Prandecka, B., ed. 1980. Polityka ochrony srodowiska w krajach socjalistycznych (The policy of environmental protection in socialist countries), p. 318. Warsaw: Panstwowe Wydawnictwo Ecknomiczne (in Polish).

Radecki, W. 1985. Odpowiedzialnosc administracyjna w ochronie srodowiska (Administrative responsibility in environmental protection), p. 1975. Warsaw: Ossolineum (in Polish).

Zagladin, V.V., and I.T. Frolov. 1986. Socialismus a globalni problemy civilizance in Socialismus a globalni problemy soucastnosti. Ceskoslovenska Akademi ved Prahe 1:31-53 (in Bohemian).

Acid Deposition: A Case Study of Scientific Uncertainty and International Decision Making

COURTNEY RIORDAN
U.S. Environmental Protection Agency

Editor's Note: *This case study of the acid deposition issue was selected and designed to illustrate some of the generic and specific difficulties of using scientific and technical information in environmental decision making—especially when the interests of more than a single nation are at stake. This chapter was prepared by a U.S. scientist who was involved in the negotiations with Canada on an air-quality treaty in 1981-83. Dr. Riordan was also responsible for management of an important part of the interagency program of research on acid deposition and its effects in the United States.*

International concern about acid deposition was raised significantly by Sweden's case study for the United Nations' Conference on the Human Environment in 1972 which was entitled "Air Pollution Across National Boundaries: The Impact on the Environment of Sulfur in Air and Precipitation" (Bolin et al., 1972). However, awareness of the "acid rain" problem in North America began in the late 1960s when scientists in Canada and the United States first began to study the changing acidity of precipitation and its effects on the continent. In both Europe and North America, scientific data pointed strongly to two important conclusions:

- Emissions of sulfur dioxide (SO_2) and nitrogen oxides (NO_x) were causing unnaturally high acidity in precipitation; and
- these increased loadings of acidic substances were leading to acidification of certain lakes and streams, and perhaps adversely affecting crops, forests, and human health.

By the late 1970s, acid deposition had become a major domestic political issue inside the United States and a major international political issue with Canada. Studies of the geographic distribution of sources of emissions, acidic precipitation, and acidic lakes indicated that lakes were acidic in the northeastern United States and southeastern Canada, and that a major contributing cause of the problem was the heavy concentration of SO_2 emissions in the midwestern United States.

In 1978, the United States Congress passed a resolution calling for bilateral discussions with Canada to "preserve and protect mutual air resources." This resolution was stimulated in part by concern about construction of a new coal-fired power plant in western Ontario near the Boundary Waters Canoe Area of northeastern Minnesota. In response to this resolution, the governments of Canada and the United States issued a joint statement in July 1979, and on August 5, 1980, signed a Memorandum of Intent (MOI) to negotiate a treaty on transboundary air pollution. The MOI noted that:

> The Government[s] share a concern about damage resulting from transboundary air pollution . . . including the already serious problem of acid rain.
>
> Are resolved to . . . improve scientific understanding of the long-range transport of air pollutants [and] develop and implement policies to combat its impact.
>
> Are convinced that the best means to protect the environment . . . is through achievement of necessary reductions in pollutant loadings.

Also in August 1979, President Jimmy Carter recommended a 10-year program of research on the causes and consequences of acid precipitation. In June 1980, the United States Congress passed the Acid Precipitation Act of 1980 (Public Law 96-294). This law added legislative authority for a 10-year research program which was later named the National Acid Precipitation Assessment Program (NAPAP). This program was to be carried out jointly by the major agencies of the U.S. government, chief among them being the Departments of Interior and Energy, the Forest Service, the National Oceanic and Atmospheric Administration, and the Environmental Protection Agency (Cowling, 1982). The annual budget for NAPAP grew from $17.4 million in 1983 to $85.6 million in 1987.

SCIENCE, POLITICS, AND RESEARCH

Although there was a certain incongruity in these two nearly simultaneous events—establishment of a major scientific research program while conducting international negotiations—there were also some benefits. The MOI called for establishment of:

> technical and scientific work groups to assist in preparations for and the conduct of negotiations on a bilateral transboundary air pollution agreement. . . . The Work Groups shall provide reports assembling and analyzing information and identifying measures which will provide the basis of proposals for inclusion in a transboundary air pollution agreement. These reports shall be provided by January 1982 and shall be based on available information.

The topics covered in the Work Group Reports included:

- the nature and extent of effects of acidic deposition (Bangay and Riordan, 1983);
- quantification of sources of emissions and costs for their reduction (Riegel and Rivers, 1982);
- the relationship between decreases in emissions and decreases in acidic deposition (Ferguson and Machta, 1982); and
- alternative strategies by which to allocate and achieve emissions decreases among geographical areas and types of pollution sources (Hawkins and Robinson, 1982).

Canadian scientists entered into the MOI work-group process with expectation that, around the beginning of 1982, an agreement would be signed by the two countries to decrease emissions of SO_2 on a prescribed schedule in order to decrease existing damage and prevent future environmental damage from acidic deposition. However, the United States had a different perspective on the work-group process. As a result, the MOI process broke down during 1982.

The Work Groups were actually able to reach agreement in a number of important areas, e.g., the amounts of emissions; the relative efficiencies and costs of technologies to decrease emissions from existing and new sources; and the relative lack of knowledge about the effects of acidic deposition on crops, forests, and materials. The Work Groups even reached agreement on the expected degree of accuracy of regional models to predict changes in annual average wet deposition in major receptor areas that would result from emissions reductions in major source regions.

It was in the aquatics-effects area, however, that the work-group process and, ultimately, the entire MOI negotiation process broke down. The scientists on both sides all agreed that total sulfur loadings from atmospheric deposition were the likely cause of long-term acidification of a number of lakes in the northeastern United States and southeastern Canada. But the U.S. and Canadian scientists disagreed on the conclusions that could be reached with respect to two basic scientific questions which were of critical importance to policy makers in both countries (Bangay and Riordan, 1983).

The first question was "What is the extent of acidification of surface waters?" As is often the case in emerging environmental problems, much of the early research on acidic deposition had concentrated on known problem areas, e.g., the Adirondacks in the United States and southern Ontario in Canada. The number of lakes observed to have an average pH below 5.0 was less than 180 in the northeastern United States and less than 100 in Canada. Further, essentially no data were available on the hydrology, soils, and geology of these lakes or how these characteristics might compare to those of the thousands of other lakes in the eastern United States and Canada.

To estimate the extent of existing acidification, Canadian scientists wanted to extrapolate from the lakes that had already been studied to the entire population of lakes. The basis for this extrapolation was their conviction of a similarity in regional scale soil type and geology. However, the U.S. scientists would not support this extrapolation for two reasons. First, they were convinced that many of the already affected lakes were impacted by local smelter emissions and not just by regionally transported deposition. Second, existing knowledge indicated that subregional soil and geology variation was an important factor in determining the response of lakes to acidic deposition.

The second major disagreement was related to the "target loading" issue. Canadian scientists wanted the Work Group on Impact Assessment to conclude that a reduction of the annual average wet deposition of sulfate to 20 kilograms per hectare per year would protect all but the most sensitive lakes from the adverse effects of acid deposition. The basis for this conclusion was empirical association. They had found acidic lakes in areas which were experiencing wet deposition greater than 25-35 kg/ha/yr and had not observed acid lakes where deposition was less than 20 kg/ha/yr.

The U.S. scientists would not agree to this view because they did not see the world so simply. Futher, in Norway and Sweden, there was evidence that some lakes could be acidic with wet deposition as low as 10 kg/ha/yr. In the opinion of the U.S. scientists, the difference between 10 and 20 kg/ha/yr could have been either the result of differences in the ratios of dry to wet deposition, or differences in the acid-neutralizing capacity of watersheds, or some combination of both factors. According to the U.S. scientists, a more defensible scientific position would be to estimate how many lakes would be acidic if annual average amounts of wet deposition were limited to a range of possible loadings, e.g., 10, 15, 20, 25, 30, 35, and 40 kg/ha/yr. In that way, policy makers would be aware that the situation was not black or white; rather, that there was a continuum on which lakes were more likely to be acidic as the amounts of annual average wet deposition increased.

With reasonable scientific confidence, both the U.S. and Canadian scientists agreed that atmospheric deposition of sulfur had caused acidification of some lakes, on the order of a few hundred in the United States and about 100 in Canada. But the U.S. scientists were not willing to extrapolate from the limited data that were then available in order to provide an estimate of damage for the entire population of lakes in the northeastern United States and southeastern Canada. Three critical elements were lacking, in their opinion. First, there was a lack of an adequate dose/response function relating acidic deposition to lake-water chemistry based on soil transport and transformation processes in watersheds; second, there was a lack of good empirical data on the relationships among geographical variation in surface-water quality, watershed soils, and hydrology; and finally, there was a lack of acceptable model estimates or measurements of dry deposition for lakes in the potentially affected regions (Bangay and Riordan, 1983).

Just as the MOI discussions on acidic deposition between the United States and Canada broke down during 1982, so did the ability of the United States Government to deal with its own domestic differences in perspectives about the acid deposition issue. Congressmen from the New England states, New York, Minnesota, and Wisconsin pushed for legislation to require decreases in SO_2 emissions from electrical utility boilers. Some support for these proposals was offered by western states which could provide low-sulfur coal if major eastern utilities were willing to shift from high- to low-sulfur coals.

However, congressmen from several midwestern states which had large deposits of high-sulfur coal, and in many cases also had large SO_2 emissions, were adamantly opposed to major new decreases in SO_2 emissions as was then-President Ronald Reagan Reagan. They argued that the new controls being called for would probably cost as much as $4 billion per year for as long as 20 years.

They believed such costs were not justified for two reasons. First, they wanted to "wait and see" what effect the 27% decreases in SO_2 emissions during the period from 1974 to 1980 would have on the problem. Second, they wanted to "do more research before making a decision;" they did not believe that existing scientific knowledge was sufficient to estimate the number of existing lakes that were acidic as a result of acid deposition, or to predict how many additional lakes might become acidic if amounts of emissions and deposition were to remain constant or to increase. They also saw gaps in knowledge of source/receptor relationships, modeling of atmospheric processes, and watershed acidification processes.

Thus, in part because of scientific uncertainties, the United States faced a political stalemate.

WHEN TO ACT: SCIENCE VERSUS POLICY

In the United States, issues such as acid deposition are hotly debated by the scientific community, elected officials, environmental groups, industry leaders, the media, and the public at large. These debates often become confused because of the seemingly unavoidable mixing of inherently distinct functions—the scientific function of discovering how nature works and how its is influenced by human activities, and the political function of deciding what values the society holds dear, and what, if anything, society ought to do about a given social, economic, environmental, or political issue.

The problem of handling this confusion of functions is particularly difficult for scientists in government and in the private sector. Too often, elected officials and the public look to scientists for answers to questions that are not scientific but political.

The problem of acidic deposition is a good example. As early as 1981, many scientists inside and outside of government had concluded that acidic deposition had, with acceptable scientific certainty, contributed in a major way to the long-term acidification of several hundred lakes in the northeastern United States and southeastern Canada. Because the lakes that had been studied were not known to be representative of the total number of lakes, there was no mutually acceptable scientific way to estimate the total number of lakes that might become acidic in the future. Nevertheless, scientists were being asked and some were providing answers to questions on how society should decrease emissions of SO_2 and thereby decrease acidic deposition.

The problem with this process is that the question of whether and how much society should pay to avoid a particular pollution effect is fundamentally not a scientific question. Rather, it is a political one, involving complex trade-offs between differing values of individuals and groups with respect to environmental quality and other activities that also affect the quality of life, e.g., food supply, education, public health, etc. In fact, even if scientists did know how many lakes were now acidic and how many more would become acidic in the future, it would still require a careful assessment of many non-scientific values to be able to determine what trade-offs, if any, society ought to make between the costs of emissions decreases and the costs of acidic lakes. Such questions are inherently political and require political judgment, not scientific judgment.

Of course, in 1982 as well as today, scientists cannot answer some of the important scientific questions about the environmental effects of acidic deposition, to say nothing of the political questions about what society ought to do about it. In view of the scientific uncertainties, many politicians in the United States have been unwilling to adopt new legislation for the purpose of decreasing SO_2 emissions below the amounts that are already being achieved in some regions under the Clean Air Act of 1970 and its amendments of 1977.

In the absence of a political solution to the acidic deposition problem, the United States conducted a major research program under the Acid Precipitation Act of 1980. President Reagan and his supporters in the United States Congress believed that a rational political decision was to invest in research that would close critical gaps in scientific knowledge. Thus, a conscious decision was made based on the following assumptions:

- There are critical gaps in policy-relevant scientific knowledge.
- Applied research can close many of these gaps during the 10-year research program that was begun in 1980.
- Widespread environmental damage at present rates of deposition is unlikely in 5-10 years.

MAJOR FEATURES OF THE NAPAP RESEARCH PROGRAM

The nature and scope of the NAPAP research program changed significantly as a result of the political stalemate that developed domestically and internationally during the period from 1982 to 1983. The extensive political and scientific debates that occurred at this time focused attention on certain gaps in scientific knowledge that were of major concern to policy makers. Those opposed to additional SO_2 controls were obliged to identify those uncertainties in the science that they believed prevented rational decisions. Those favoring additional SO_2 controls insisted on three basic conditions for their willingness to support research as opposed to action:

- First, those opposed to controls had to demonstrate why a gap in knowledge or uncertainty was critical to policy;
- Second, NAPAP research plans and approaches had to provide reasonable assurance that major decreases in scientific uncertainties could be achieved over a period of 5-10 years; and
- Third, the research had to be affordable and actually funded.

During the period from 1983 to 1984, the researchers in the NAPAP program worked closely with policy makers to develop a major expansion in the applied research program to address significant gaps in knowledge with projects that met the three conditions listed above.

In some ways, the timing of these discussions could not have been better. Research had already provided insights about some of the causes and effects of acidic deposition. These early results provided the foundation for a larger scale applied-research effort. Thus, the bulk of the new resources in NAPAP were directed to provide better answers to policy-relevant scientific questions in the following categories:

Aquatic Effects

- To what extent have surface waters been acidified by acidic deposition?
- How many more lakes and streams are likely to be acidified if deposition rates remain constant or increase?

• What is the dose/response function that relates acidic deposition to surface water acidification?

Forest Effects

• Is acidic deposition alone or in combination with other factors responsible for observed growth reductions and damage to selected forests in the eastern United States?

• What is the extent of damage in these forests that might be attributable to air pollution?

• What is the dose/response function that relates acidic deposition to growth declines and/or damage in these forests?

Emissions

• What are the historical and present amounts of emissions of acidic deposition precursors?

• What techniques are available for decreasing these emissions and at what cost?

Emission/Deposition Relationships

• What are the present patterns of dry deposition?

• What changes in patterns of wet and dry deposition of sulfur and nitrogen compounds would result from a change in the pattern of emissions?

The research approaches used in pursuing these several applied research questions are summarized briefly below:

The *National Surface Water Survey* (NSWS) was initiated in 1984 to provide a statistically based estimate of the number of acidified lakes and streams in various parts of the United States. In Phase I, samples were taken in regions of the country that were known to contain a significant percentage of lakes and streams with alkalinity less than 400 microequivalents per liter. In Phase II, representative subsets of lakes with alkalinity less than 200 microequivalents per liter were sampled to determine spatial and temporal variations of acidity in each lake over the spring, summer, and fall seasons. This study was also designed to determine the presence or absence of various fish species in some subregions. Phase III is a long-term monitoring program for a set of lakes with alkalinity less than 200 microequivalents per liter in areas with different acid deposition loads.

The *Direct/Delayed Response Program* (DDRP) was designed to supplement the results of the NSWS by providing detailed information on the dynamic responses of watersheds and lakewater chemistry to acid inputs. The DDRP funded the development and application of models that can use data on vegetation, soils, and hydrogeology of watersheds to predict future changes in lakewater chemistry which may occur under a variety of future acid deposition scenarios. Three different models were run using the detailed data on soil type and watershed characteristics gathered for 145 watersheds in the northeastern and the southeastern sections of the United States. The three models were the so-called ILWAS (Integrated Lake-Water Acidification Study), TRICKLE DOWN, and MAGIC (Model of Acidification of Groundwater In Catchments) models.

The *Watershed Response Program* is a watershed manipulation program designed to test critical features of the three models listed above in the field at the plot and catchment level. Simulated acidic deposition is applied to watersheds and then the response of vegetation, soils, and surface waters is observed. The data generated by these watershed-manipulation studies should provide a definitive test of the power and utility of the watershed models for predicting lakewater responses to possible future changes in acidic deposition loadings.

The *Forest Response Program* was designed to determine the possible effects of acid deposition and other airborne pollutant chemicals on forests. During the late 1970s, acid deposition was considered a major contributing cause of damage to forests in certain areas of Germany. During the early 1980s, two sources of data for tree injury and decline that were not explained by natural causes began to appear in the United States. Many scientists were concerned that these initial reports of changes in the condition of forest trees in the United States might become comparable in magnitude and extent to those observed in Europe.

A four-part program was initiated to include:

- field studies to identify and quantify changes in forest health;
- controlled exposure/response experiments to determine the impact of acidic deposition on tree seedling growth;
- research on physiological processes to identify cause-and-effect mechanisms; and
- development of models to predict tree and forest response to acidic deposition.

After much debate about alternative approaches to the study of emissions/deposition relationships, NAPAP decided to develop the *Regional Acid Deposition Model* (RADM). This model is a six-elevation Eulerian Model with horizontal grids that are 80 kilometers on a side. RADM 1 contains first generation descriptions of transport, clean-air chemistry, wet scavenging, and deposition.

A preliminary evaluation of RADM 1 was completed during 1986 using two limited data sets: the Oxidant and Scavenging Characterization of April Rains (OSCAR) and the Cross-Appalachian Tracer Experiment

(CAPTEX). OSCAR measured wet deposition amounts and chemistry within certain specific weather events. CAPTEX measured plumes of inert tracer material across the northeastern United States and Canada also during selected meteorological events. The preliminary evaluation results are being used to revise RADM.

NAPAP also decided that some measure of dry deposition and trends was essential. Such measures were needed to improve estimates of total deposition of acidic materials in receptor areas and as a means of evaluating RADM. Unfortunately, dry deposition was a case where the need outpaced the feasibility of science and the availability of funding in the NAPAP research program. As a result, a decision was made to employ an indirect air concentration/deposition velocity approach at monitoring sites. This decision was made in the absence of demonstration that an appropriate deposition velocity algorithm could be developed for operational sites based on actual flux measurements developed at a limited number of core research sites. The rationale for this decision was that even if the technique failed, air quality information would still be available for model evaluation.

NAPAP has not funded research on new combustion technologies or new post-combustion clean-up technologies. However, a great deal of research is being carried out by other programs in the U.S. government and by the private sector.

NAPAP INTERIM ASSESSMENT

In September 1987, NAPAP issued an Interim Assessment of its research program findings. The report was expected to be used by policy makers in the Executive and Legislative branches of the U.S. government in their reassessment of acid deposition policy. Although the scientific chapters of this Interim Assessment (NAPAP 1987b,c,d) provided a valuable summary of both NAPAP-sponsored and non-NAPAP-sponsored research findings, substantial controversy resulted from the disparities in substance and tone between the Executive Summary (NAPAP, 1987a) and the scientific chapters (LeFohn and Krupa, 1988).

From a policy perspective, the most important and most controversial conclusions related to the probability of future adverse environmental effects if current rates of acidic deposition were maintained in the future. The Executive Summary of the Interim Assessment emphasized that:

- Available observations and current theory suggest that there will not be an abrupt change in aquatic systems, crops, or forests at present levels of air pollution.
- Some lakes and streams in sensitive regions appear to have been acidified by atmospheric deposition at some point in the last 50 years.

Available data suggest that most watersheds in the glaciated northeast are at or near steady state with respect to sulfur deposition, and that further significant surface water acidification is unlikely to occur rapidly at current deposition levels. Although no lakes and streams with a pH of less than 5.0 have been found in the Southern Blue Ridge Province, water bodies in this region are generally not at steady state with respect to sulfur deposition, and gradual increases in surface water sulfate and decreases in acid neutralizing capacity (ANC) may occur as the sulfur absorption capacity of the soil decreases.

- At current levels of acidic deposition, short-term direct foliar effects on crops or healthy forests are unlikely. Acidic deposition may have a cumulative effect on trees growing on certain low-nutrient soils, but this effect is expected to be gradual and has not been reported in the United States at current levels. It is unlikely that regional sulfur dioxide concentrations are causing damage to crops or forests. Trees and crops can exhibit severe damage and even mortality from high concentrations of ozone and sulfur dioxide. Such occurrences are rare today because of emission controls on most major point sources. At the more typical ambient chronic concentrations of ozone, some crop damage is observed. For many tree species in low-elevation forests, growth reduction may be occurring at ambient ozone concentrations. With the possible exception of above-cloud-base forests where high mortality has occurred from unknown causes, most U.S. forests are not expected to show an abrupt change in health at current ambient air pollutant concentration levels and deposition rates.

Perhaps as important were the Interim Assessments' enumeration of scientific uncertainties which NAPAP hopes to reduce by 1990:

- the sources, quantities, and reactivities of natural emissions of sulfur dioxide, nitrogen oxides, volatile organic compounds, methane, and alkaline substances (current emissions of these substances are uncertain by about a factor of about 3);
- the origin and distribution of hydrogen peroxide, a primary oxidizing agent in clouds;
- the influence of urban emissions on deposition locally (<30 km) and in the mesoscale (30 km to 200 km) downwind;
- the source/receptor relationship resolved to a state level on a seasonal and annual basis;
- the current spatial and seasonal distribution of dry deposition of sulfur dioxide and nitric acid;
- identification of forest soils which are potentially sensitive to change by ambient acidic deposition and which might affect tree health;

- the relative contribution of acidic deposition, ozone, hydrogen peroxide, nitrate, and natural stresses to the decline of above-cloud-base forests in the Appalachians;
- methods to extrapolate results of dose/response experiments of pollutants on seedlings and saplings to mature trees;
- methods to estimate change in the regional distribution of surface water chemistry (lakes and streams) over the next half-century at present or changed rates of acidic deposition; and
- the effect of episodic acidic events on the health and reproduction of fish in streams and lakes.

The NAPAP Final Assessment is due to be completed in 1990 (Mahoney et al., 1989).

SUMMARY

In recent years, progress has been slow in resolving many scientific and public-policy questions about the causes, consequences, and management of acid deposition in North America. Part of the reason for this slow progress has been uncertainties about the science involved. But equally important has been the absence of a public consensus between Canada and the United States as well as among the several states within the United States about what, if anything, should be done about acidic deposition. It appears that the degree of scientific certainty that is required to reach a decision about such a complex issue of science and public policy is an inverse function of the degree of public consensus about the same issue. Acid deposition is certainly an example of this generalization.

REFERENCES

Bolin, B., et al. 1972. Air pollution across national boundaries: The impact of sulfur in air and precipitation. Case study prepared by Sweden for the United Nations Conference on the Human Environment. Stockholm, Sweden: Norstedt and Sons, p. 97.

Bangay, G.E., and C. Riordan. 1983. United States-Canada Memorandum of Intent on Transboundary Air Pollution. Final Report prepared by the Impact Assessment Work Group I. U.S. Department of State and the Embassy of Canada, Washington, DC.

Cowling, E.B. 1982. Acid precipitation in historical perspective. Environmental Science and Technology 16:110A-123A.

Hawkins, D.G., and R. Robinson. 1982. United States-Canada Memorandum of Intent on Transboundary Air Pollution. Final Report prepared by the Strategies Development and Implementation Work Group 3A. U.S. Department of State and the Embassy of Canada, Washington, DC.

Ferguson, H., and L. Machta. 1982. United States-Canada Memorandum of Intent on Transboundary Air Pollution. Final Report prepared by the Atmospheric Modeling Work Group 2. U.S. Department of State and the Embassy of Canada, Washington, DC.

LeFohn, A., and S. Krupa. 1988. A technical amplification of NAPAP's Interim Assessment. Air Pollution Control Association, Washington DC.

Mahoney, J.R., P.M. Irving, and J.L. Malanchuk. 1989. Plan and schedule for NAPAP's 1989 and 1990 assessment reports. Journal of the Air Pollution Association 38:1489-1496.

NAPAP. 1987a. Executive summary. The Causes and Effects of Acidic Deposition: Interim Assessment, Volume I National Acid Precipitation Assessment Program. Washington, DC.

NAPAP. 1987b. Emissions and control. The Causes and Effects of Acidic Deposition: Interim Assessment, Volume II. National Acid Precipitation Assessment Program. Washington, DC.

NAPAP. 1987c. Atmospheric processes. The Causes and Effects of Acidic Deposition: Interim Assessment, Volume III. National Acid Precipitation Assessment Program. Washington, DC.

NAPAP. 1987d. Effects of acidic deposition. The Causes and Effects of Acidic Deposition: Interim Assessment, Volume IV. National Acid Precipitation Assessment Program. Washington, DC.

Riegel, K.W., and M.E. Rivers. 1982. United States-Canada Memorandum of Intent on Transboundary Air Pollution. Final Report prepared by the Emissions, Costs, and Engineering Assessment Work Group 3B. U.S. Department of State and the Embassy of Canada, Washington, DC.

U.S. Department of State and the Embassy of Canada. 1980. Memorandum of Intent between the Government of the United States of America and the Government of Canada concerning transboundary air pollution. Washington, DC.

Diagnosis of Environmental Protection Problems in Poland

ANDRZEJ T. KASSENBERG
Institute of Geography and Spatial Organization
Polish Academy of Sciences

The natural environment defines the potential for each country's development; however, it also creates barriers and restrictions to that development. Studies of the state of the environment provide a basis for long-term development plans when they recognize this dual role. Such studies aim at gathering information on environmental conditions and the processes at work; evaluating the backlog in environmental protection activities; and presenting the consequences of a situation in ecological, social, and economic terms, ultimately addressing the whole issue in the form of a spatial synthesis.

Such diagnostic studies require a comprehensive approach that takes into account the existing relationship between environmental factors and socioeconomic development. This can be approached from four viewpoints (Kassenberg and Rolewicz, 1985):

- *Effectiveness*, i.e., the conservation as well as the rational use of resources and environmental "goods";
- *Consequences*, i.e., social, economic, and ecological profits and losses related to the state of the environment;
- *Hazards*, i.e., the results of a scale of development which are potentially harmful to social and economic activities;
- *Safeguards*, i.e., the application of suitable measures to protect or reconstruct a degraded environment.

In Poland, the first studies of this type were undertaken at the end of the 1960s and beginning of the 1970s by the Institute of Geography of Jagiellonian University and by the Committee for Protection of Nature and Natural Resources of the Polish Academy of Sciences (Brykowicz and Waksmundzki, 1972; Leszczynski, 1974). In the first study, the extent of the concentration of harmful phenomena in one *voivodship* (administrative district) of Poland was determined on the basis of qualitative and quantitative analysis of distortions and damage in the environment. However, due to lack of information for the whole country, it was impossible to apply the methods broadly. Furthermore, it did not address the relationships among the source of pollution, the state of the environment, and proposed protection and recultivation efforts.

Parallel with previous studies, pre-plan spatial diagnoses of the state of environmental protection in Poland were undertaken by the Planning Commission of the Council of Ministers. Four generations of studies were carried out: the 1970 and 1975 versions were highly simplified, while the 1980 and 1985 versions constituted full-scale assessments. In principle, the studies were designed to develop a spatial synthesis through the diagnosis of individual environmental problems. Though some differences resulted from their specific character, each of these studies followed the sequence:

1. analysis of the state of the environment;
2. type and features of sources of environmental threat;
3. effectiveness of protection and recultivation efforts;
4. volume and type of pollution emissions;
5. consequences of pollution for the economy, society, and nature.

The following four fundamental environmental issues were studied in particular detail and will be discussed in subsequent sections:

- air quality;
- water quality;
- condition of land; and
- conservation of nature.

AIR QUALITY

Maintaining the purity of air is a crucial problem in Poland, as air quality has been deteriorating year by year. In 1975, the area of measured air quality exceeding Polish pollution standards was 8,400 km^2; by 1980, this area had grown by 27% and continues to grow, albeit at a slower rate due to decreased industrial production. It is estimated that almost 20% of Poland's population lives in these areas. Inhabitants of the Katowice district in southern Poland are the most exposed to these hazards in terms of both the concentration and range of excessive pollutants. The number of people threatened to varying degrees by the impact of air pollution on human health is approaching 3 million.

Gaseous air pollution, particularly sulfur dioxide, is the most serious problem. In about two-thirds of the country's area, the yearly average of SO_2 concentration exceeds 20 $\mu g/m^3$—a level which may cause first-degree damage to coniferous forests. This high concentration of SO_2 is paralleled by deposition of 8 tons of sulfur compounds yearly on each square kilometer of land in Poland. In 10% of the area, the level exceeds 50 t/km^2 (Kassenberg and Rolewicz, 1985).

Apart from SO_2, the most dangerous air pollutants are:

- particulates, nitrogen oxides, and carbon dioxide (commonly occurring, high concentrations);
- lead, cadmium, arsenic, and mercury (quite common, highly toxic, and durable in the environment); and
- carcinogenic hydrocarbons (commonly occurring, particularly dangerous to health).

Extremely high concentrations of air pollution containing such compounds as carbon disulfide, fluorine, and heavy metal dusts occur in localized sites which are often near chemical and non-ferrous metallurgy plants.

Estimates of total emissions of major air pollutants in Poland in the period 1985-1987 are shown in Table 1. According to the table, emissions of pollution from harmful sources included in 1987 official statistics amount to 1.8 million metric tons of particulates and 5.4 million metric tons of gases. Of the particulate emissions, 22% occurred in the Katowice district, 7% in Jelenia Gora, 6% in Krakow, and 5% each in Konin and Warsaw. Of the gaseous emissions, 28% occurred in Katowice, 11% in Krakow, 6% in Piortrkow Trybunalski, 5% each in Legnica and Jelenia Gora, and 4% in Tarnobrzeg. Increasing pollution is of concern particularly in districts regarded as "clean" and having great touristic and recreational value, such as the districts of Bialystok, Koszalin, Krosno, Lomza, Ostroleka, and Zamosc.

Most sources of air pollution in Poland are concentrated in a few dozen cities and localities. The largest is a cluster of eight towns in the Upper Silesian industrial district (i.e., Dabrowa, Gornicza, Jaworzno, Laziska Gorne, Trzebinia, Chorzow, Bedzin, and Bytom). Other polluted areas include Krakow, Bogatynia, Konin, Warsaw, Rybnik, Polaniec, Ostroleka, Plock, Kedzierzyn-Kozle, Skawina, and Oswiecim. Statistics for 1987 show that of 1,342 industrial plants emitting particulates, 89% have facilities for pollution reduction, but only 211 of these (i.e., 16%) reduce pollution satisfactorily (i.e., minimum of 90% reduction). With regard to gaseous pollution, of 1,362 industrial plants, 91% have no purifying facilities and only 17 plants (i.e., 1.2%) have produced satisfactory results (i.e., minimum 50% reduction) (Main Statistical Office, 1988).

The state of air quality in Poland is also influenced by pollution coming from other countries, so-called transboundary air pollution. Although there is an approximate balance between the import and export of pollution to and

TABLE 1 Emissions of air pollutants in Poland, 1985-1987 (in thousands of metric tons per year).

EMISSION SOURCES	PARTICULATES	GASES SO_2	NO_x	CO	Hydrocarbons
TOTAL	2,980	4,200	1,530	3,770	510
Power stations	865	2,050	410	70	5
Industrial heat stations	100	730	215		
				1,680	205
Industrial technologies	1,380	390	400		
District heating and commercial/ residential	630	930	145	1,270	200
Motor vehicles	5	100	360	750	100

SOURCE: Ministry of Environmental Protection and Natural Resources, Warsaw, 1988.

TABLE 2 Water quality in Poland, 1964-1986 (as percentages).

Classes of water purity	1964 - 1967	1971 - 1973	1978 - 1983	1986	Required purity*
Class I	31.6	23.4	6.8	4.2	53.0
Class II	25.6	32.2	27.9	26.7	40.0
Class III	14.0	18.0	29.0	27.3	7.0
Below any standards	28.8	26.4	36.3	41.3	--

* According to classification made in view of present or planned utilization of water (M. Roman)

SOURCE: Main Statistical Office, Warsaw, 1988, 1989.

from Poland as a whole, the environmental situation in border regions of the German Democratic Republic and Czechoslovakia is very unfavorable for Poland (see Chapter 23, this volume). Pollution coming from the west has a great impact on the degradation of forests in the Sudety area. Prevailing southwesterly winds bring pollution from the Karvina-Ostrava region of Czechoslovakia to the Rybnik coal district and the Upper Silesian industrial district.

WATER QUALITY

Water purity in Poland is a problem that parallels air pollution. When comparing the present situation with that of 20 years ago, one can notice a continuous deteriorating trend. For example, the amount of river water of highest quality has decreased dramatically from 32% to 5%. Over the same period, the amount of river water which is not useful for any purpose grew from 29% to 42% (Main Statistical Office, 1988). This is illustrated in Table 2; for detailed descriptions of classes of water purity, see Chapter 19 (this volume).

Many rivers consist of water that cannot be used along their entire lengths, e.g., Bzura, Krzna, and Ner. Apart from these, the most polluted large rivers are the Brda, Bobr, Bug, Drweca, Ina, Kamienna, Kwisa, Liwiec, Mala Panew, Notec, Nysa Luzycka, Oder, Prosna, Skrwa Prawa, Vistula, Warta, Widawa, Wierzyca, and Wkra.

In the upper parts of the Vistula and Oder drainage areas, excessive salinization of water as the result of mining operations is a serious problem, amounting to 6-7,000 metric tons per day. Salt concentrations exceeding by several times permitted values can be found in the Olza, Ruda, Bierawka, Klodnica, Ilownica, Przemsza, and other rivers, but particularly in long stretches of the Vistula and Oder.

A similar problem exists with respect to lake purity. Out of 500 large lakes, about 300 have been endangered as a result of pollution, mainly municipal and agricultural wastes (Chapter 17, this volume). The largest number of degraded lakes occur in the districts of Koszalin, Olsztyn, Slupsk, Suwalki, and Szczecin. Unfortunately, included among them are some of the Great Masurian Lakes, e.g., Mikolajskie, Niegocin, and Talty.

The fundamental reason that surface-water quality has deteriorated so greatly in Poland is that these waters are very convenient waste receivers. The growing volume of liquid wastes being discharged into rivers has not been addressed by the construction of additional sewage treatment plants. In addition, the operation and functioning of existing plants is for the most part inefficient and could benefit from systemic improvements. In 1987, the annual volume of industrial and municipal liquid wastes requiring purification was 4.5 billion m^3, including 2.6 billion m^3 of industrial wastes. Almost 62% were treated in some way, but only 50% with highly effective chemical and biological methods.

Of the 4,732 industrial plants with high impact on water resource management in Poland, more than 2,900 plants discharge their liquid wastes directly into surface water. Of these, almost 16% have no treatment facilities and another 9% have treatment plants with insufficient capacity. More than 1,800 remaining plants discharge their liquid wastes directly into municipal sewers.

Of the 818 cities in Poland, only 436 have sewage treatment plants, and of these, only 60% are capable of providing adequate mechanical-biological treatment of the wastewaters delivered to the plants. Many of these existing plants have insufficient capacities. In addition, there are many district capitals which do not have municipal sewage treatment plants, e.g., Warsaw, Bialystok, Elblag, Kalisz, Lodz, Radom, and Rzeszow. In the Lodz district, no single town has a municipal sewage treatment plant (Main Statistical Office, 1988).

The pollution of the Baltic Sea, which has recently shown a growing trend, has international as well as regional and national significance. Two of Poland's largest rivers—the Vistula and Oder—carry enormous quantities of organic and toxic pollution into the Baltic. This causes marked changes in the chemical composition of the sea water, particularly reduced oxygen content in the bottom water. Hydrogen sulfide occurs periodically and causes the disappearance of life in the most polluted areas, i.e., the Bays of Gdansk and Puck, the Vistula Lagoon, the Pomeranian Bay, and the Szczecin Lagoon. Some of the pollution comes from towns and industrial plants located in coastal areas that discharge their liquid wastes directly into the sea. The purity of sea water is also much affected by the ships calling at Polish ports.

Bacteriological pollution of the coastal zone has caused the closing of practically all beaches on the Gdansk and Puck Bays as well as on the Vistula and Szczecin Lagoons. Also, some beaches have been closed in the central seaside areas of Kolobrzeg, Dzwirzyno, Ustka, and Leba.

CONDITION OF LAND

Degradation of land in Poland resulting from intensive mining operations is another serious problem. Such degraded areas occur mainly in the mining regions of Upper Silesia, Legnica and Glogow, Tarnobrzeg, Konin, Turoszow, Kielce, Chelm, and Belchatow. It should be stressed that current recultivation operations proceed too slowly. In 1987, only 4,273 hectares were recultivated, whereas the area where mining operations were completed covered 100,000 hectares (Main Statistical Office, 1988).

Erosion and the subsequent acquisition of steppe characteristics cause the deterioration of productive qualities of soil. The area threatened or already damaged by erosion in Poland is estimated to be 2 million hectares. The intensity of the process is illustrated by fact that land with about 10% slope loses a layer of 5 mm of soil per year. Erosion is most hazardous to the districts of Zamosc, Lublin, Bydgoszcz, Krakow, Chelm, Torun, Wroclaw, and Zielona Gora. Areas at risk of becoming steppes cover about 5 million hectares, including 0.6 million hectares of critically threatened land located mainly in central Poland.

Another problem connected with land surface protection is incorrect management of industrial and municipal wastes. At the end of 1988, industrial plants producing over 1,000 tons of environmentally harmful waste accumulated over 1.5 billion tons of waste (2.5 times more than in 1975) in an area of over 10,600 hectares. Almost 90% of the wastes were collected in the ten districts of Katowice, Legnica, Walbrzych, Krakow, Kielce, Tarnobrzeg, Szczecin, Bydgoszcz, and Konin. The current situation regarding the re-use of industrial wastes in connection with their production is not satisfactory. Of the 186 million tons of industrial waste produced in 1988, only 106 million tons were utilized for economic purposes, i.e., approximately 57%. The sources of collection and production of industrial wastes are concentrated in about a dozen towns, particularly five towns in the Upper Silesian industrial district (Gliwice, Piekary Slaskie, Knurow, Ruda Slaska, Zabrze), as well as in Polkowice, Jastrzebie-Zdroj, Walbrzych, Lubin, Krakow, Wodzislaw Slaski, and Rybnik (Main Statistical Office, 1988).

Municipal wastes are also hazardous to the environment. In 1987, the total accumulated municipal wastes for Poland was almost 46 million m^3. The largest concentrations of municipal wastes are in big cities, especially in Katowice, Lodz, and Warsaw. Municipal wastes are generally not utilized. The unsanitary condition and location of most dumping grounds are very hazardous to the environment. Composting and burning of wastes are practically unheard of in Poland. Under such circumstances, a growing quantity of waste is stored in forests, water reservoirs, and streams, making recreation impossible. Also, municipal dumping grounds pose a problem as they are operated without any regulation of their location, use, or capacity. They are usually small in size (up to 0.5 hectares), are unenclosed, and lack preventive measures against spread of pollution.

CONSERVATION OF NATURE

Forests perform a special role in the environment in addition to serving tourism and recreation purposes. The forested area in Poland is 8.7 million hectares, almost 28% of the total surface. Forests are not evenly distributed over Poland's area; the index of afforestation ranges from 11.9% in the Plock district to 48.3% in the Zielona Gora district. Limited variety in the types of timber stands is a problem. There are vast areas of coniferous monocultures, which are less immune to many biotic, abiotic, and man-made stress factors. Monoculture and even-aged coniferous timber stands often lead to degradation of habitat quality for wildlife and recreational uses. A majority of Polish forests are young; the average age of state-owned forests is 49 years. The shortage of timber stands aged 80 and above is estimated at 0.7 million hectares, which means that the required increase in such timber stands is 70%.

The sanitary and health situation of forests is the fundamental problem in forest management. About two-thirds of the forested area is in zones of permanent or periodic hazard by harmful insects, parasitic fungi, emissions of air pollutants, or unfavorable weather conditions. At present, the area of standing timber damaged as a result of air pollution is estimated to be almost 800,000 hectares, i.e., over 9% of Polish forests. The threat of damage to forests affects as much as 75% of the country's area. The process of deforestation has begun in Poland, primarily in the Katowice district where almost 100% of the forests have been damaged and where there are frequent cases of deforested land. Similar areas exist in the Izery and Karkonosze Mountains. Apart from Katowice, areas where a substantial share of forests are damaged are the districts of Lodz (42%), Krakow (41%), Jelenia Gora (40%), Tarnobrzeg and Legnica (28% each), Bielsko, Czestochowa, and Tarnow (26% each), and Wroclaw (23%). This great threat from air pollution is due to the structure of industry, outdated technologies, lack of gas-purifying facilities, high concentration of emission sources, location of plants close to forests, and the predominance of species sensitive to toxic gases and particulates (Main Statistical Office, 1988).

In addition, the following factors have an impact on the sanitary condition of forests:

- overpopulation of game in relation to ecological capacity, resulting in damages to an area of 140,000 hectares annually;
- forest fires, as 30% of Poland's forested area is in the highest class of fire hazard; and
- weather conditions, particularly winds and heavy snowfalls as well as freezing temperatures (Kassenberg and Rolewicz, 1985).

Apart from forest protection, which is a crucial issue of environmental protection in Poland, attention should also be focused on the larger issue of nature and landscape protection. Legally protected areas in Poland are divided into four distinct categories, which are listed below in order of the stringency of the regulations which govern them:

- national parks;
- nature reserves;
- landscape parks; and
- areas of protected landscape.

These protected areas have been continually expanding and at the end of 1988 amounted to almost 4.5 million hectares, i.e., over 14% of the country's total area. The largest share of these areas occurs in the following districts: Przemysl (50%), Bialystok and Gorzow (39%), Konin

(36%), Skierniewice (33%), Chelm (32%), Krosno (29%), and Zielona Gora (28%). The ultimate goal for such protected areas is to cover 30-40% of the area in Poland (Main Statistical Office, 1988).

Among natural resources subject to conservation protection, *national parks* are the most important. There are 14 national parks in Poland covering a total area of 127,000 hectares, including over 70% of Polish forests. National parks in Poland are characterized by a large variety of bioconeoses, types of plants and animals, as well as forms of landscape. These parks cover about 0.4% of the country's total area, which is a relatively low proportion in comparison with other countries. For instance, in Czechoslovakia national parks cover 1.25% of the country's area; in Sweden, 2.41%; in Great Britain, 5.38%; and in Japan, 6.4%.

Another important form of conservation protection are *nature reserves*, which are areas where nature as a whole or an individual element (e.g., groups of plants or animals or fragments of landscape) are under protection. At present, there are almost 1,000 nature reserves in Poland covering over 114,000 hectares, i.e., about 0.3% of the country's surface. Most of these nature reserves provide forest and floristic protection; there are only a few nature reserves for the non-living environment (e.g., land forms and geological phenomena). Protection of water complexes is also insufficient. The average area for nature reserves in Poland (i.e., about 115 hectares) is among the smallest in Europe—in Czechoslovakia, it is 140 hectares; in Holland, 145 hectares; and in the FRG, 290 hectares. This relative lack of nature reserves has a negative effect on environmental quality, as does the lack of adequate areas of national parks mentioned above.

There are also two distinct categories of landscape protection. First, *landscape parks* are protected areas because of their extraordinary environmental, aesthetic, or touristic value. There are 43 landscape parks in Poland, covering over 1,000,000 hectares, i.e., 3% of the country's area. The goal is to increase this proportion to 4% in the future. The second category includes *areas of protected landscape*, which provide potential areas for tourism and recreation. The goal is for such areas to cover at least 20-30% of the country in the future and be developed into an interconnected system based on natural conditions. By the end of 1987, areas of protected landscape approved by resolutions of district councils covered over 3 million hectares, i.e., 10% of the country's area (Main Statistical Office, 1988). Recognition of elements of nature as valuable and unique also plays a significant role in nature protection and enrichment of landscape values. Almost 17,000 objects have been designated as natural monuments in Poland.

However, each form of protection mentioned above is greatly threatened by development of industry, expansion of cities, mining, intensification of agriculture and forestry, expanded motor traffic, and tourism. The extent of hazards to individual national parks differs, but there is no national park free of the negative effects of human activity (Chapter 14, this volume). Among the most threatened parks is the Ojcow National Park near Krakow. Many nature reserves and landscape parks are also located in hazardous zones.

Changes occurring in the environment are followed by destruction of flora and fauna. The transformation of vegetation in Poland can be summarized as follows. Only 8% of the country's area has retained its natural character. Another 19% is made up of areas in the transition stage between the natural environment and an environment changed by human impact. The remaining 73% are areas dominated by economic activities with a visibly changed environment, including 10% of degraded areas (Falinski, 1975).

Many species from Poland's original fauna are dying out or are extinct. Aurochs, tarpans, and saiga antelopes are now extinct; rare and endangered species include bear, elk, wildcat, beaver, mink, marten, vulture, and eagle. Great changes have occurred in quantitative relations and distribution of animals. Excluding introduced and migrating types, there are currently 430 vertebrate species in Poland, of which only 10% are not endangered (Glowacinski et al., 1980).

THE STATE OF ENVIRONMENTAL PROTECTION

Poland is not a large country. In fact, it is relatively small in terms of area, although it has a very differentiated natural environment. Every 1,000 km^2 is characterized by different natural conditions. In the north is the Baltic Sea; southward is a stretch of lake districts, followed by a zone of lowlands and then a zone of uplands; and in the extreme south are lower and higher mountains.

The 1970s and 1980s have been characterized by increased hazards to the environment. As ecological factors have had no bearing on the strategy of the country's development, and efforts for environment protection have had little effect, the condition of the environment has greatly deteriorated. An ecological crisis which annually increases in severity and affects the entire economy as well as every citizen is manifested primarily though:

- wasteful exploitation of mineral, water, and forest resources;
- rapidly deteriorating conditions in air, water, and soil pollution;
- overall degradation of natural potential which threatens further development of agricultural production and forest production, as well as efficient management of water resources;
- deterioration of landscape;
- increasing contamination of foodstuffs;

TABLE 3 Environmental losses in Poland (in millions of zlotys per year).

Agriculture	150,000
Forestry	50,000
Water resources	65,000
Corrosion	215,000
Minerals (due to inefficient extraction)	130,000
Raw materials (in liquid wastes and air pollutants)	50,000
Health	115,000
TOTAL	775,000

SOURCE: Main Statistical Office, Warsaw, 1988.

- rapidly increasing hazards to human health; and
- general deterioration of living conditions and quality of life.

The present condition of the environment contributes to measurable and immeasurable losses. At present, minimum losses due to pollution of the environment are estimated at about 800 billion zlotys, i.e., over 10% of the annual national economic product (Table 3); however, some experts estimate the losses to be twice as much (Main Statistical Office, 1988).

In accordance with methodological assumptions of this diagnosis, a spatial synthesis of the state of environmental protection has been created. Four categories have been distinguished where conditions are highly unsatisfactory for a variety of reasons (Kassenberg and Rolewicz, 1985). These are:

- *areas of ecological hazard*;
- *cities and towns* with major sources of pollution which are not located in areas of ecological hazard;
- *nature and landscape conservation areas* with unstable natural balance;
- *spas* endangered with losing their healing values due to unstable natural balance.

The first category includes 27 areas where the natural balance has been completely broken, manifested by loss of immunity, elimination systems, as well as intensified hazards to human health. These areas cover about 35,000 km^2 and are inhabited by over one-third of Poland's population.

The basic criteria used to distinguish an area of ecological hazard are:

- violation of permitted standards or strong degradation (pollution) in at least two environmental aspects; and

TABLE 4 Characteristics of Areas of Ecological Hazard in Poland.

CHARACTERISTIC	UNIT OF MEASURE	AREAS OF ECOLOGICAL HAZARD	POLAND
Area			
size	km^2	35,220	312,683
share of the country	%	11.3	---
Population (1987)			
number	thousand	13,329	37,664
share of country	%	35.4	---
population density	pop/km^2	381	120
Source of hazard (1982)			
a) towns with over 100,000 inhabitants			
number		28	38
population:			
total	thousand	7,674	10,813
share of country	%	71.0	---
b) towns with 50-100,000 inhabitants			
number		24	42
population:			
total	thousand	1,679	2,860
share of country	%	58.7	---
c) towns with 20-50,000 inhabitants			
number		29	114
population:			
total	thousand	939	3,537
share of country	%	26.5	---
d) towns with under 20,000 inhabitants			
number		80	611
population:			
total	thousand	710	4,446
share of country	%	16.0	---
e) industrial plants in particularly hazardous branches of industry*			
number		1,143	2,607
employment:			
total	thousand	896	1,110
share of country	%	80.7	---
f) industrial plants in particularly hazardous branches of industry**			
number		4,228	16,072
employment:			
total	thousand	345	1,000
share of country	%	34.5	---

(continued...)

TABLE 4. Continued

CHARACTERISTIC	UNIT OF MEASURE	AREAS OF ECOLOGICAL HAZARD	POLAND
g) industrial plants in branches of industry with low risk***			
number		7,658	19,988
employment:			
total	thousand	1,061	2,364
share of country	%	44.9	---
Municipal and industrial liquid wastes requiring purification (1987)			
share of country	%	62.4	---
degree of concentration	thous m^3/km^2	80.1	14.4
total treated waste	%	64.1	61.7
share of waste treated biologically/chemically	%	25.6	27.5
Emission of particulate pollution (1987)			
share of country	%	76.8	---
degree of concentration	$tons/km^2$	39.5	5.8
degree of reduction	%	94.7	94.3
Emission of gas pollution (1987)			
share of country	%	81.2	---
degree of concentration	$tons/km^2$	125.2	17.3
degree of reduction	%	13.6	12.4
Degraded soil (1987)			
share of country	%	34.7	---
Endangered forests (1987)			
share of country (area)	%	64.4	---
share of country (mass)	%	61.7	---
Industrial wastes (1987)			
share of total accumulated waste in country	%	92.6	---
degree of generation of accumulated wastes	$tons/km^2$	38.5	4.7
share of wastes produced	%	89.5	---
degree of concentration	$tons/km^2$	4,598.1	575.6
degree of economic utilization	%	54.0	54.4

* fuel, power engineering, metallurgy, chemical
** wood/paper, mineral, food processing
*** light, electroengineering, etc.

SOURCES: Main Statistical Office, Warsaw, 1986; Kassenberg, 1986.

TABLE 5 Environmental risk in areas of ecological hazard in 1987.

Areas of ecological hazard	Industrial and municipal liquid wastes requiring purification discharged to surface waters: Total		Treated		Index of water intakes closed cycles in %	Industrial air pollution: particulates			gases		
	hm^3	$dam^3/1km^2$	in % of the total	biologically and chemically in % of the total		produced in tons/km^2	emitted in thousands tons	reduction degree in %	produced in tons/km^2	emitted in thousands of tons	reduction degree in %
Belchatow	7	30	100	21	73	17,186	32	99	1,427	315	--
Bydgoszcz-Torun	117	61	16	9	4	164	39	88	33	57	10
Chelm	10	7	95	66	1	66	27	70	33	45	--
Czestochowa	43	106	75	63	71	526	34	81	147	60	--
Gdansk	121	38	79	22	1	152	54	89	25	72	12
Upper Silesia	866	279	58	15	35	1,730	386	93	487	1,307	14
Inowroclaw	42	55	48	2	7	351	23	92	44	33	1
Jelenia Gora	32	60	76	609	1	35	7	64	29	15	2
Kielce	32	31	52	48	10	969	19	98	20	21	1
Konin	130	128	100	3	1	1,225	92	93	164	166	0
Krakow	190	80	45	2	12	464	111	90	252	591	0
Legnica-Glogow	122	30	95	31	65	83	19	94	146	267	55
Lodz	157	190	4	4	7	571	39	92	72	58	2
Myszkow - Zawiercie	16	73	94	89	69	280	27	60	38	7	0
Opole	111	110	81	84	44	1,519	67	96	178	150	16
Plock	40	159	98	93	35	9	1	43	695	129	24
Poznan	86	86	38	8	3	114	10	91	22	19	12
Pulawy	18	25	100	71	15	142	11	90	52	39	15
Rybnik	65	62	84	39	40	1,681	41	98	202	209	0
Szczecin	129	38	61	34	0	411	76	94	59	178	10
Tarnobrzeg	132	52	89	54	1	725	58	97	79	216	3
Tarnow	73	65	62	52	50	198	35	84	73	59	27
Tomaszow	31	74	99	90	2	35	3	82	45	11	42
Turoszow	39	79	98	30	66	7,854	109	97	489	244	0
Walbrzych	39	80	87	42	9	239	23	80	53	24	8
Wloclawek	47	75	76	34	17	29	7	60	35	22	--
Wroclaw	109	121	94	74	23	494	36	93	78	68	3

Areas of ecological hazard	Industrial wastes: accumulated in mio tons	Industrial wastes: produced in mio tons	Industrial wastes: produced in tons/km^2	Industrial wastes: proportion of economic utilization in %	Soil degraded as result of industrial activity and peat mining in hectares (1986)	Forest area in zones hazardous to air pollution emission: in thousands hectares	Forest area in zones hazardous to air pollution emission: in % of forest area	Greenery areas in towns in m^2 per inhabitant (1986)
Belchatow	11	5	22,157	18	1,722	8	58	--
Bydgoszcz-Torun	2	1	307	36	373	44	57	10.5
Chelm	0	0	13	20	1,443	0	2	9.7
Czestochowa	3	0	1,700	89	606	28	83	5.6
Gdansk	13	1	397	18	1,784	6	6	8.0
Upper Silesia	348	57	18,338	78	8,759	103	99	13.3
Inowroclaw	42	2	3,025	38	489	1	4	12.0
Jelenia Gora	0	0	174	76	138	25	60	18.1
Kielce	33	2	2,304	31	1,182	6	30	9.3
Konin	33	2	1,830	4	4,615	5	18	8.3
Krakow	75	5	2,314	72	1,192	19	64	9.1
Legnica-Glogow	303	28	6,991	29	2,369	30	23	12.2
Lodz	1	1	852	90	185	--	--	8.8
Myszkow - Zawiercie	3	0	1,743	39	85	--	--	8.8
Opole	22	1	836	23	3,139	38	61	11.1
Plock	0	0	606	16	45	3	23	4.3
Poznan	0	0	308	49	396	2	4	18.4
Pulawy	3	0	468	6	120	8	58	5.5
Rybnik	241	34	33,207	66	2,177	15	100	9.2
Szczecin	56	5	1,550	9	577	10	41	9.2
Tarnobrzeg	53	4	1,707	11	810	47	58	6.0
Tarnow	3	1	522	28	540	15	60	8.0
Tomaszow	1	0	291	83	358	3	15	5.8
Turoszow	0	4	8,308	3	2,764	15	100	9.5
Walbrzych	94	4	8,581	35	589	5	21	13.3
Wloclawek	0	0	193	44	162	14	54	6.9
Wroclaw	7	1	807	20	385	4	13	14.1

SOURCE: Main Statistical Office, Warsaw, 1987, 1988.

- multiple violation of permitted standards or very strong degradation (pollution) in one environmental aspect, or violation of permitted standards by a very toxic substance.

Specific characteristics of areas of ecological hazard are presented in Tables 4 and 5 (preceding pages). The data presented show that the selected areas are not homogeneous with respect to the quantity and type of pollution. Taking this into account, they were divided into the following subgroups:

- areas of ecological disaster (13,783 km^2, population 6.2 million);
- areas of extensive pollution (19,400 km^2, population 6.2 million);

and

- areas with serious air pollution (2,037 km^2, population 0.5 million).

The second category (cities and towns) includes localities apart from the above-mentioned areas which also contain hazardous sources of pollution. They are characterized by pollution of the environment and degradation surpassing the overall character of the region. Sixty such localities are inhabited by about 5 million people.

The third category (conservation areas) refers to national and landscape parks. In these areas, there is decreased natural resistance to environmental stresses and weakened self-regulating processes due to pollution of water and air, even from distant sources of emission, as well as excessive tourist traffic, urbanization pressure, industrialization, and incorrect irrigation procedures. These problems have resulted in the extinction of many less resistant or uncommon types of vegetation and animals. In all of Poland, there are 15 such areas, of which the most endangered national parks are Ojcow, Babia Gora, Karkonosze, Swietokrzyski, Kampinos, Wielkopolski, and Pieniny.

The fourth category (endangered spas) includes 23 spas threatened with the loss of their natural healing values due to incorrect management of water resources and liquid wastes, excessive air pollution, and heavy traffic patterns which alter the local climate.

CONCLUSION

Based on the diagnosis of the condition of the environment presented here, as well as on efforts undertaken for its protection, four zones can be distinguished in Poland which require different environmental policies (Figure 1, at end of chapter). These are:

- *Zone I*: characterized by geographic integration of areas of ecological hazard. These areas require activities to restore natural living conditions, including conduct of appropriate economic activities.

- *Zone II*: where there exists a potential danger that two coastal areas of ecological hazard may be connected in a continuous land/sea stretch of degraded environment. These areas require a reduction of population and use pressure to a level determined by the natural capacity of the area.
- *Zone III*: where there are regional environmental problems connected mostly with dispersed areas of ecological hazard. These areas require that natural barriers be overcome through regional development.
- *Zone IV*: a relatively clean environment, without areas of ecological hazard beyond local environmental problems. These areas require efforts to prevent the spread of new pollution, as well as the implementation of appropriate economic development principles in order to maintain the quality of the region.

REFERENCES

Brykowicz, K., and K. Waksmundzki. 1972. Complete map of degradation and pollution in geographical environment. In Man and knowledge. Weidza Powszechna. Warsaw.

Falinski, J. 1975. Anthropogenic transformation of vegetation in Poland. Phytocenosis 4,2.

Glowancinski, Z., et al. 1980. State of vertebrates and selected invertebrates in Poland: List of species, their disposal, danger, and protection. Panstowowe Wydawnictwa Naukowe, Warsaw-Krakow.

Kassenberg, A. 1986. Areas of ecological hazard: New planning category. Cosmos 1.

Kassenberg, A., and C. Rolewicz. 1985. Spatial diagnosis of environmental protection in Poland. Studies of Committee of Spatial Development, Polish Academy of Sciences, volume LXXXIX. Panstwowe Wydawnictwa Naukowe, Warsaw.

Kozlowski, S. 1985. Ecodevelopment: Concept of development. Man and World Outlook, May 5, p. 232

Leszczynski, S. 1974. Problems of protection of human environment. Geographical Papers 108.

Main Statistical Office, Warsaw. 1986. Environmental protection and water economy.

Main Statistical Office, Warsaw. 1987. Environmental protection and water economy.

Main Statistical Office, Warsaw. 1988. Environmental protection and water economy.

Main Statistical Office, Warsaw. 1989. Information about the state, risk, and protection of the environment in 1988.

Ministry of Environmental Protection and Natural Resources, 1988. National Program of Environmental Protection Through 2010 (draft). Warsaw.

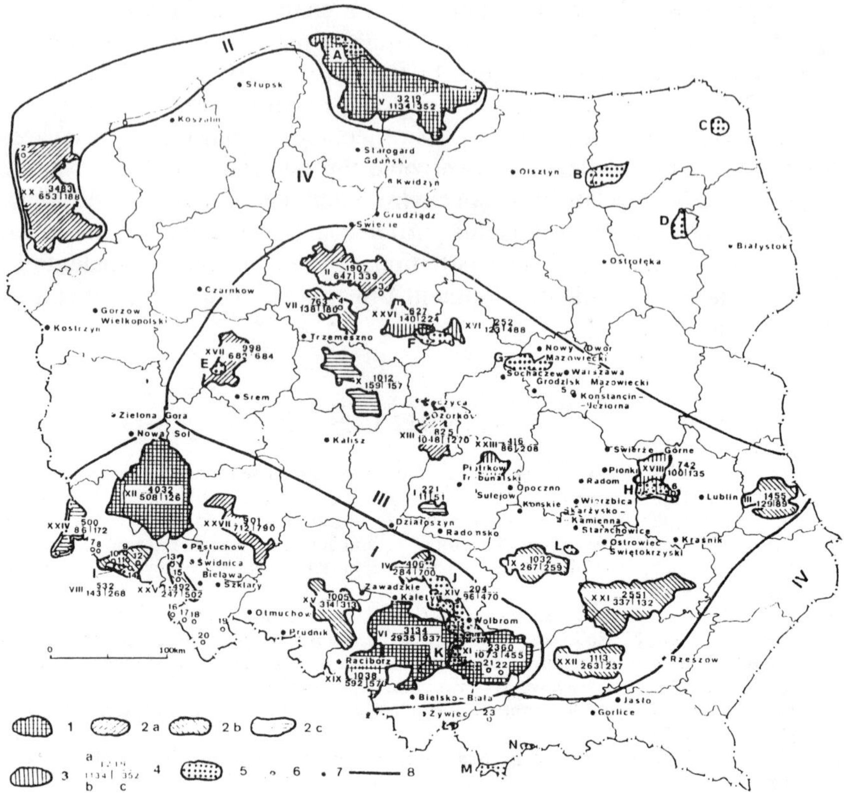

FIGURE 1 Zones of environmental quality in Poland (Kassenberg, 1986).

Areas of Ecological Hazard: I = Belchatow; II = Bydgoszcz-Torun; III = Chelm; IV = Czestochowa; V = Gdansk; VI = Upper Silesia; VII = Inowroclaw; VIII = Jelenia Gora; IX = Kielce; X = Konin; XI = Krakow; XII = Legnica Glogow; XIII = Lodz; XIV = Myszkow-Zawiercie; XV = Opole; XVI = Plock; XVII = Poznan; XVIII =Pulawy; XIX = Rybnik; XX = Szczecin; XXI = Tarnobrzeg; XXII = Tarnow; XXIII = Tomaszow; XXIV = Turoszow; XXV = Walbrzych; XXVI = Wloclawek; XXVII = Wroclaw.

1 **Areas of ecological disaster**
2a **Metropolitan areas with extensive pollution**
2b **Industrial districts with extensive pollution**
2c **Mining and energy producing districts with extensive pollution**
3 **Areas of predominant air pollution**
4 **(a) total area in km^2**
(b) population in thousands
(c) density persons/km^2

5 **Nature and landscape conservation areas with unstable natural balance:** A = Nadmorski Landscape Park; B = Mazurian Landscape Park; C = Wigry National Park; D = Biebrza Wetland (proposed national park); E = Wielkopolksi National Park; F = Gostynin-Wloclawek Landscape Park; G = Kampinos National Park; H = Kazimierz Landscape Park; I = Karkonosze National Park; J = Jura Landscape Park; K = Ojcow National Park; L = Swietokrzyski National Park; L/ = Babia-Gora National Park; M = Tatry National Park; N = Pieniny National Park.

6 **Spas in danger of losing their healing values due to unstable natural balance:** 1 = Kolobrzeg; 2 = Swinoujscie; 3 = Ciechocinek; 4 = Inowroclaw; 5 = Konstancin-Jeziorna; 6 = Naleczow; 7 = Czerniawa Zdroj; 8 = Swieradow Zdroj; 9 = Cieplice; 10 = Szklarska Poreba; 11 = Sosnowka; 12 = Kowary; 13 = Szczawno Zdroj; 14 = Karpacz; 15 = Jedlina Zdroj; 16 = Kudowa Zdroj; 17 = Duszniki Zdroj; 18 = Polanica Zdroj; 19 = Ladek Zdroj; 20 = Dlugopole Zdroj; 21 = Swoszowice; 22 = Wieliczka; 23 = Rabka.

7 **Cities and towns with major sources of pollution not located in areas of ecological hazard**

8 **Boundaries of zones of environmental quality**

Energy Use and Environmental Consequences in Poland

JAN H. JUDA
KAROL BUDZINSKI
Institute of Environmental Engineering
Warsaw Technical University

The idealistic goal of most social systems and their economic activities is to ensure optimum living conditions and human well-being both now and in the foreseeable future. It is becoming increasingly apparent that this goal will not be achieved unless greater attention is given to the proper management and protection of the environment. This concept is gaining support in all countries regardless of their differing political ideologies and economic systems. Unfortunately, difficulties arise with the practical implementation of this worthwhile principle.

The prevailing general opinion, in Poland as well as internationally, is that the best means of analyzing problems of environmental protection is through use of system analysis, an approach which takes into account economic and sociological as well as ecological factors. These relationships and interdependencies are illustrated in Figure 1.

The analysis of the interdependencies occuring in large and complex systems should begin with the selection of optimum consumption models. However, the model of consumption which has developed spontaneously in the United States and is now considered as a target model for many developing countries does not seem to be the optimum solution, particularly from the point of view of rational management of the environment. Without going into further philosophical considerations, it should be pointed out that rational change in a society's consumption model is one of the most significant factors affecting environmental quality over the long term.

It is impossible, even using advanced computer modeling, to analyze the large system presented in Figure 1 to include all the consumption patterns and resulting interrelationships between social and economic activity and the environment. For this reason, it is necessary to divide the large system into subsystems which comprise selected segments of social

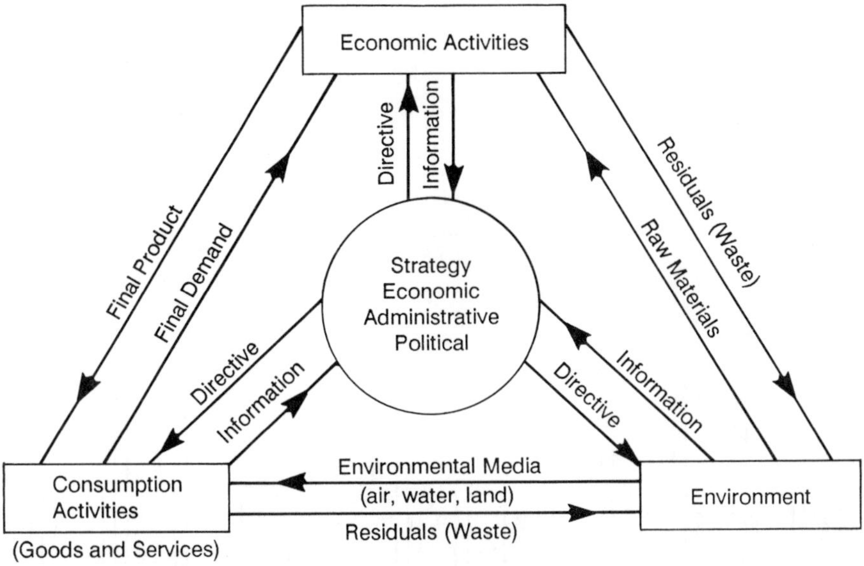

FIGURE 1 Schematic system analysis approach.

demands. In Poland, large research programs have been devoted to the study of the subsystem of supply and its effect on the environment. The following discussion presents some initial results and conclusions of this program.

ENERGY SUPPLY

From the viewpoint of system analysis, Poland should first of all consider possibilities of reducing gross energy consumption (Figure 2). It is regrettable that in Poland energy consumption per unit of national income according to estimates by the World Bank is nearly twice as high as that in the western developed countries. It is desirable to investigate all possible cause-and-effect relationships. Planned changes in the Polish economic structure and the gradual implementation of low and non-waste technologies should lead to a decrease of energy and raw material input into the Gross National Product.

Table 1 presents current data on gross energy consumption (GEC) in tons of coal equivalent in Poland, the United States, and the world, in total, per capita, and per km^2. GEC per capita in Poland is 0.65 times lower then it is in the United States. Keeping in mind that environmental risk depends on GEC per area unit of the country, it appears that in Poland the potential risk is over three times as great as it is in the United States.

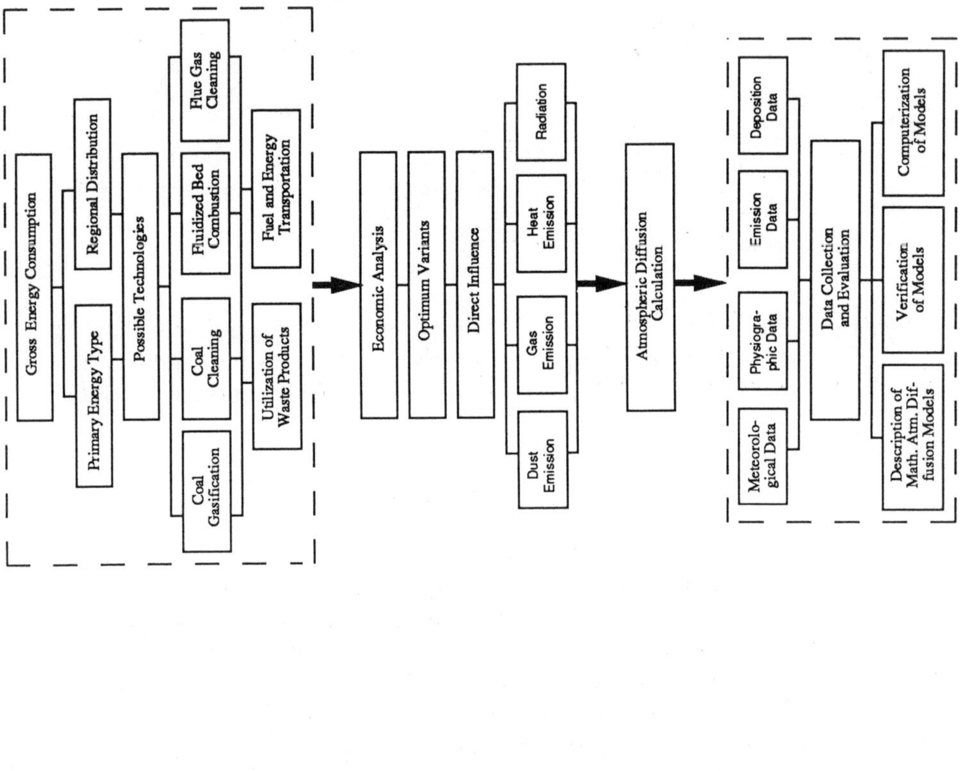
Gross Energy Consumption
Primary Energy Type
Regional Distribution
Possible Technologies
Coal Gasification
Coal Cleaning
Fluidized Bed Combustion
Flue Gas Cleaning
Utilization of Waste Products
Fuel and Energy Transportation
Economic Analysis
Optimum Variants
Direct Influence
Dust Emission
Gas Emission
Heat Emission
Radiation
Atmospheric Diffusion Calculation
Meteorological Data
Physiographic Data
Emission Data
Deposition Data
Data Collection and Evaluation
Description of Math. Atm. Diffusion Models
Verification of Models
Computerization of Models

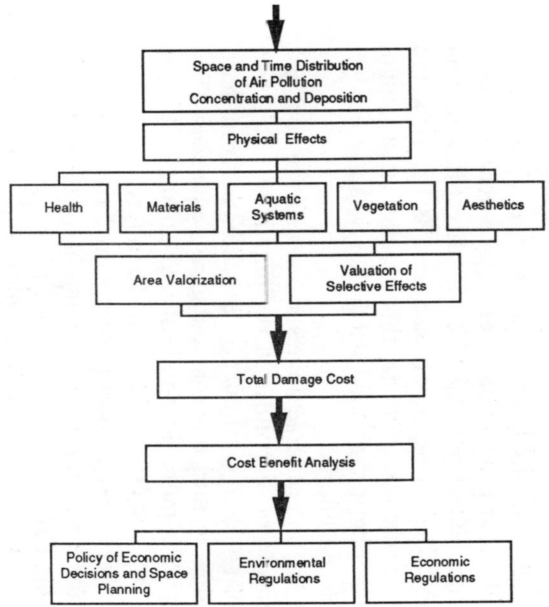

FIGURE 2 Evaluation and assessment of energy consumption and environmental impact.

TABLE 1 Gross energy consumption in 1985, in millions of tons coal equivalent (tce).

	Unit	Poland	USA	World
Gross Energy Consumption (GEC)	10^6 tce	178	1,742	9,400
GEC per inhabitant	10^6 tce	4.81	7.38	1.97
GEC per km^2	10^6 tce	570	186	---
Energy Mix (percent of GEC)				
Solid fuels	%	79	26	
Liquid fuels	%	13	39	
Gaseous fuels	%	7	23	
Hydro and geo	%	1	5	
Nuclear	%	0	8	

Another significant issue concerns types of energy use. The United States uses 19% of the total world production of energy—a very high proportion. Although GEC per capita or per km^2 offers no direct information on environmental problems, what is of importance here is the structure of Primary Energy Supply. Generally speaking, this includes solid, liquid, and gas fuels; hydro- and geothermic energy; wind energy; solar energy; and—last but not least—nuclear energy.

As a result of the combination of large domestic resources of coal and economic problems resulting from difficulties in the import of liquid and gas fuels in Poland, 80% of all energy production is generated from hard coal and brown coal, causing serious environmental consequences. The situation in the United States is more advantageous, as solid-fuel energy amounts to no more then 26% of GEC (Table 1). Therefore, the use of hard and brown coal for energy generation purposes in Poland may be taken as the starting point for discussion of ways to minimize the ill effects of energy use on the environment.

Investigations carried out in Poland concern the following technologies:

- coal gasification;
- coal cleaning;
- fluidized bed combustion; and
- flue gas cleaning.

Research indicates that coal gasification technology will not be applied on a large scale in Poland. The significant technological achievements which have already been implemented on an industrial scale are obtained in solid fuel cleaning, including elimination of pyrite. However, since the large quantities of waste produced by coal cleaning still possess thermic energy, the storage of waste products causes problems because of the danger of spontaneous combustion.

Fluidized bed technology for solid fuel combustion has been developed on a semi-technical scale. Here, the combustion of waste products brings

TABLE 2 National emission of SO_2 and NO_x, 1985.

	Unit	Poland	USA	West Germany
Total SO_2 emission	10^6 t/year	4,300	20,800	2,400
Total NO_x emission	10^6 t/year	1,500	19,400	2,900
SO_2 emission per inhabitant	kg	116	83	40
NO_x emission per inhabitant	kg	40	80	49
SO_2 emission/km^2	kg	14	2.2	9.6
NO_x emission/km^2	kg	4.8	2.0	11.7

about a significant energetic gain. The Polish government program outlining national coal usage calls for solid fuel cleaning and the utilization of waste products in fluidized bed combustion.

In relation to flue gas cleaning technology, it should be noted that Poland has fully implemented all existing techniques of dust separation. Progress in constuction of new flue gas desulfurization installations currently appears highly unsatisfactory. There are only a few pilot installations of a semi-technical scale in Poland. Generally speaking, while the techniques of coal cleaning and dust separation from flue gas are beginning to be widely implemented in Poland, installation of fluidized bed combustion and gas desulfurization processes lags even further behind. Today, the prevailing opinion in Poland is that technological progress in this field can only be achieved through international cooperation.

A complex approach to these technical problems makes possible the development of introductory economic analyses for the purpose of estimating the costs of investments and exploitation of a given technology. Next, the optimum variants can be established. For each variant we can determine the direct effects of energy production technologies on the environment through dust emissions, gas emissions, heat emissions, and radiation.

Table 2 illustrates gas emissions in Poland, West Germany, and the United States. The major risks in Poland are caused by SO_2 emissions, in a relative sense. NO_x emissions do not have as serious an effect on the environment when compared to the situation in developed western countries (e.g., West Germany).

AIR POLLUTION MODELING

There is often no basis to measure directly the environmental impacts of an emission source; therefore, dispersion modeling techniques are used. The range of problems related to air pollution modeling are being investigated through numerous research projects in Poland. Developing air pollution models calls for data as presented in Figure 2: meteorological and physiographic data, as well as data concerning emissions and depositions.

With collected and processed data, mathematical models of atmospheric dispersion can be designed. These models are then computerized and verified. In Poland, a whole "family" of air pollution models have been designed for various time and space scales, among which there are urban, regional, and country models, as well as models for long-range transport.

Let us first turn to the long-range transport models. Due to its central geographic location in Europe and the fact that the majority of regional winds blow westward, Poland is in a particularly disadvantageous position. Figure 3 shows mean sulfur depositions in Europe. Dispersion calculations have been made within the cooperative program for monitoring and evaluation of the long-range transmission of air pollutants in Europe. The highest streams of sulfur compounds in Europe are found in Poland, East Germany, and Czechoslovakia. When considering the source and balance of these pollutants, it appears that 52% of dry and wet deposition of sulfur compounds in Poland may be attributed to foreign sources, and only 48% comes from domestic sources. Figure 4 shows the calculated amounts of sulfur compounds imported to and exported from Poland. A conclusion derived directly from these data is that close international cooperation in the field of air pollution limitation is called for if we are to solve the problem.

Special attention in Poland is being given to verification of models, particularly those on urban and regional scales. To design a mathematical model is relatively easy; it should precisely describe the present state and confirm measurement data. A variety of research projects have been carried out in this field in Poland, beginning with a monitoring project conducted in Krakow in February 1984. Emissions from 283 surface and point sources located in various parts of the city were monitored simultaneously, according to meteorological parameters measured at 15 weather stations in the area. In addition, air pollution concentrations were measured at 24 points in Krakow. Since the calculations concerned only SO_2 levels, it was necessary to determine contributions to emission levels from non-Krakow sources.

The data obtained formed the basis for verification of several dispersion models. The one-level Gaussian model did not yield a satisfactory description of the actual dispersion pattern. The highest compatibility of results was found in a three-level numerical model. Further information of the experimental results can be found in Juda (1986).

In this type of experiment, which concerns concentrations of air pollutants like SO_2, it is difficult to determine precisely the pollution inflow from sources located outside the target area. For this reason a technique involving pollutant tracers and plume dyes in model verification was employed.

Figure 5 presents the experimental design of this technique. A tracer

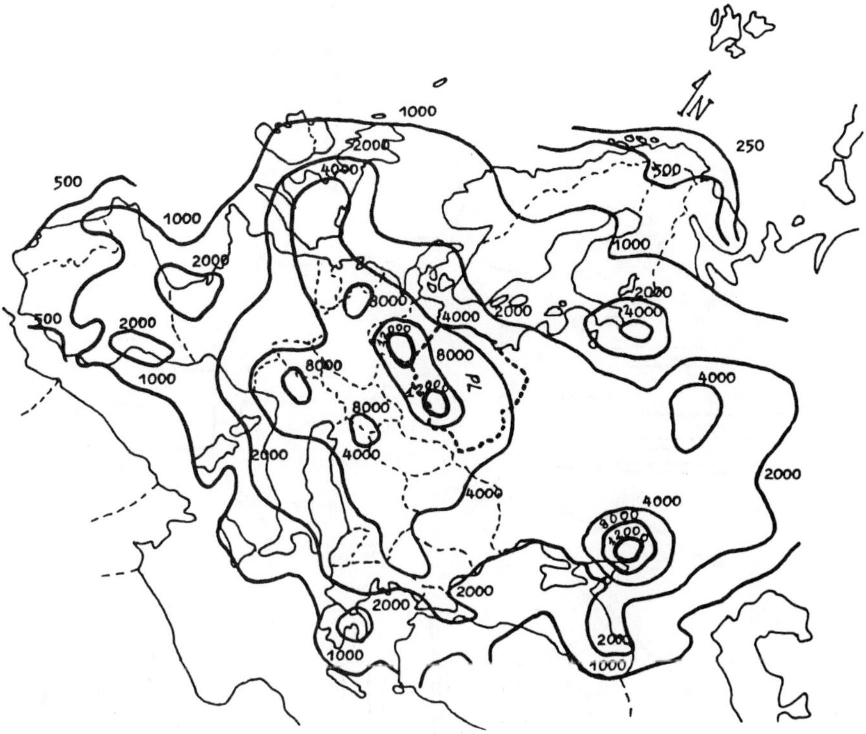

FIGURE 3 Mean annual dry and wet deposition of sulfur compounds in MG-s/m^2 (EMEP/MSCW Report 1/85).

(SF6) and plume dye is inserted into chimneys 100 m, 160 m, and 300 m in height. In the vicinity of the experimental installation, meteorological measurements are carried out in the following ranges:

- ground level up to 18 m with the use of a mast;
- atmospheric soundings with the use of captive balloon up to 500 m; and
- higher stratum examination with the use of a free balloon.

Depending on wind directions within a radius of several kilometers, air samples are collected using specially designed injectors. The samples are then chromatographed in order to determine SF6 concentrations. The plume is examined with three theodolites associated with a film camera so that all parameters of dispersion can be defined. In some cases, additional measurements are carried out with the use of a plane. Initial results indicate that dispersion model parameters established for stack hights to 120 m can not be extrapolated to greater stack heights (e.g., 300 m).

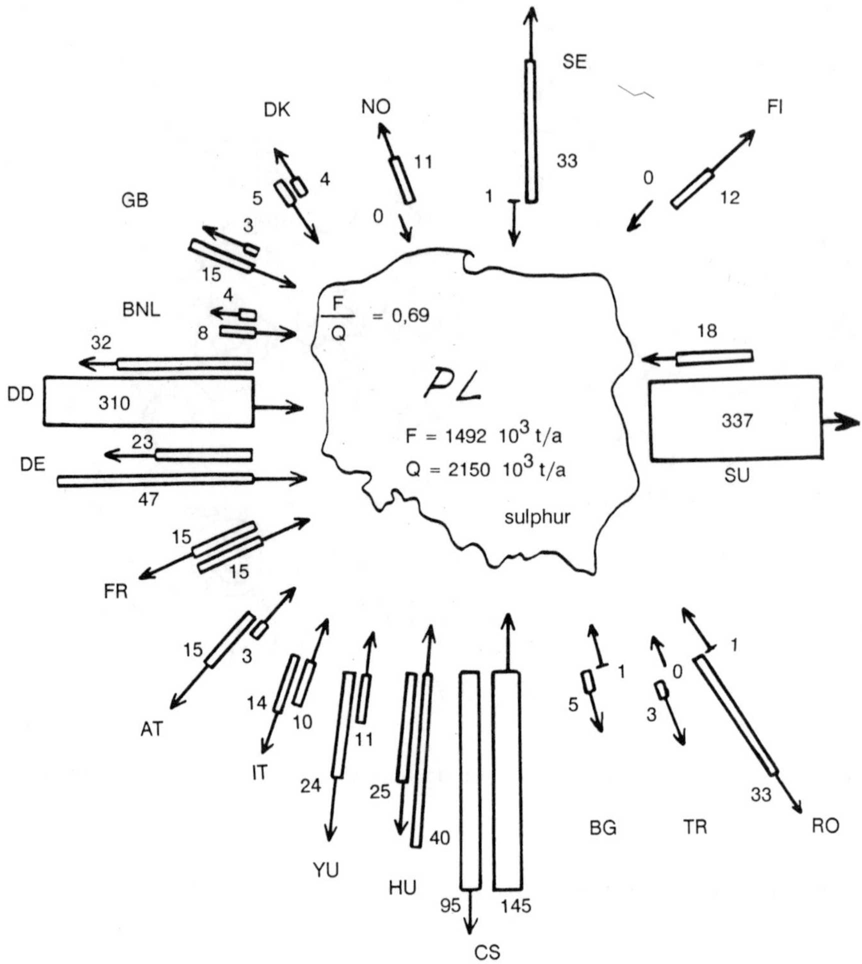

FIGURE 4 Calculated sulfur budget "imported" to and "exported" from Poland. Unit=10^3 tons of sulfur per annum (EME:/MSCW Report 1/88).

Models already verified provide information necessary for mapping of air pollution in various scales in urban areas, regions, or the whole country. The existing maps illustrate only sulfur pollutants, since the data on sulfur emission sources are available in Poland. The maps thus developed are used in comprehensive planning as well as in the development of environmental protection programs.

Currently, a program of this kind is being prepared which aims at the limitation of SO_2 emissions in Poland through 2010. For research purposes, maps of air pollution distribution are made for regions where

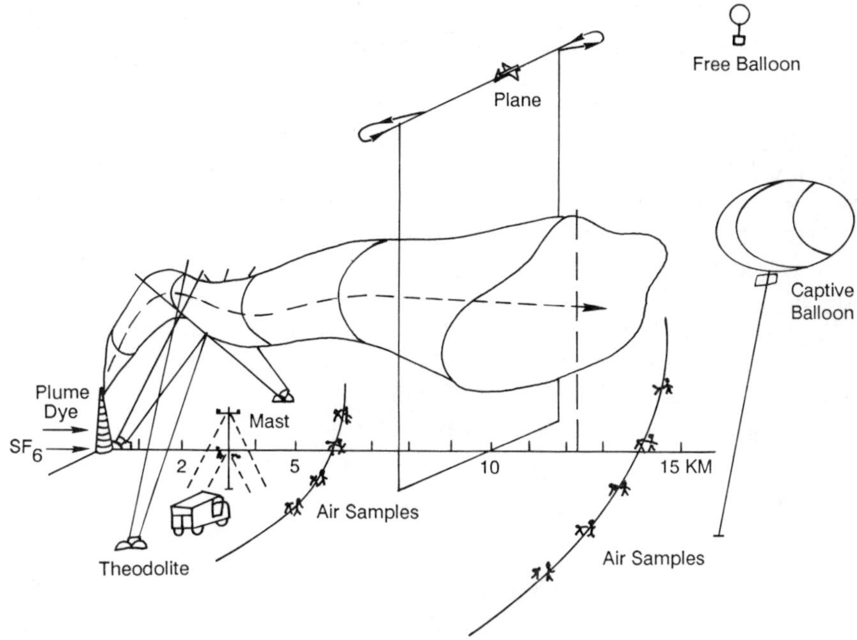

FIGURE 5 Experimental air pollution model verification.

ecological monitoring is done, in order to determine correlations between the pollutant concentrations and ecological effects.

Although a large amount of modeling of air pollution dispersion has been carried out both in Poland and elswhere in the world, among the issues that require further research are the following:

- physical and chemical reactions of pollutants in the atmosphere;
- pollutant wash-out by atmospheric fallout;
- pollutant transportation inside the clouds; and
- ground absorption.

Description of pollution dispersion in the atmosphere as well as description of ground deposition of pollutants provide the basis for further description of circulation of pollutants in other elements of the environment, i.e., water, soil, and plants.

COST/BENEFIT ANALYSIS

The next step in the system analysis process is to examine the effects of pollution on human health, materials, water systems, vegetation, and aesthetics. Although a great deal of research has been carried out in

Poland and elsewhere, we are still not able to define absolute relationships between environmental pollution and measurable economic losses due to the multiplicity of interdependencies which occur in all ecosystems.

Nevertheless, some attempts have been made to measure losses caused by particular air pollutants. Information is available mainly for SO_2 and its secondary effects, such as acid rain. In Europe, for example, estimates of damages resulting from the emission of one ton of SO_2 range from $1,000 to $3,000. Thus, with more precise knowledge of environmental damage, it is possible to undertake a cost/benefit analysis. Briefly, the choice of optimum variants for rational management of the environment boils down to direct comparisons of global costs of environmental protection on a scale of a region (*CE*) and loss costs (*CL*) caused by exploitation of natural resources as well as environmental pollution.

The optimum solution, which takes a form

$$Dmin = f(CE, CL)$$

means, in this case, the minimum sum of environmental protection costs and losses.

Because of the "uncertainty factor," attempts at loss estimation have been made in Poland to develop new methods of solving the problem. A notion of area valorization has been introduced which evaluates area sensivity to environmental pollution since, to date, investigations have been limited to air pollution. Considering parameters of the area such as agricultural uses, forestation, water systems, population density, and fixed capital invested, an area value index can be derived in a 0-1 or 1-10 scale. The higher the area value, the higher the potential environmental damage caused by a definite concentration of pollutants, a concept which can be presented in the mathematical formula:

$$Dmin = \frac{1}{T} \int_0^T \int_\Omega Vi(x, y) Ci(X, y, t) d\Omega dt$$

where *D* equals nondimension loss function. Summing the products of area value indices (*Vi*) and pollution concentration indices (*Ci*) in time and space, the goal is the minimum value of the dependent variable *D*.

Omitting damage estimates in rational value, the procedure allows for optimum comprehensive planning and helps identify those emission sources in which investments will generate the greatest gain in pollution reductions. This is the theoretical basis of optimum cost estimates for environmental protection.

CONCLUSION AND RECOMMENDATIONS

The above review of research on environmental protection may be summarized in the following propositions:

- Issues of environmental protection should be viewed in a complex manner through a systems analysis approach that encompasses all interdependencies among social needs, economic activity, and environmental conditions.
- For technical reasons, subsystems should serve as the scale of analysis. One of the most significant subsystems is energy supply and its environmental effects.
- A description of an analyzed subsystem should enable both simulation and optimalization models to be developed for a given developmental variant.

In the "energy environment" subsystem, there still exist areas requiring further research. They are:

- the improvement and verification of air pollution dispersion models;
- the construction and improvement of models of pollutant circulation in biogeochemical cycles; and
- the estimation of damage to environment caused by emission of specific pollutants, a process which should consider all existing (particularly additive) interdependencies of various pollutants.

REFERENCE

Juda, K. 1986. Modeling of air pollution in the Krakow area. Atmospheric Environment, Vol. 20, No. 12, pp. 2449-2458.

Recommendations

Recommendations for a Science-Based Program of Ecological Risk Assessment and Environmental Protection

STANLEY I. AUERBACH
Oak Ridge National Laboratory

ALAN W. MAKI
Exxon Company, USA

Threats to the life-support systems of the planet Earth need to become the concern of the social and political systems of every nation. These threats result from many deliberate and inadvertent activities of human beings in every society. While there is growing international recognition of environmental problems, preventive and corrective actions and policies have not been put forth quickly for a number of reasons. These reasons include ignorance, differing priorities among social groups within the same society, political or social apathy, lack of financial resources, and lack of qualified personnel. In this book we have tried to meld the ecological perspectives and ideals of two differing societies in the hope of identifying common goals and general procedures by which to improve the assessment and management of ecological risks.

Some readers may not be aware of the official policies on matters of environment in the East European countries. All of the participants in these U.S.-Polish workshops were surprised to learn how similar the official policies of East European countries were to those of the United States and other western countries. For this reason, we recommend Chapter 20 where Trojan has summarized the official policies of East European countries in a way that provides a useful frame of reference for the recommendations that have been developed in this chapter and some of the other chapters of this book.

These official policies reflect a combination of political, economic, and social ideals which may affect the degree and intensity of implementation of action programs. We hope the recommendations for research discussed in this book will be consistent with governmental policies. Both publicly and privately supported research is needed on the concepts of ecological risk assessment and the management procedures that are necessary to develop

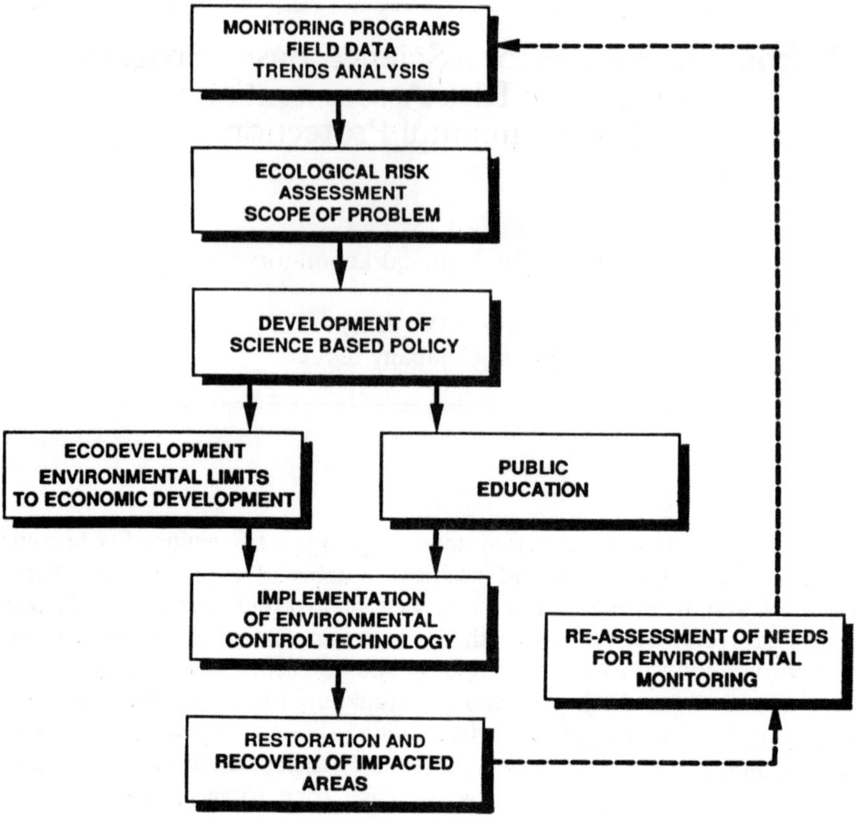

FIGURE 1 Conceptual framework for a research-based environmental policy and restoration program for pollution-impacted ecosystems.

comprehensive programs to assess ecosystem exposures, quantify ecological impacts, and develop policies for management of ecological risks.

Figure 1 provides a conceptual framework for a research-based environmental policy and restoration program for pollution-impacted ecosystems. It begins with the enhancement of ecological awareness and ends ideally with the prevention, protection, restoration, and/or recovery of impacted ecosystems. Thus, we believe it is a blueprint for development of enlightened environmental policy not only in Poland and the United States but in many other countries of the world.

Figure 1 also underscores the fact that carefully designed environmental monitoring programs are essential for the identification and quantification of ecological trends, such as changing forest productivity or buildup of heavy metals in river sediments. With such data in hand, the scope of existing or potential problems can be identified and the magnitude of environmental risks can be assessed. Environmental scientists, leaders in industry, and regulatory officials can use the results of these assessments to develop science-based recommendations for voluntary and/or regulatory actions to prevent, decrease, or ameliorate key environmental problems. Public education and recognition of environmental limits to economic development are essential no matter which preventive or ameliorative actions may be required. This education ideally should include all segments of society, from school children to leaders in industry, regulatory agencies, and the public at large. All groups must recognize that the capacity of the environment to assimilate wastes is not infinite, and that breakdown of ecological structure and function will occur if the limits of ecosystem resiliency are not understood and respected.

Most detrimental impacts of changes in air or water quality can be ameliorated by control technologies which can lead to the eventual restoration and recovery of impacted ecosystems. Clearly, continued economic development will require regular reassessment of needs for environmental monitoring. Data from such long-term monitoring can then be used to feed back to the beginning of the sequence for reanalysis of trends and reassessment of incremental environmental risks resulting from altered plans for industrial, residential, commercial, or recreational development (Figure 1).

SYNTHESIS AND INTEGRATION

Air Pollution Impacts—Monitoring and Trends Analysis

Very large parts of the environmental research programs in both the United States and Poland are devoted to the development of monitoring data for air, water quality, and ecological trends. Indeed, the majority of data discussed during the workshops regarding terrestrial and aquatic ecosystems was the direct result of ongoing monitoring programs. Monitoring is an essential first step in ecological risk assessment.

One of the projects recommended for future collaboration between the United States and Poland is a cooperative research activity aimed at characterizing effects on forest ecosystems from air pollution and other future environmental stresses. An important outgrowth of the first NAS-PAN workshop was a follow-up planning meeting for a joint U.S./Polish program subtitled "Long-term Ecological Monitoring of Forests." A group of American and Polish forest scientists met in Poland in 1988 and made the following recommendations:

- All ecosystem parameters are characterized by natural trends and variability; we must be able to recognize if observed variability exceeds the range of normal variability.

- Long-term measurements are needed—long-term being defined as more than 30 years.
- Measurements of greatest potential are those that enable pre-visual detection of change and integration of the effects of several stress factors.
- Monitoring should not be static. The linkages between research on field monitoring and mathematical modeling of ecosystem processes must be interactive. Monitoring programs must be designed to include changes in methods that research identifies as being useful to meet the monitoring objectives.
- Because of the diversity of ecological responses to environmental stresses, decisions should be made early in a monitoring program as to what constitutes a significant change. Just because something is detectable does not mean that it is important. On the other hand, many changes in an ecosystem that ultimately are important may be subtle and difficult to detect.
- The monitoring approach must identify those ecosystem values that are important and develop indicator measurements that can be used to assess changes in those values.
- The first goal of the proposed monitoring system should be to evaluate changes caused by air pollution. Later, however, other environmental stress factors should be included.
- A pilot monitoring project focusing on the United States and Poland should be implemented with collaboration also in other European countries. Emphasis should be on intensity of effort rather than extent. If all parameters can not be monitored, an attempt should be made to include as many integrative factors as possible.

Issues of environmental protection should be viewed in a complex way through a systems-analysis approach that encompasses all interdependencies among social needs, economic activity, and environmental conditions. For ease of analysis as well as tractability, subsystems should serve as the scale of analysis. One of the most significant subsystems is energy supply and its environmental effects. As noted in Chapter 23, within the selected subsystem of energy and environment interactions there are a number of areas requiring further research. These include:

- improvement and verification of air pollution dispersion models;
- construction and improvement of models of pollutant circulation in biogeochemical cycles; and
- estimation of damage to the environment caused by emission of specific pollutants—a process which should consider all existing interdependencies of various pollutants.

Ecological Risk Assessment

Once environmental monitoring data have been developed for pollutants and ecosystems of interest, decisions regarding significance of the observed trends and concentrations of pollutants must be made. All too frequently, monitoring data are developed at considerable expense with no clear concept of how the data are to be used and, more importantly, how the significance of the observed values will be determined.

In its simplest form, risk assessment involves comparisons between known or estimated concentrations of particular pollutants and measurements of ecological effects or toxicological data for indicator species. This comparison serves as the basis for definition of "concentrations of concern" and for deciding whether environmental or human-health risks are acceptable or unacceptable. When monitoring shows that concentrations of a particular pollutant in the environment exceed known toxic concentrations, an appropriate control strategy is needed to limit discharge of this material into the environment. Conversely, when biological effects are known to occur only at concentrations in excess of those monitored in the environment, concerns are minimized. Thus, a procedure for regular comparison of environmental monitoring data with biological effects data yields an objective data base with which to prioritize environmental problems and manage ecological risks, i.e., a science-based policy.

Development of Science-Based Policy

In a very real sense, essentially all regulatory initiatives and much current environmental decision making have a basis in the risk-assessment approach, i.e., comparison of environmental-fate and environmental-effects information. Air and water quality standards are developed to protect human health and the environment. These standards recognize the importance of relating the known biological effects of particular pollutants to real-world exposures. Similarly, criteria for toxic pollutants are developed from comparisons and predictions of biological effects with resultant workplace or other environmental exposures.

Much current land-use planning and assessment of potential impacts from industrial development also involve comparisons of ecological effects with predicted impacts of the proposed activity. For example, the National Environmental Protection Act (NEPA) in the United States requires that an Environmental Impact Statement (EIS) be assembled for new developments which may have substantial impacts on the environment. The EIS describes the biological structure and function of the existing ecosystem as a baseline and then predicts how this baseline is expected to be altered by the proposed development in a risk assessment context. Decisions on the acceptability of a proposed project, suggested alternatives, and mitigative measures are all made based on the severity of changes from the baseline functions of the ecosystem. Thus, scientific information gathered from carefully designed and conducted field and laboratory programs provides the foundation for current decision making and the development of future environmental policy.

Ecodevelopment: Environmental Limits to Economic Development

In Chapter 4, Marek and Kassenberg discuss the concept of ecodevelopment, recognizing that ecological factors can place finite limits on the extent of economic development. This concept recognizes that humans have had an impact on the environment as a result of day-to-day activities since before recorded history. As human populations have increased and associated economic development has proceeded, impacts have become progressively more severe. Contemporary ecological stresses are demonstrated by chronically degraded air and water quality, as well as the breakdown of ecological structure and function in areas that are most severely impacted. In these heavily impacted areas of Poland and the United States, we have come to realize that the capacity of the environment to assimilate wastes is not infinite, and that degradation of environmental values is the inevitable result of over-stressed ecosystems.

Guidelines for Sustainable Agricultural Development

In Chapter 15, Ryszkowski proposes a series of guidelines for sustainable development related to sound ecological management of agricultural areas:

- Relate the structure and magnitude of agricultural production to the natural environment.
- Introduce new technologies to ensure that agricultural enterprises internalize the costs of restoring impacted ecosystems.
- Introduce landscape-level planning to ensure that ecological structure and function are maintained in areas under increasing pressures to intensify agricultural production.

Public Education

There is a continuing need for environmental scientists to present the results of their research with clear and concise conclusions for leaders in industry and government and the public at large. Ambiguous and inconclusive data, or data from poorly designed programs, should be avoided—they serve only to obscure public understanding of complex ecological issues. In

Chapter 20, Trojan points to the "collision of interests" between industrial development and environmental concerns. He cites the need for enhanced education programs to produce good researchers and to keep the public fully informed of environmental consequences of planned activities.

IMPLEMENTATION OF ENVIRONMENTAL CONTROL TECHNOLOGY

In closing this summary chapter, we refer again to the "Conceptual Framework for a Research-Based Environmental Policy and Restoration Program for Pollution Impacted Ecosystems" shown in Figure 1. In the course of this chapter, we have progressed from trends analysis and risk assessment to development of science-based policy and public education. The key to success in the assessment and management of ecological risk is acceptance and implementation of environmental control technologies that bring about ecological improvements and permit sustained development with minimal further degradation of the environment.

Clearly, the control technologies needed to achieve many contemporary environmental goals are already available. Equally clear are the high costs of implementation. Although the costs of not implementing necessary environmental management are less widely recognized, they are no less real. The ecological situation in many areas of the United States and Poland are critical; both environmental concerns and human health are at serious risk in the most heavily developed areas of both countries. If economic development and the quality of life in our societies are to be improved, then the costs of environmental control technologies such as wastewater treatment, clean-coal technologies, use of air scrubbers and precipitators, and more ecologically sound methods for agriculture and forest management must be factored into development decisions.

We believe that Figure 1 contains the keys to success in the assessment and management of ecological risks, not only in the United States and Poland but in many other countries of the world as well.

REFERENCES

Auerbach, S.I. 1989. Monitoring the environment in the 21st century. In the Proceedings of the Second U.S.-USSR Symposium on Air Pollution Effects on Vegetation, R.D. Noble, J.L. Martin, and K.F. Jensen, eds. U.S. Department of Agriculture, Forest Service—Northeastern Forest Experiment Station.

Barnthouse, L.W., and G. W. Suter II, eds. 1986. Users manual for ecological risk assessment. Oak Ridge National Laboratory Report ORNL-6251, Oak Ridge, Tenn.

Clarke, R., ed. 1986. The handbook of ecological monitoring. New York: Clarendon Press.

Cohrssen, J.J., and V.T. Covello. 1989. Risk analysis: A guide to principles and methods for analyzing health and environmental risks. Council on Environmental Quality NTIS #PB89-137772. U.S. Department of Commerce, Springfield, Va.

National Research Council. 1983. Risk assessment in the federal government: Managing the process. Washington, D.C.: National Academy Press.
Ruckelshaus, W.D. 1983. Science, risk, and public policy. Science 221:1026-1028.
Ruckelshaus, W.D. 1984. How to better assess and manage risk. Governance (Spring):3-5.
Schweitzer, G.E., and A.S. Phillips, eds. 1986. Monitoring and managing envinronmental impact: American and soviet perspectives. Washington, D.C.: National Academy Press.
Suter, G.W. II, L.W. Barnthouse, and R.V. O'Neill. 1987. Treatment of risk in environmental impact assessment. Environmental Management 11:295-303.

Appendices

Appendix 1
List of Participants

STANLEY I. AUERBACH
Environmental Sciences Division
Oak Ridge National Laboratory
Building 1505
Oak Ridge, Tennessee 37831
USA

ANN M. BARTUSKA
U.S. Forest Service
U.S. Department of Agriculture
1509 Varsity Drive
Raleigh, North Carolina 27606
USA

ALICJA I. BREYMEYER (*PAN Chairman, 1988*)
Institute of Geography and Spatial Organization
Polish Academy of Sciences
Krakowskie Przedmiescie 30
00-927 Warsaw
POLAND

KAROL BUDZINSKI
Institute of Environmental Engineering
Warsaw Technical University
Nowowiejska 20
00-653 Warsaw
POLAND

WILLIAM E. COOPER
Center for Environmental Toxicology
Department of Zoology
Michigan State University
East Lansing, Michigan 48824
USA

ELLIS B. COWLING (*NAS Chairman*)
School of Forest Resources
North Carolina State University
Raleigh, North Carolina 27695
USA

STEFAN GODZIK
Institute of Environmental Engineering
Polish Academy of Sciences
Marii Sklodowskiej-Curie 34
41-800 Zabrze
POLAND

KRYSTYNA GRODZINSKA
Institute of Botany
Polish Academy of Sciences
Lubicz 46
31-512 Krakow
POLAND

WLADYSLAW GRODZINSKI* (*PAN Chairman, 1987*)
Department of Ecosystem Studies
Jagiellonian University
M. Karasia 6
30-060 Krakow
POLAND

MAREK J. GROMIEC
Institute of Meteorology and Water Management
Department of Water Management
Podlesna 61
01-673 Warsaw
POLAND

* *Deceased*

MARK A. HARWELL
Ecosystems Research Center
Cornell University
Ithaca, New York 14853
USA

WALTER W. HECK
Agricultural Research Service
U.S. Department of Agriculture
1509 Varsity Drive
Raleigh, North Carolina 27606
USA

ANNA HILLBRICHT-ILKOWSKA
Institute of Ecology
Polish Academy of Sciences
Dziekanow Lesny
05-092 Lomianki
POLAND

WARREN E. JOHNSTON
Department of Agricultural Economics
University of California
Davis, California 95616
USA

JAN JUDA
Institute of Environmental Engineering
Warsaw Technical University
Nowowiejska 20
00-653 Warsaw
POLAND

STANLEY J. KABALA
University of Pittsburgh
2016 Sampson Street
Pittsburgh, Pennsylvania 15221
USA

ANDRZEJ T. KASSENBERG
Institute of Geography and Spatial Organization
Polish Academy of Sciences
Warsaw
POLAND

ALAN W. MAKI
Exxon Company, USA
3301 C Street, P.O. Box 196601
Anchorage, Alaska 99519
USA

MICHAL J. MAREK
Institute of Public Administration and Management
Wawelska 56
02-067 Warsaw
POLAND

BOGUSLAW MOLSKI*
Botanical Garden
Polish Academy of Sciences
Warsaw
POLAND

HALINA OBUCHOWICZ
Office of International Relations
Polish Academy of Sciences
Palac Kultury i Nauki
Warsaw
POLAND

ANNA S. PHILLIPS
Soviet and East European Affairs
National Academy of Sciences
2101 Constitution Avenue
Washington, D.C. 20418
USA

COURTNEY RIORDAN
Office of Research and Development
U.S. Environmental Protection Agency
401 M Street S.W. (RD-680)
Washington, D.C. 20460
USA

LECH RYSZKOWSKI
Department of Agrobiology and Forestry
Polish Academy of Sciences
Swierczewskiego 19
60-809 Poznan
POLAND

* *Deceased*

MILTON RUSSELL
Energy, Environment, and Resources Center
University of Tennessee
322 South Stadium Hall
Knoxville, Tennessee 37996
USA

GLENN E. SCHWEITZER
Soviet and East European Affairs
National Academy of Sciences
2101 Constitution Avenue
Washington, D.C. 20418
USA

MICHAEL W. SLIMAK
Office of Research and Development
U.S. Environmental Protection Agency
401 M Street S.W. (RD-680)
Washington, D.C. 20460
USA

SARAH M. TERRY
Department of Political Science
Tufts University
Medford, Massachusetts 02155
USA

PRZEMYSLAW TROJAN
Institute of Ecology
Polish Academy of Sciences
Dziekanow Lesny
05-092 Lomianki
POLAND

GARY R. WAXMONSKY
Office of International Activities
U.S. Environmental Protection Agency
401 M Street S.W. (A-106)
Washington, D.C. 20460
USA

Science Attache
U.S. Embassy, Warsaw (1987-89)

Appendix 2
Summary Memoranda

SUMMARY REPORT OF NAS-PAN WORKSHOP ON ECOLOGICAL RESEARCH AND ENVIRONMENTAL PROTECTION

October 1987

A workshop on "Ecological Research and Environmental Protection" was jointly sponsored by the National Academy of Sciences (NAS) of the United States and the Polish Academy of Sciences (PAN). This was the first workshop held within the framework of the current NAS-PAN interacademy exchange program. Dr. Ellis B. Cowling, Associate Dean for Research at the School of Forest Resources, North Carolina State University, and Prof. Wladyslaw Grodzinski, Head of the Department of Ecosystem Studies, Jagiellonian University, and Director of the PAN Institute of Freshwater Biology, served as workshop co-chairmen.

In addition to the workshop sessions, the participants visited a number of relevant governmental organizations and scientific institutes in central and southern Poland. These visits provided opportunities for the U.S. delegation to gain insight into the organization of research activities in the fields of ecology and environmental protection. Field trips included the Nowa Huta Steel Mill, Niepolomice Forest, Dobczyce Reservoir, and various industrial areas in Upper Silesia.

During the workshop sessions, the participants presented 22 papers on various topics. The discussion of these papers was so lively and productive that the participants decided to prepare manuscripts for a joint publication entitled "Ecological Risks: Perspectives from Poland and the United States."

Participants discussed a variety of proposals for cooperative activities in areas of mutual interest, including workshops, seminars, exchanges of literature, and research projects.

Participants developed plans for a reciprocal meeting in the U.S. in the second or third quarter of 1988 to finalize draft texts for the joint publication and to further elaborate cooperative research proposals.

The NAS delegation expressed its gratitude to Prof. Grodzinski, the PAN delegation, and staff for making their visit to Poland a great success. The support of the Rockefeller Brothers Fund for the workshop was also gratefully acknowledged.

For NAS
Ellis B. Cowling

For PAN
Wladyslaw Grodzinski

Mogilany, Poland
October 23, 1987

MEMORANDUM

ON THE RESULTS OF THE NAS-PAN WORKSHOPS ON ECOLOGICAL RESEARCH AND ENVIRONMENTAL PROTECTION AND ON FURTHER COOPERATION IN THE PROTECTION OF ECOLOGICAL RESOURCES

November, 1988

Two bilateral workshops were held within the framework of the Memorandum of Understanding on Scientific Exchanges between the National Academy of Sciences (NAS) and the Polish Academy of Sciences (PAN). The first workshop, together with associated visits to organizations and research institutions concerned with environmental protection and ecological research, took place in Poland October 11-23, 1987, and the second in the United States November 6-19, 1988.

The main objective of the two workshops was to discuss research activities in the following fields:

- Concepts regarding the relationships among ecosystem health, pollutant exposure, economic development, and environmental management;
- Case studies of specific changes in ecosystem health and productivity in relation to air and water pollution stresses;
- Recommendations for improvement of research and education relating to environmental protection policy;
- Development of cooperative activities directed to the protection of ecological resources.

Manuscripts that were presented at the workshops by the Polish and American participants will be published by NAS as an edited volume. One hundred copies will be provided to PAN and 100 copies will be distributed by NAS. Arrangements are being explored by which additional copies can be made available through NAS and PAN.

The participants considered the workshops and associated visits very important to the development of contacts between Polish and American institutions interested in the specific topics that were discussed. NAS and PAN will encourage research institutions in the two countries to develop cooperative projects involving exchange visits and joint research activities. These projects will be implemented outside the exchange quota provided by the NAS-PAN agreement with the participating research institutions assuming responsibility for financing the projects.

These collaborative projects will be designed to complement and not duplicate the many other cooperative activities between American and Polish scientists in related scientific fields. The projects will be consistent with the objectives of bilateral agreements between the American and Polish Governments in the fields of environmental protection and ecological research.

From the initial 19 projects proposed at the first workshop in Poland, five are being implemented under the auspices of the U.S. Environmental Protection Agency (EPA) and the U.S. Department of Agriculture (USDA). An additional seven projects are currently being developed.

Among the areas identified by the participants as particularly promising for future cooperation are:

- Biogeochemical cycles in ecological systems; models of accumulation, flows, and budgets of pollutants and important nutrients in ecosystems;
- Ecological boundaries, functions, and management of landscapes;
- Methods for assessing ecological risks and their implications for environmental policy and decision making;
- Systems for monitoring geosphere/biosphere changes across different time and space scales.

The Polish participants expressed their gratitude to the NAS Office of Soviet and East European Affairs for the efficient organization of the workshop, particularly the extensive program of associated visits.

All participants expressed their appreciation to the Rockefeller Brothers Fund and the Ford Foundation for providing financial support for the workshops.

The participants were deeply saddened by the death of Academician Wladyslaw Grodzinski during the time of the workshop in Washington. As co-chairman of these NAS-PAN workshops, Ladd provided vital leadership to assure their success. But Ladd's contribution was much more profound, giving inspiration to us all through his intellectual creativity, his spirit of camaraderie, and his remarkable courage to the end. We dedicate our continuing collaborative efforts to his memory.

Prepared in Washington, D.C., on November 11, 1988, in English.

Alicja I. Breymeyer
Workshop Co-Chair
Polish Academy of Sciences

Ellis B. Cowling
Workshop Co-Chair
National Academy of Sciences

Appendix 3
Cooperative Projects Suggested at 1987 Workshop

1. **Managing Aquatic Ecotones for the *In Situ* Treatment of Nutrients and Toxic Substances**

   A river-lake ecotone will be managed to develop patterns of macrophyte populations that will process nutrients and toxic substances by active biological uptake and passive sedimentation.

2. **American-Polish Development and Training Program for Water Quality Modeling and Environmental Risk Assessment**

   Water quality models for transport, fate, and exposure processes will be coupled with risk assessment models. These will be utilized both as a training methodology and in a direct validation experiment under field conditions in Poland.

3. **Prototype Application of the American Environmental Protection Experience to Polish Regional and Local Decision Making**

   The project will include an evaluation of the U.S. environmental impact statement and risk assessment methodology within the context of the Polish social and economic system. A modified version will be tested at the community level in Poland.

4. **Mechanisms of Foliar Leaching Due to Air Pollution**

   Post-doctoral support for a researcher from the PAN Institute of Environmental Engineering will be provided to collaborate with the EPA laboratory in Corvallis, Oregon, to pursue research on cuticular leaching associated with air pollution. Funding through the EPA post-doctoral program will be pursued.

5. **Ambient Ozone Monitoring in Southern Poland**

The absence of ozone date for the forested regions of southern Poland presents a major uncertainty in understanding forest response to air pollution. Surplus ozone monitors from U.S. scientists will be lent to PAN scientists to obtain these essential data.

6. **Deposition in Forest Canopies**

Scientists at the PAN Institute of Ecology (Dziekanow Lesny) have examined wet, dry, and aerosol deposition to forested watersheds under heavy impact from air pollution. A post-doctoral opportunity for a young scientist from the Institute of Ecology to a major watershed research site in the eastern United States is proposed. EPRI-Integrated Forest Study Sites are the preferred locations, particularly Coweeta or Whiteface.

7. **Mycorrhizal Response to Air Pollution in Polish Forests**

The research efforts in the Niepolomice Forest are expanding to include an assessment of mycorrhizal health and function, although mycorrhizal expertise is not readily available within the local scientific community. We propose to send a U.S. scientist who works with mycorrhizal and air pollution to Poland for 6-18 month research and training opportunity.

8. **Impacts of Regional and Local Air Pollution on Agricultural and Horticultural Crops in Southern Poland**

Impacts of SO_2, metals, and other particulates on agricultural crops in southern Poland are probable, although no data are available for response under clean air conditions. Open-top chambers will be obtained from U.S. scientists and sent on loan to Poland to evaluate agricultural response under filtered air and non-filtered conditions. Data obtained can be used to assess changes in crop quality and quantity, with and without air pollutant stress, for crops grown in pollution inputs greater than currently found in the United States.

9. **Joint U.S.-Polish Studies on Characterizing the Effects of Air Pollutants on Forest Ecosystems**

The purpose of this project is to establish continuing, multi-level research on forest ecosystem indicators of air pollution stress, beginning with a technical workshop on indicators of response. A pilot field program for identifying and evaluating appropriate indicators of ecosystem response, and the establishment of a long-term monitoring and modeling program focused on selected indicators of response of forest ecosystems in the United States and Poland representing a gradient of air pollution stress are also proposed. This program will contribute significantly to a comparative understanding of biogeochemical cycles and ecosystem responses to stress, and will be directly relevant to the International Geosphere Biosphere Program (IGBP).

10. **Comparative Analysis of Biogeochemical Cycling in Gradients of Forests**

Using watersheds approach, we would like to estimate and compare the input/output balance of heavy metals, sulfur components, etc. in various ecosystems. Special attention will be paid to *accumulation* and *bioelimination* of pollutants on the ecosystem level.

11. **Biogeochemistry of Metals in Aquatic and Terrestrial Systems**

High levels of metals have been added to both aquatic and terrestrial systems as a result of heavy industrial air and wastewater effluent discharges. The cycling, mobilization, organic complexation, and bioavailability of these metals at the high ambient levels found in Upper Silesia provide an excellent research opportunity.

12. **Air Quality Impacts on Forest Production Ecology**

The high impact areas of southern Poland offer an excellent laboratory to study the ongoing effects of air quality impairment on forest production ecology. Evolving methods for this assessment should be refined and applied to specific impact areas in southern Polish forests.

13. **Adaptations of Terrestrial Ecosystems to Air Pollution**

This project will focus on the successional changes that occur in stressed ecosystems, e.g., the shift from a coniferous forest to a deciduous one in southern Poland. An important component of this study will be the study of ecosystem function (e.g., biogeochemical cycling) that changes with succession. This project will also investigate the importance of mitigation techniques such as fertilization.

14. Evaluation of Regional Air Pollution Models

Cooperation in three areas is proposed: (1) the quantification of uncertainty in the structure and parameters of models; (2) the design of field studies to collect data to evaluate the performance of models; and (3) to compare the performance of regional models given the same emissions, meteorology, and terrain.

15. Use of Bioindicators to Quantify Dry Deposition of Sulfur in Forested Areas

This project will be a joint evaluation of Polish results on monitoring sulfur content of pine needles to determine a quantitative relationship to deposition type (total, wet, dry) both seasonally and annually. We will explore possible U.S. collection of similar data to compare with Polish results and to observe values of deposition using direct wet and indirect dry measurement methods.

16. Methodologies for Assessing Human Exposure to Toxic Metals

Cooperative efforts will be directed to improving approaches to estimating human exposure to toxic metals such as lead and cadmium in urban areas, drawing on expertise in Chattanooga, Tennessee, and Katowice, Poland. Particular attention will be given to environmental monitoring, biological monitoring, and associated quality assurance procedures.

IDEAS FOR FUTURE DISCUSSION

17. Educational Needs in Environmental Protection

Wise decisions leading to improved protection of the environment require knowledgeable technical personnel, prudent leaders in industry and government, and well-informed citizens. Increased investments in environmental education are expected to lead to improved protection of the environment in both Poland and the United States.

18. Ecological Dimensions of Spatial Planning in Metropolitan Areas

Cooperative efforts should build on Polish successes in developing an ecological infrastructure within the country and U.S. efforts to improve planning in areas surrounding urban development.

19. Control Technology to Reduce Emissions of Pollutants in Stack Gases

This project will exchange data on performance of emerging technologies and will involve cooperation in the identification and performance of joint projects to develop technologies for the control of specific pollutants.

20. Exploration of the Applicability of American Remote Sensing Technologies to Solution of Polish Environmental Protection Problems

This project will determine the utility of using U.S. data for purposes of meeting specific analytical needs of the Polish environmental situation.

21. Development and Testing of Biological Markers

This project will involve development of bioindicators as early indicators of ecological change at the population level.

Appendix 4
Cooperative Projects Suggested at 1988 Workshop

1. **Monitoring of Forest Response to Environmental Health**

   A workshop was held in Poland in October, 1988, to identify key issues and parameters for forest monitoring (see Appendix 3, item 9).

2. **Use of Bioindicators to Quantify Dry Deposition of Sulfur in Forested Areas**

   (See Appendix 3, item 15)

3. **Applicability of Remote Sensing Technologies to Solution of Polish Environmental Problems**

   (See Appendix 3, item 20)

4. **Development and Use of Water Quality Models with Emphases on Microcomputer Applications and Innovations**

5. **Information Systems and Computer Models for Water Management in Large River Basins**

6. **River Basin Environmental Management**

   The maintenance of water quality in lakes and rivers requires management of the watersheds. The TVA model will constitute an initial hypothesis for evaluation. A river-lake ecotone will be managed to develop *in situ* controls of toxic substances and nutrients. The experimental watersheds will be located in the "Lake District" in northeastern Poland where extensive baseline data already exist.

7. U.S.-Polish Development and Training Program for Water Quality Modeling and Risk Assessment

Water quality management in Poland requires new, innovative technologies that do not require massive commitments of money and energy. Water quality models for transport, fate, and exposure processes will be coupled with risk assessment methodologies. These will be utilized both as a training methodology and as a direct validation experiment under field conditions in Poland.

8. Function and Management of Ecological Boundaries Within Landscapes

Ecotones between various ecosystems as well as shelterbelts, narrow stretches of meadows, and other biological filters have great influence on many processes basic for ecosystem functioning such as matter and energy fluxes, distribution of biota, etc. Ecological boundaries have substantial influence on so-called "self-purification environmental processes" and play an important role in the protection of ecosystems against pollutants.

The object of this proposal is to compare and summarize the results obtained in the United States and Poland and draw general conclusions for environmental conservancy and sustainable development of concerned countries. There will be a mutual exchange of investigators for cooperative work. The project will include organization of a workshop to summarize results.

9. Evaluation of Ecological Processes by Remote Sensing Methods: Litter Fall as an Index of Organic Matter Input to Ecosystems

The program will involve continuous registration of litter fall on the ground and on satellite/aircraft photography. Broadleaved forests throughout Poland are proposed as the main ecosystem type to be studied. Collection and elaboration of ground data will be done at Polish site, and remote sensing and modeling will be done at U.S. site.

10. Development and Testing of Biological Markers

DNA adducts, isoenzymes, and biochemical changes at the subcellular level have considerable potential for use in ecological, epidemiological studies in contaminated ecosystems. Ultimately, these kinds of studies may also be useful for prediction of threats to human health.

11. Methodologies for Assessing Human Exposure to Toxic Metals

(See Appendix 3, item 16)

12. Prototype Application of the U.S. Environmental Protection Experience to Polish Regional and Local Decision Making

(See Appendix 3, item 3)